TECHNOLOGY MANAGEMENT

1997

Robert Szakonyi, Editor

Auerbach

Copyright © 1997 RIA Group

ISBN 0-7913-3001-X

All rights reserved. No part of this work covered by the copyright hereon may be reproduced or used in any form or by any means—graphic, electronic, or mechanical, including photocopying, recording, taping, or information storage and retrieval systems—without the written permission of the publisher.

Copyright to "Transforming the Software Environment at Applicon" held by Barbara Purchia.

Copyright to "Core Technological Competencies," by William J. Walsh, held by Eaton Corp.

David P. Sorensen, Kerry S. Nelson, John P. Tomsyck, *Evaluation Review*, (vol. 18, no. 1), pp. 52-64, copyright © 1994 by Sage Publications, Inc.
Reprinted by permission of Sage Permissions, Inc.

Printed in the United States of America

Auerbach Publications
31 St. James Avenue
Boston MA 02116 USA

CONTRIBUTORS

Timothy H. Bohrer
Vice-President, Technology, James River Corp., Cincinnati OH

J.R. Champion
Consultant, Denver

Roger W. Cohen
Senior Planning Advisor, Exxon Research and Engineering Company, Florham Park NJ

Glenn Dugan
Chief Engineer, Sears Manufacturing Co., Davenport IA

Manek R. Dustoor
Director for Technologies and Concepts, Haworth Inc., Holland MI

Thomas P. Fidelle
Manager, Applications Research, Great Lakes Chemical Corp., West Lafayette IN

Stephen J. Fraenkel
President, Technology Services Inc., Northfield Il

Horst Geschka
Founder, Geschka & Partner Management Consultants

William J. Guman
Director, Contract R&D Technology Development, Grumman Corp., Bethpage NY

Gary Hanbaum
Financial Planner for Individuals and Small Businesses, Newark OH

Mark T. Hehnen
Director of Technology Commercialization, Weyerhaeuser Co., Tacoma WA

Stephen Hellman
Director of Technology Planning, Clairol Corp., Stamford CT

Russell Horres
President, CyberRx, Inc.

Norman E. Johnson
Vice-President of Corporate Research, Engineering, and Technology Commercialization, Weyerhaeuser Co., Tacoma, WA

James R. Key
Assistant to the Director for Industrial Program Development, Engineering Research Center, University of Maryland, College Park

Timothy Kriewall
Technical Manager, Sarns 3M Health Care, Ann Arbor MI

Henry LaMuth
President, AlphaComm, Inc.

Colin MacPhee
Principal, Albion Enterprises, Scarborough, Ontario, Canada

Gerald L. Majewski
Vice-President, Research and Development, INCSTAR Corp.

George Manners
Director of Services and Facilities, James River Corp., Neenah WI

Ronald A. Martin
Vice-President, Drug Development, Parke-Davis Pharmaceutical Research Division, Warner-Lambert Co., Ann Arbor MI

Keith E. McKee
Director, Manufacturing Productivity Center, Illinois Institute of Technology Research Institute, Chicago

John J. Moran
Certified Management Consultant and a Principal, J.J. Moran & Associates, Santa Monica CA

Sakharam Patil
Vice-President of Marketing and Commercial Development, American Maize, Hammond IN

John Peterson
Manager, Technology Strategy for a Fortune 500 Company

Peter S. Petrunich
Independent Technomarketing Consultant, Simpson, SC

Barbara Purchia
Senior Manager, Development Process Improvement Group, Lotus Development Corp.

Scott Schaefer
Vice-President, Disk Drive Products Development, Hutchinson Technology, Hutchinson MN

John L. Schlafer
Manager of Product Engineering, EcoWater Systems, Inc., Woodbury MN

Michael W. Schoonover
Senior Manager of R&D, VOP, Des Plaines IL

David P. Sorensen
Executive Director, Corporate Technical Planning and Coordination Department, 3M Company

Edward L. Soule
Vice-President of Corporate Research and Engineering, Weyerhaeuser Co., Tacoma WA

Stewart L. Stokes, Jr.
Senior Vice-President, QED Information Sciences, Inc., Wellesley MA

Lorraine Y. Stroumtsos
Project Leader, Innovation and Quality, Corporate Research, Exxon Research and Engineering Company, Clinton NJ

Robert Szakonyi
Director, Center on Technology Management, Illinois Institute of Technology Research Institute, Chicago

Philip R. Taylor
Business Systems Consultant, Fort Wayne IN

Don Tijunelis
Vice-President of R&D, Viskase Corp., Chicago

Adrian Timms
Associate Manager, Hershey Foods Corp. Technical Center, Hershey PA

William D. Torregrossa
Director of R&D Planning and Analysis, Hershey Foods Corp. Technical Center, Hershey PA

Thomas B. Turner
Retired Director of Department of Defense Manufacturing Technology Information Analysis Center, Towers Perrin

Monique Verhagen
Geschka & Partner Management Consultants

Harold G. Wakeley
Principal Scientist, Human Factors Research Group, Winnebago IL

William J. Walsh
Director of Engineering, Eaton Corp., Carol Stream IL

James A. Ward
Independent Consultant, Mays Landing NJ

Barbara Winckler-Rub
Geschka & Partner Management Consultants

CONTENTS

Introduction xi

SECTION 1
Linking Technology to Business Goals 1

  1 Aligning R&D Efforts with Business Objectives
     William J. Guman 3

  2 The Importance of Developing a Technology Plan
     Robert Szakonyi 17

  3 Creating a Strategic Technology Plan
     Adrian Timms and William D. Torregrossa 25

  4 Integrating Technology Planning with Business Planning
     Robert Szakonyi 35

  5 Corporate Strategic Technology Planning: A Case Study
     J.R. Champion 45

  6 One Corporation's Strategy and Technology Alignment: A Case Study
     John Peterson 63

  7 Core Technological Competencies
     William J. Walsh 85

  8 Conducting a Corporate R&D Retrenchment
     Mark T. Hehnen, Norman E. Johnson, and Edward L. Soule ... 99

SECTION 2
Management of R&D 105

  9 Developing an R&D Strategy and Strengthening R&D Administration
     Robert Szakonyi 107

  10 R&D Management: A Balancing Act
     Manek R. Dustoor 121

  11 Setting R&D Priorities: A Customer-Driven Process
     John J. Moran 133

  12 R&D Strategy and Coordination in Multinational Settings
     Stephen Hellman 143

13 Evaluating R&D Productivity
    Harold G. Wakeley 149

14 R&D Cost Accounting
    Stephen Hellman 173

15 Recognizing and Rewarding R&D Personnel
    Timothy J. Kriewall 181

16 Guidelines for Motivating Scientists and Engineers
    Stephen J. Fraenkel 195

SECTION 3
Project Management and Teamwork 205

17 Establishing Discipline in the Selection, Planning, and Carrying Out of R&D Projects
    Robert Szakonyi 207

18 Controlling Project Variables
    James A. Ward 223

19 An Action Plan for Self-Managed Teams
    Stewart L. Stokes, Jr. 231

20 Establishing a Cross-Disciplinary Team for Developing a New Product
    Robert Szakonyi 241

21 Motivating R&D Teams
    Ronald A. Martin 249

22 Air Drops and Total Quality
    George Manners 259

SECTION 4
New Product Development 271

23 Organizing for Effective Product Development
    Scott Schaefer 273

24 Product Development Effectiveness: Execution Is Key
    Peter S. Petrunich 281

25 Industrial R&D Program Evaluation Techniques
    David P. Sorensen, Kerry S. Nelson, and John P. Tomsyck 289

26 A Strategy for Product Design Evolution
   John L. Schlafer 299

27 Reducing New Product Development Risks
   Colin MacPhee 311

28 The New Product Idea Paradox
   Timothy H. Bohrer 321

29 Using Focus Groups to Capture the Voice of the High-Technology Customer
   John J. Moran 333

30 The Role of Applied Research in Corporate Development
   Thomas P. Fidelle 341

31 Transforming the Software Environment at Applicon
   Barbara Purchia 353

32 Making Concurrent Engineering Happen
   Philip R. Taylor 365

SECTION 5
Technology Transfer 381

33 Reengineering the Technology Delivery Process
   Michael W. Schoonover 383

34 The R&D Project: A Feedback Process
   Don Tijunelis 401

35 Technology Transfer During a Project
   Don Tijunelis 411

36 Improving the Engineering-Manufacturing Interface
   Thomas B. Turner 423

37 Transferring New Technology into the Manufacturing Plant
   Keith E. McKee 433

38 Laying the Groundwork for Technology Transfer
   James R. Key 453

39 Using Cross-Functional Teams for Technology Transfer and New Product Development
   Sakharam Patil 465

40 Technology Transfer for Small Companies
   Glenn Dugan 475

SECTION 6
Nurturing Innovation 483

41 Establishing a Scientific Network
   Gerald L. Majewski 485

42 A New Approach to Successful Technological Innovation
   L.Y. Stroumtsos and R.W. Cohen 497

43 Acquiring and Using Technology Competitively
   Henry LaMuth 511

44 Managing Innovation in Start-Up Versus Established Environments
   Russell Horres 519

45 Scenarios for Innovation Strategies
   Horst Geschka, Monique Verhagen, and Barbara Winckler-Rub 525

Index 543

INTRODUCTION

In its fifth volume, *Technology Management* continues in its mission of telling R&D managers how their colleagues are handling some of the most important problems and opportunities of their profession. It offers this insight in a collection of case studies and analyses informed by personal experience in technology management. Like the previous four volumes, this one is divided into six sections, each of which addresses issues of concern to technology managers.

The first section explores how companies can better link their R&D and business planning. For more than a decade, US industry has been struggling to compete with foreign manufacturers that employ cheaper labor and operate in less regulated environments. Many observers have suggested that the solution for these US companies is to exploit more effectively their one of their greatest advantages—their expertise in science and technology. They should leverage their R&D to develop new products that outperform the foreign competition. Although this sounds like a great idea, it is difficult to put into practice. Most US companies can not leverage their R&D more effectively because they have not coordinated it with their business goals. The contributors to this section have drawn on their own experiences to advise others how they can coordinate R&D goals and operations with those of their companies.

Section two addresses the nuts and bolts of R&D management. No matter how well a R&D department's goals compliments its business strategy, this function will fail if it is not effectively managed. This section discusses such essential management issues as developing a R&D strategy, improving R&D administration, managing R&D in multinational settings, and motivating R&D personnel.

Section three addresses a topic of great current interest—teamwork. It describes how to reward and motivate team members, how to work with cross-functional teams, and how self-managed teams function best. This section also explores many important issues of project management, such as common project pitfalls and controlling project variables.

The fourth section discusses new product development. It describes how to organize for effective product development, reduce new product development risks, and search for those all important new product ideas.

Section five addresses technology transfer. Its chapters describe how this takes place in small companies, how to plan for technology transfer, and what R&D can learn from a transfer.

Finally, section six explores how organizations can nurture innovation. Its contributors address such issues as acquiring and using technology competitively and the advantages and disadvantages of managing innovation in start-up firms.

SECTION 1

LINKING TECHNOLOGY TO BUSINESS GOALS

Linking technology to business goals is the subject most discussed by technical and nontechnical managers. Every company can cite R&D projects that were directed toward the wrong commercial goals, or technical goals that were not supported sufficiently to accomplish the desired business goals. In this section, the process of linking technology to business goals is examined from various viewpoints.

William J. Guman describes how project management within an R&D department fits in with its company's operations. Robert Szakonyi explains how developing a technology plan ensures that R&D goals support corporate ones.

Adrian Timms and William D. Torregrossa describe the evolution of their approach to technology planning. They discuss such matters as getting the support of both senior management and R&D personnel for technology planning, developing a balance between qualitative and quantitative analysis, and formulating principles for technology planning. Robert Szakonyi suggests methods that other organizations have used to integrate business and technology planning.

J.R. Champion describes the system of strategic planning that he helped develop and explains how to implement it step by step—from technology assessment, to program development, documentation, review, and approval.

John Peterson describes how a floundering company reversed its sliding fortunes through the strategic selection of R&D projects.

William J. Walsh shows how his company used core technological competencies to provide productivity and quality. Similarly, Mark T. Hehnen, Normal E. Johnson, and Edward L. Soule describe Weyerhaeuser Co.'s efforts to commercialize technology from the corporate R&D department, explaining the rationale behind their company's methodology of identifying, developing, and exploiting its core technologies.

ALIGNING R&D EFFORTS WITH BUSINESS OBJECTIVES

Managing in an R&D-driven industry is often a balancing act in which trade-offs must be made between creativity and control. Similarly, managers are often confronted with situations in which their short-term objectives seem to be in conflict with their long-term goals.

William J. Guman

Senior managers of large, multidivision corporations use various methods, structures, and controls to ensure that the efforts of all divisions are directed toward achieving the strategic goals of the corporation. In some cases, executives have adopted a philosophy of empowerment, seeking to tap the creativity and entrepreneurial spirit of employees and achieve levels of innovation typically associated with small, start-up companies. In other cases, senior management has placed extremely tight controls on the activities of all divisions in their company—that is, they have created a microcontrolled environment—to ensure that all business units are working toward a common set of objectives. This chapter details the experiences of an organization in which senior corporate management created a microcontrolled R&D environment to ensure that the R&D activities of each division supported the corporation's overall business objectives.

SETTING THE STAGE

Understanding how the company described in this case study microcontrolled R&D to achieve its business objectives entails a review of some company background. The following sections of the case study describe the company's primary business, its R&D program and investment strategy, its organizational structure, and the parent company's strategic plan for R&D, control and administration of R&D resources, and documentation of R&D plans.

The Company's Business

The Alpha Co was a major profitable division within Beta Corp. (Both Alpha and Beta are fictitious names). Beta's corporatewide business was diversified and extended into the following major market segments: government (defense and nondefense) aerospace; communications and electronics; space; commercial aviation; and commercial industrial products.

WILLIAM J. GUMAN, PhD, is involved with corporate strategic planning at Grumman Corp. He was director of R&D in Grumman's Corporate Technology organization and held several R&D management positions with a major aerospace company, including director of R&D, director of Advanced Product Development, and director of Technology & Operations. He has published more than 60 technical papers in various professional journals and is the author and coauthor of numerous company and in-house R&D contract reports.

Approximately 85% of Alpha's business was related to defense aerospace. This portion of the business provided between $100 million per year and $570 million per year of Alpha's total annual sales from 1972 to 1986. The peak occurred during 1981. These sales came from the defense-related product lines within each of its four major strategic business units (SBUs). Beta dissolved Alpha as a division in late 1987, and Alpha's military aircraft program and associated assets were acquired by another major aerospace corporation.

The Company's R&D Program

In the defense industry, R&D is called independent research & development (IR&D), which refers to contractor-initiated efforts relating to basic and applied research, product development, or systems and concept formulation studies. IR&D is distinguished from R&D in that it is independent of any contractual or grant commitments to the government. In essence, a company uses IR&D to technically and competitively position itself in the defense industry. The major characteristic of the bid-and-proposal (B&P) effort in defense work is the intent to use the results of IR&D directly for preparing a specific bid or proposal to the government.

Both IR&D and B&P are indirect costs carried in the overhead account of a contractor's accounting system. IR&D and B&P are the only overhead costs, however, that are required by law to have a separately negotiated advance agreement. This places a ceiling on the annual amount of such costs that may be included in the overhead allocated to all work performed for the Department of Defense (DoD). To determine these IR&D/B&P cost ceilings, the government uses the following evaluation factors:

- The contractor's commercial and government business activity.
- The technical quality and the potential military relationship of the planned IR&D projects.
- Past contractor performance.
- Business base projection.
- Other forecast business and financial data.

The IR&D/B&P Investment Strategy. As a major defense aerospace contractor, Alpha negotiated annual IR&D/B&P advance agreements with the DoD's Tri-Service Negotiating Group. Alpha's annually planned IR&D/B&P budget and program were heavily influenced by this annual cost recovery process. For annual planning purposes, the company's IR&D/B&P work focused on sustaining and generating new aerospace defense business. The IR&D/B&P effort was to be performed within the limits of the annually negotiated cost ceilings. These annual ceilings were approximately 2% to 3% of Alpha's annual sales to the DoD. Supplementary R&D funding was provided by Beta management and was generally allocated to Alpha for new non-defense-related business.

A primary aim of Alpha's senior management and Beta's management was to maintain the IR&D/B&P program within the negotiated ceiling. In effect, this ceiling produced an implicit management objective and also defined Alpha's operational environment for the IR&D/B&P program. The management strategy that was used to achieve this implicit objective was to microcontrol the IR&D/B&P program and its expenditures.

Deviations from the business plan were permitted only after Alpha received approval from Beta's corporate vice-president for aerospace development.

Under the terms of the agreement with Tri-Services, Alpha could fully interchange the individual IR&D and B&P ceilings that had been negotiated. In other words, the company was legally permitted to increase recovery of costs for either IR&D or B&P above the individually negotiated ceilings, provided that recovery of costs for the other would be decreased below its ceiling by a similar amount and provided that the total IR&D and B&P amount would not exceed the amount negotiated. During the 16 years that Alpha negotiated IR&D/B&P ceilings, the company overspent the annual IR&D/B&P ceiling approximately 50% of the time. Such overrun expenditures came out of profit. Over the same period, the company underspent the IR&D portion of the total ceiling dollars 75% of the time. Alpha senior management elected to invoke this interchangeability option and to use the larger amount of B&P funding thus made available to enhance the company's position to win major, new short-term defense business. In short, Alpha gave B&P higher priority than it did IR&D.

The Company's Flexible Matrix Organization

For the specific purpose of managing the company's IR&D/B&P program, the company was viewed as a flexible matrix organization in pursuit of its key business (i.e., product) areas, supported by its functional directorates. The company set up key functional directorates for: business development; engineering; manufacturing; materials; finance, contracts, and legal; and integrated logistics support. These directorates supported the IR&D/B&P program directors and managers in the matrix.

Corporate Control over the IR&D/B&P Program

Beta had a corporate strategic plan that defined the direction the corporation was taking to seek new business, retain and increase current market shares, and increase technical skills. The plan was coordinated annually with the senior management of all of Beta's divisions and subsidiaries, after Beta received suggested programs from them.

The business opportunities assigned to Alpha were an outgrowth of agreements between Alpha and Beta senior management. The programs to be pursued were designed to permit Alpha to meet its goals, which supported Beta's goals for profit and growth. Deviations from the plan by Alpha were permitted only after Alpha received approval from Beta's corporate vice-president for aerospace development.

Beta had a documented corporate policy entitled "Bid & Proposal and Independent Research & Development Programs" that strongly affected how Alpha's IR&D/B&P program was managed. As part of Beta's strategic planning effort, each division and subsidiary within Beta formulated its own long-range plan for IR&D/B&P, consistent with its own strategic plan. The corporate vice-president of product development approved all such IR&D/B&P plans and changes. Furthermore, if planned annual expenditure for any IR&D/B&P project at Alpha would exceed $25,000, Beta's approval of detailed project documentation was required before the project could be initiated. It was also Beta's policy that the following written reports be submitted to Beta:

- Monthly reports, presenting the status of resources expended on IR&D/B&P.
- Quarterly reports, with a brief narrative description and a milestone chart detailing the progress made on each IR&D/B&P project.
- A detailed year-end IR&D/B&P progress report. Furthermore, Beta conducted quarterly oral reviews of the IR&D/B&P program.

Control and Administration of IR&D/B&P Resources

Within Alpha, the engineering directorate administered and monitored the efforts, expenditures, and progress of the company's entire IR&D/B&P program. However, the business development directorate controlled how the B&P budget was used and applied for all proposal activities that focused on winning major new defense aerospace business. Specific IR&D projects were dedicated to developing and providing the requisite focused technology for the product lines within each SBU. The engineering directorate, however, controlled the use and application of a smaller percentage of the total IR&D budget, to develop the technology base in the technical disciplines that were applicable to all SBUs in the immediate future (i.e., in less than two years). Such technologies included manufacturing technology, advanced concepts, and operations analysis and simulation.

Furthermore, the engineering directorate was also allocated a B&P budget of approximately 2% to 5% of the company's total IR&D/B&P budget. This budgeted B&P activity was dedicated to winning government-funded R&D contracts in newly emerging technologies or new system-enabling technologies.

Company Leadership Styles

Alpha had five company presidents during a nine-year period, and each president's management approach and leadership style varied somewhat with respect to the IR&D/B&P program. However, Alpha's IR&D/B&P program

Exhibit 1
Generic Outline of Alpha's Annual Strategic Plan

1. Where are we today?
   A. Mission statement
   B. Strategic issues
      - Impact
      - Actions required
   C. Situation analysis
      - External environment and competition:
         — Products
         — Companies
      - Internal environment and capabilities:
         — Strengths
         — Weaknesses
         — Threats

2. Where do we want to be . . . when?
   A. Objectives (five-year)
      - Industry position
      - New business development
   B. Assumptions
      - External environment
      - Internal environment

3. How are we going to get there?
   A. Analysis of market opportunities
   B. Strategies, programs, and goals
   C. Financial measurement
   D. Resource deployment
      - Facility requirements
      - IR&D/B&P
      - Labor resources
   E. Implementation (i.e., action) plans
      - Strategy, interim goal, and action:
         — Individual, due date, and expected result

was always microcontrolled and consistent with Beta's published policy on IR&D and B&P programs.

Planning Documents for the Annual IR&D Program

During preparation of the company's annual strategic plan, each division within Beta Corp followed a generic outline provided by Beta (see Exhibit 1). Specific details of the company's plan were based on its business assignments. The IR&D/B&P program was one of the company's implementation plans in its overall strategic plan. The IR&D/B&P effort addressed the strategic issue of how Alpha would attain its objectives while staying below the IR&D/B&P cost ceiling.

Alpha's requirements for short-term and intermediate-term development of technology under IR&D were linked directly to its specific defense aerospace product lines. Much of the associated IR&D planning documentation pertaining to the technical aspects and resource requirements was generated annually by Alpha's IR&D project managers and middle management. Various versions of this documentation were required by Alpha management, Beta management, and the DoD Tri-Services. The rationale underlying this documentation is depicted in Exhibit 2.

Implementation Plan for Technology R&D Contracts

In addition to the IR&D plans that were generated annually, planning and implementation documentation was generated for technology-developing R&D

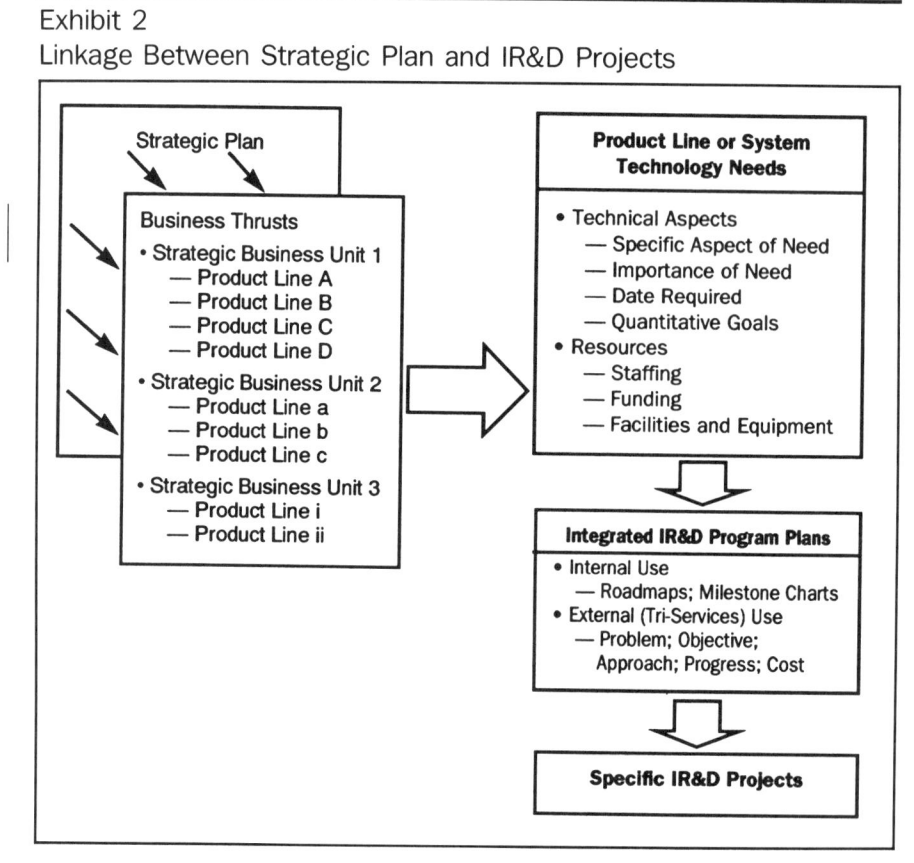

Exhibit 2
Linkage Between Strategic Plan and IR&D Projects

Strategic Plan

Business Thrusts
- Strategic Business Unit 1
  — Product Line A
  — Product Line B
  — Product Line C
  — Product Line D
- Strategic Business Unit 2
  — Product Line a
  — Product Line b
  — Product Line c
- Strategic Business Unit 3
  — Product Line i
  — Product Line ii

Product Line or System Technology Needs
- Technical Aspects
  — Specific Aspect of Need
  — Importance of Need
  — Date Required
  — Quantitative Goals
- Resources
  — Staffing
  — Funding
  — Facilities and Equipment

Integrated IR&D Program Plans
- Internal Use
  — Roadmaps; Milestone Charts
- External (Tri-Services) Use
  — Problem; Objective; Approach; Progress; Cost

Specific IR&D Projects

contract orders. These contracts were to be won using the small budget (approximately 2% to 5% of Alpha's total annual IR&D/B&P budget) that had been allocated to Alpha's engineering directorate. In general, these government-funded R&D contracts entailed conducting R&D in newly emerging or new system-enabling technologies. Technology-developing contracts in manufacturing technology, advanced composites, and operations analysis and simulation were also pursued under this contract program. The annual goals and five-year goals for this program were always expressed in terms of dollars. Senior management closely tracked the annual goal of contract R&D dollars won.

> **A microcontrolled environment merely defines a scenario in which a program manager must produce results despite the environmentally imposed constraints.**

A formal status review of each contract R&D project was conducted once a month at a meeting with the principal investigator of that project. During this meeting, a small review group—with representatives from the engineering, contracts, and finance departments—completed a one-page status report of each project. The report included assessments of technical performance, financial status, scheduling, staffing, and any other relevant factors. Contract R&D projects that had a definite problem were brought to the attention of the company's president by the finance department during monthly management review meetings.

ANALYSIS OF THE STRATEGY OF MICROCONTROLLING R&D

The experiences of Beta and Alpha can help technology managers who are considering the use of a microcontrolled R&D environment in their own organizations. The following sections review the factors that contributed to the success of Alpha's R&D environment and the lessons learned by Alpha's management.

Success Factors

The microcontrolled R&D environment at Alpha was merely part of Beta's corporate culture. In this environment, several factors were important to achieving the company's short-term objectives. The presence of one or a few of these factors within the company did not necessarily guarantee success; however, when most of the factors were present in a given situation, their cumulative effect significantly enhanced the probability of success.

Strong Program Leadership. A microcontrolled R&D environment merely defines a scenario in which a program manager must perform and produce expected results despite the environmentally imposed constraints. R&D program managers who performed well in this environment were able to advocate and successfully defend their programs to senior management. Their charisma and ability to be perceived as winners usually enhanced their teams' morale and motivation to succeed. Certainly, a strong ego drive for achieving success contributed significantly to the performance of program leaders in the microcontrolled R&D environment.

In addition, such strong leaders sought, hired, and retained the best-qualified people to work on their programs. The presence of several equally capable program managers led to strong internal competition for priority in the ongoing use of the limited number of key technical specialists in Alpha's matrix organization. In conjunction with this personnel competition, these program leaders also emphatically sought to obtain and control the amount of the IR&D/B&P budget they perceived necessary to do their job, regardless of the effect this action would have on other ongoing or planned R&D projects.

Despite the microcontrolled environment, the successful leaders usually also ensured that their program area could draw readily available resources for use during unforeseen contingencies. This resource reserve was usually either a management reserve budget they established or a funded special project they initiated and controlled.

Support from Senior Management. R&D projects that had either the CEO's or company president's personal attention and endorsement were always allocated an operating budget. However, some endorsements caused discomfort for R&D middle managers when they discovered that a visibly endorsed program was technically tenuous but knew that it had been sold to the highest levels of management. The microcontrolled R&D environment could not prevent the occurrence of such an anomalous project; fortunately, in the few isolated instances encountered, such projects eventually tended to fade away.

Senior management's attention to contract R&D projects also contributed significantly to the progress of these projects. When technology-developing R&D contracts were pursued, won, and performed specifically to implement some aspect of Alpha's short-term strategic plan to enter a new business, senior management followed the progress being made by the supporting IR&D/B&P contract R&D effort.

Adequate Resources. In general, the R&D projects that achieved their objectives received the budget, technical personnel, and capital assets consistent with the results expected of the program by senior management. At Alpha, these projects always directly supported short-term business plans. It was not uncommon for IR&D/B&P funds to be reallocated from several lower-priority projects to a high-priority project within a matter of days. The microcontrolled R&D environment at Alpha provided management with the timely information needed to assess the likely impact of such program interruptions and resource reallocations.

Contract-Funded Technology Development. By law, IR&D/B&P funds cannot be used to support contractually required work on R&D contracts awarded by the government. Because they provided Alpha with R&D funding above the level of the negotiated IR&D/B&P ceiling, management usually viewed these R&D contracts as a way to augment its R&D budget for developing technology. Seeking to obtain this larger R&D budget, senior management established annual goals to win a specified amount of contract dollars provided by these R&D contracts. To meet this financial goal, technical personnel could pursue, win, and perform R&D contracts if they met the following three conditions:

> **Contract R&D programs provided the opportunity to establish formal relationships with customers in anticipation of future R&D business.**

- R&D management assessed the probability of winning the contract to be relatively high.
- The B&P funding required to win the R&D contract was not excessive.
- The B&P funding could be provided by the relatively small B&P budget for contract R&D projects (i.e., 2% to 5% of Alpha's total IR&D/B&P budget).

Contract R&D activity provided an outlet for technical specialists to conduct the type of research they preferred. The program also provided the opportunity to establish formal relationships with customers in anticipation of future R&D business. Alpha's contract R&D program successfully established several technology-developing programs in addition to those funded by the IR&D program. Alpha did not develop a formal strategy, however, regarding the potential of the contract R&D program to enhance the process of developing future new major business, nor did it fully exploit its potential. New enabling technology that was developed for the customer by these contracts was rarely used to strengthen a particular long-range business plan. In addition, the monthly review process of all contract R&D programs caused such discomfort to some talented technologists that they preferred not to pursue and win R&D contracts in the microcontrolled R&D environment that prevailed at Alpha.

Assignment and Tracking of Employees' Annual Performance Goals.

Alpha conducted formal annual performance appraisals for exempt personnel. (*Exempt* legally signifies a job function that is excluded from the application of the Fair Labor Standards Act and the Walsh Heally Public Contracts Act, primarily as they relate to payment for overtime.)

Individuals being appraised would meet with their immediate manager. During the review, the individual's performance, potential, career interests, and development needs would be discussed. The performance of the individual's major goals and special assignments would be rated numerically. Furthermore, the employee and manager would jointly agree on a set of major goals and tasks with measurable criteria for the next appraisal period. These were usually stretch goals, instead of goals that could be easily reached. The scored assessment of how well the goals were met became a factor during the annual salary review process.

R&D personnel usually had goals related to an identifiable improvement of some specific operational or technical aspect of Alpha's technical capability and competitiveness. Another goal commonly set by an R&D program manager was to win a specified dollar amount of R&D contracts with a specific customer in a particular area of technology. Sometimes Alpha would set a goal to formally establish a team arrangement with another company for a future joint effort. Frequently, an individual's goals would be a subset of a major goal of the company. By setting and tracking goals at various levels of the R&D organization, management had a means to anticipate the possible results of R&D projects if all individual goals were met. Furthermore, tying the goals to the annual salary review process motivated R&D personnel to achieve their goals.

Dedication of IR&D Projects to the Short-Term Technical Needs of Specific SBUs.
At Alpha, IR&D projects were dedicated to help the business development directorate meet its short-term business objectives. The focus of this R&D was systems oriented, and it had several enduring objectives.

One objective was to provide the technical foundation for the task of verifying that significant modification of one of the company's major products (e.g., military aircraft) would result in a product variant that would either perform a new military mission or significantly expand its present capabilities. Although the R&D organization successfully accomplished this objective, the business development directorate failed to obtain government funding for producing any of these derivative military aircraft, despite intensive marketing efforts.

A second objective of the systems-oriented R&D endeavor was to provide the technical foundation for winning major new system business in the SBUs. The R&D program was performed to provide the technical foundation to win defense business that was planned for and budgeted by the government. In general, this R&D activity successfully supported the business development directorate in winning such new short-term business.

A third objective of Alpha's R&D program was to establish the feasibility of some special technical enhancement of an existing product. In this case, the research activities were primarily part of Alpha's engineering directorate's IR&D program. The engineering directorate also successfully developed new analysis tools and simulation capabilities to support the feasibility-verifying efforts of the systems-oriented IR&D mentioned in the discussion of the first objective. The engineering directorate was also responsible for developing new manufacturing technology along with assessing the applicability of new structural material to existing and future company products. Senior management's interest contributed to the success of this latter R&D effort to enhance Alpha's competitive manufacturing capability.

Results of the Microcontrolled R&D Environment

The overwhelming drive to win major new short-term business caused a greatly unbalanced distribution of the type of technology that was developed under IR&D. As can be expected, people worked within the company rules and were productive. However, they also learned how to play the games deemed necessary for them to function in the microcontrolled environment. The following sections examine the results from Alpha's strategy of using a microcontrolled R&D environment to achieve business objectives.

> Alpha's R&D personnel used their creative talent to seek evolutionary modifications and improvements to existing products.

Technical Innovation Was Not Smothered. Senior management required that the main thrust of Alpha's IR&D be linked to its short-term business objectives. The net effect of this focus, along with the planning and tracking of results, was that the IR&D program became primarily development oriented; however, technical innovation was not smothered. Instead, Alpha's R&D personnel used their creative talent to seek evolutionary modifications, improvements, and derivatives to existing Alpha products or processes.

In addition, the microcontrolled environment did not prevent R&D personnel from seeking out entirely new applications of Alpha's technical expertise. In one case, this particular type of lateral technology transfer allowed Alpha to rapidly establish itself at the forefront of an emerging area of business within the defense industry. This lateral transfer of technology was facilitated by the attitudes of technical personnel and their interaction with potential customers.

Two additional outlets that fostered technical innovation were acceptable to Beta and Alpha's senior management. One was the technology-developing R&D contracts awarded by the government. The other was the highly focused corporate-funded and directed R&D program (i.e., new aerospace defense business).

Management Awareness Was Enhanced. Frequent reporting provided management with timely project status information for both the IR&D/B&P contract R&D projects and the Beta-funded R&D projects. Two types of reports were generated and issued: numerical data on the financial and human resources used during the reporting period and progress reports based on the input received from each project manager. In addition to those reports that were used within Alpha, Beta management received monthly, quarterly, and year-end written reports for all Alpha's IR&D/B&P projects with annual expenditures that would exceed $25,000.

The usefulness of the weekly financial expenditure and monthly open financial commitments data was significantly enhanced after it was graphically displayed against time. These graphs were generated for each IR&D/B&P

project as well as for the entire IR&D/B&P program. Extrapolating and analyzing the trend of expenditures at a central point within the company usually caused potential problems to surface sufficiently early to initiate inquiries and, if necessary, corrective action.

In the microcontrolled environment, project management was characterized by minimal communication. The written progress reports and the associated milestone charts did not necessarily reveal what was really happening in a project. Therefore, some Alpha presidents conducted monthly oral progress reviews of major IR&D/B&P projects. Dialogues initiated between the president and the project managers sometimes uncovered issues not discernible from the printed status reports.

Beta managers also conducted quarterly oral reviews of Alpha's major IR&D/B&P programs and issued the results of these reviews in a memo to the CEO of Beta Corp. In addition, permanent consultants to Alpha occasionally worked with IR&D/B&P program managers and provided assessments and recommendations.

The microcontrolled environment did reduce passive or static performance of R&D personnel. However, surprises occurred. For example, a vendor persistently delayed delivery of an essential item and the delivery milestone was not being tracked by the formal system. This situation eventually produced the surprise that prevented the project from being completed on schedule—despite the project manager's diligent efforts.

In addition, management became aware that inappropriate goals were often set and tracked. New technology-developing R&D contracts were planned for and tracked by Alpha primarily for the dollars they added to the IR&D/B&P budget. Thus, contracts that cost little to win and that were relatively easy to win were pursued, instead of contracts that would enhance the company's intermediate- to long-term strategic technical position.

Finally, management realized that the true intent of a goal can be misunderstood. Project managers and R&D personnel must have an unambiguous understanding with their supervisors of the results that are expected of them. The dialogues that took place during employees' annual goal-setting sessions and performance reviews could occasionally fall short of communicating what was really expected from each employee.

Two cases can be cited to illustrate this point. In one case, the goal was to develop a specific new capability for the company in the short term. However, the R&D program manager who was responsible for the project did not know that this new capability also had to be transferred out of R&D and put into practical use within the company. In another case, the individual's goal was to win an R&D contract from a particular customer and for that contract to have a targeted contract dollar value. When that planned procurement was canceled by the government, the R&D program manager was initially unaware that the targeted dollar value of the contract being pursued would remain a goal. The R&D program manager subsequently learned that receiving a contract from the originally targeted customer was not considered the real pertinent issue of the goal.

The Microcontrolled Environment Linked the Potential of New Managers. In addition to being exposed to the different leadership styles of various presidents, Alpha's engineering and R&D personnel were also exposed to the different leadership styles of various vice-presidents and directors who were hired and brought into the company from outside Beta Corp. The professional experience of these individuals augmented the in-house experience, and this new source of experience introduced different insights and approaches to R&D planning and problem solving.

However, their talent and this additional outside experience focused primarily on meeting short-term business objectives and achieving victories soon after they joined the company. Microcontrolling R&D primarily for this purpose essentially prevented the full potential of this new technical management—and of the company's inherent R&D talent—from being used to prepare Alpha to win new business that would use emerging technologies.

> Project managers and R&D personnel must have an unambiguous understanding of the results expected of them.

CONCLUSION

Microcontrolling R&D was an effective implicit strategy for realizing the particular business objectives to which this strategy was applied. At Alpha, the primary thrust of the IR&D/B&P contract R&D efforts to win short-term profitable major new business was too sharply focused and unbalanced in scope. Despite excellent R&D management, high morale, and select islands of technical expertise, the secondary R&D activities did not position the company to win contracts for the technically more-sophisticated, next-generation products that eventually replaced the company's existing products. Indeed, of the last three major prime military aircraft programs that Alpha won over a period of approximately 20 years, each new program was for a technically less-sophisticated Alpha product line than the one it replaced.

What cannot be determined is the percentage of the total annual IR&D/B&P budget a company must dedicate and manage strategically to prevent its product line from becoming technologically obsolete. This problem is further complicated by the need to manage technology-developing contract R&D projects and to sustain interest and motivation from senior management. The threats and opportunities of emerging technologies must be integrated carefully into the overall long-range business development strategy and must be an integral part of the scope of a microcontrolled R&D environment.

THE IMPORTANCE OF DEVELOPING A TECHNOLOGY PLAN

*Most US companies use little more than hunches and educated guesses to choose their **R&D** projects. This chapter explains why these companies should instead develop a formal technology plan.*

Robert Szakonyi

Most US companies do not make the fullest use of their R&D organizations. Because this function is poorly integrated into these companies' business planning processes, their R&D organizations often choose to pursue projects that make sense technologically but little sense from a business standpoint.

Edward D. Weil and Robert R. Cangemi asked R&D managers at 55 US companies how good the business/R&D coordination was in their organizations. Sixty-five percent responded that their companies needed improvement or had a severe problem in this area. Forty-two percent indicated that their R&D organizations provided no significant input to corporate planning, and 91% believed that their entire research staff was not informed about corporate goals and strategies. Consequently, Weil and Cangemi concluded that "there is a substantial problem of linking long-range planning and long-range research in technologically oriented companies."[1]

Technology planning—that is, planning an entire R&D organization's entire output—is essential to integrating R&D into business plans. Unfortunately, US companies are also weak in this area. Lowell Steele, in describing how to do technology planning (or as he called it, "strategic planning for R&D") stated that:

> This kind of planning has rarely been undertaken in American industry, despite the fact that a great deal of attention has been devoted to the subject of planning research and development projects . . . (The application of R&D project planning) assumes an unsupported conclusion—that the objective has already been established. The goal of these planning techniques is to increase the efficiency with which objectives are delivered. On the other hand, the specification of the objectives is the crucial feature of strategic planning (for R&D).[2]

ROBERT SZAKONYI, PhD, is the director of the Center on Technology Management of IIT Research Institute, Chicago. He has performed consulting work for many companies in various industries and written numerous books and articles on technology management.

DEVELOPING A TECHNOLOGY PLAN

A technology plan consists of decisions about which technologies a company will invest in, which it will monitor, and which it will not support at all. Such a plan should be based on an assessment of an R&D organization's capabilities; the immediate and strategic business needs that the R&D organization must serve; the future technical potential of various technologies; and the leverage that those technologies could provide in business applications.

The greatest benefit of a technology plan is the clear direction it provides an R&D organization. By evaluating systematically all the key issues that affect an R&D organization's output, R&D managers learn where they need to go and what their priorities should be. A technology plan gives R&D managers a method for selecting projects that are linked to the company's objectives.

A technology plan (if put together well) provides these benefits because it helps an R&D organization do three things that many others neglect to do: take an overall perspective; establish clear priorities with business managers about which technical programs will be pursued; and forecast the development of various technologies and the impact of economic and social trends on this development.

Many R&D organizations do not take such a systematic view of the way they generate proposals and select projects. Typically, an R&D organization relies solely on a piecemeal and bottoms-up approach for generating R&D proposals. That is, with little guidance from the R&D managers, individual researchers and engineers propose what they find technically promising. The R&D managers then select projects to support based on their intuition about each proposal's worth.

A technology plan forces R&D managers to decide which technical programs are important. It enables R&D managers to tell researchers and engineers what kind of R&D is needed, and it provides R&D managers with criteria for evaluating and selecting from the proposals that are submitted.

Many R&D organizations do not establish clear priorities with business managers about which technical programs will be pursued. Consequently, many R&D organizations are forced to alter their project priorities according to which R&D projects are the "hottest" any given week or month. A technology plan provides R&D managers a tool for resisting company pressures to constantly change project priorities. It gives R&D managers a way to link R&D projects to the technical objectives that were already agreed on by business managers.

Many R&D organizations do not systematically evaluate the potential of technologies through forecasts of technological, economic, and social trends. In many organizations, one person is responsible for monitoring new technologies. Usually this person is also given other assignments, such as developing a project management system, doing R&D budgeting, and serving as a liaison with the business units. This person spends at best one or two days a month monitoring new technologies.

In addition, this person's findings are rarely expressed in the context of technological, economic, and social trends. To forecast well, an R&D organization must evaluate its own technical capabilities and what they will lead to, its competitors' capabilities, the potential of various relevant technologies (some of which perhaps no company is currently developing), future company and customer needs, and potential changes in economic and social environment that may affect technology. The real benefit of making such forecasts is that R&D managers must extrapolate their intuitive evaluations about future technology developments and applications.

Many R&D organizations do not take a systematic view of the way they generate proposals and select projects.

Forecasts related to technology, however, are not a technology plan. James Quinn wrote that:

> Forecasts assess future environments and the mutual impact of these environments on the organization and the organization on the environments. Planning occurs later when the organization takes forecast information and converts it into goals, policies, programs, and procedures which guide action. Many executives forget this distinction, and mistakenly look at forecasts as generating specific research plans. Forecasting can be delegated to staff groups. But planning from these forecasts requires decisions from major line executives.[3]

Furthermore, as Quinn points out, decisions about long-term R&D do not require precise forecasting. All that is necessary is general accuracy, for example about the direction of future events and the probable range of those events.

On the other hand, when R&D managers have not put a technology plan together well—or when they neglect to develop one—an R&D organization usually operates with blind spots. Blind spots can undermine a major technical program. Some typical blind spots of R&D organizations are:

- Inadequate attention to competitors' actions or to new technological developments.
- An unaggressive strategy in licensing technologies that the company or its competitors may use.
- Preservation of the status quo in how technical programs are budgeted (e.g., existing businesses keep their share of technical resources while few resources are available to counter potential threats or create new opportunities.
- A narrow idea of how technology can create competitive advantages (e.g., to erect barriers to entry by raising or lowering economies by scale, keeping learning curves proprietary, altering capital requirements, affecting access to conventional distribution channels, changing switching costs, or keeping outside firms from overcoming entry barriers).
- Overloading an R&D portfolio with low-risk R&D projects.

Management Considerations

Developing a technology plan is almost always a far greater challenge than expected. Two of the biggest hurdles to overcome are the resistance to planning of R&D personnel and the lack of a company language for discussing technology planning.

Many R&D personnel (and managers) resist technology planning because it can upset existing arrangements for selecting R&D projects. After technology planning is carried out, interesting R&D proposals with excellent technical merits and good potential commercial payoffs are no longer automatically supported, for R&D proposals are no longer evaluated in isolation. Instead, they are evaluated in terms of a clear set of priorities, and under these conditions, many pet projects are turned down.

To counter the resistance of R&D personnel to technology planning, senior R&D managers should solidly support the process of technology planning. They should also assign R&D staff to participate in technology planning who are highly respected by their peers, as well as imaginative and adept at getting things done in their companies.

The importance of having very competent R&D planners is illustrated by the following two case histories. In one company, the R&D planners were the alter ego of the senior technical manager, providing him with analyses that addressed issues that were neglected. Because these R&D planners were closely tied to senior R&D management, their analyses eventually provided the framework for reallocating technical resources to address neglected issues. In the other company, the R&D planners were isolated from both the senior R&D management and in their R&D departments. These R&D planners spent most of their time collecting data that the rest of the R&D organization seldom used. Eventually, the R&D organization grew so disenchanted with their contribution that the entire six-person planing group was disbanded.

R&D planners must establish a process for technology planning. An important element in establishing this process involves developing a company language, or classification system, that an R&D organization can use to analyze its technologies and then communicate its findings to business managers. Steele points out that "this classification system may be structured by discipline, by product, by elements of the production process, by market, or by industry."[4] Whatever the classification system, this company language must provide a method for looking sensibly at technical programs and informing business managers about the probable impact of technical programs on their businesses.

Developing a company language is very difficult because the language should not only make analytical sense but it should also be acceptable to R&D staff and business managers. R&D planners must work closely with both the R&D staff and business managers in developing it. The process is slow but indispensable to creating a planning methodology.

After R&D planners develop a company language, they must forge an agreement on what their company's key technologies are. At a defense electronics company, R&D planners started this process by establishing various technical councils composed of senior technical managers, scientists, and engineers from the operating units. These councils were supposed to advise business managers about the key technologies in a given area. To do this, they had to evaluate the strengths and weaknesses of many technologies and identify their company's key technologies. At a natural-resources company, R&D planners took a quite different approach. They conducted case studies of the technologies in each of their company's major manufacturing plants. They reviewed the historical development of these technologies, looking at successive improvements and the ways in which they were applied. With this overarching perspective, the R&D planners—and the plant manager—then identified their organization's key technologies.

Many R&D personnel (and managers) resist technology planning because it can upset existing arrangements for selecting R&D projects.

To elicit support for the technology plan, R&D planners must get others involved in the planning process. One way to do this is to ask researchers and engineers to evaluate the potential of various technologies. Another way is to ask business managers to evaluate current technical programs.

Tools for Analyzing Technology

One of the most important tools for analyzing technology is technological forecasting. This is a general approach for estimating advances of various technologies and for understanding the critical forces behind and implications of these advances. It can also refer to a specific methodology developed during the 1960s and 1970s by James R. Bright. This methodology is used to predict quantitatively when various technological advances will occur and how significant they will be.

This methodology has fallen into disfavor in most US companies, probably because its proponents promised too many benefits and industrial managers expected too much from it. Whatever the case, the methodology is seldom used in the US. Its various analytical techniques, such as morphological analysis, simulation, scenario-building, and cross-impact analysis, have unfortunately made many industrial managers skeptical about its value.

Nevertheless, the more general approach to technological forecasting can help R&D and business managers answer many important questions, such as:

- For roughly how long will company's core technologies provide competitive advantages?
- What technologies will most likely replace these technologies after they have matured?
- Are technologies being developed in other industries that might be used in competition with the company's present or future technologies?
- How will different technologies affect each other?

- What kinds of social changes, government actions, or international events are likely to affect the development and use of various technologies?
- Are there relatively unknown technological developments that will become vastly more important because of, for example, new government regulations?
- What are the competition's most likely future technological advances?
- About when will these technological advances occur?
- How might future economic conditions affect the development and use of certain key technologies?
- What kinds of scientific advances might university researchers be expected to make in the most relevant technical disciplines in the next five to ten years?
- How might customers' future needs affect the development of technology?
- What kind of competitive advantages might a certain technology provide at different stages in its life cycle?

The answers, of course, need not be precise to be illuminating. Company managers need only to gain an improved perspective on their technologies to make better decisions about technological priorities. There are four methods for evaluating technologies.

Method 1. One method consists of evaluating how a company's technologies make the company's product acceptable to customers.[5] Seven aspects of product acceptability can be used in evaluating technologies:

- Performance
- Acquisition cost.
- Ease of use.
- Operating cost.
- Reliability.
- Serviceability.
- Compatibility with other parts of a system.

Method 2. Harris, Shaw, and Summers identified two aspects of technology evaluation: the importance of the technology and the company's relative position on it. According to these authors, if a company has a technology of high importance and it also has a high position on it, the company should "bet" (i.e., support the technology well). If a company has a technology of high importance but has only a low position on, it should "draw" (i.e., either invest more in the technology or abandon it). If a company has a technology of low importance that it has a high position on, it should "cash in" (i.e., abandon the technology). Finally, if a company has a technology of low importance that it has a low position on, it should "fold in" (i.e., pull out).[6]

Method 3. A third method is to use profiles of technologies drawn through case studies of their historical development. (The previously mentioned natural-resources company that analyzed the technologies employed in its manufacturing plants used this approach.) The historical development of a company's technologies can provide the basis for forecasting its technological advances and thus as a method for setting priorities.

> One of the most important tools for analyzing technology is technological forecasting.

Method 4. A fourth method is to assess technologies' potential leverage and the likelihood of success in gaining leverage. Steele identified criteria for evaluating technologies: the size of the market that would be affected, this market's projected rate of growth, and the degree of market penetration a company might expect to achieve.[7] Consequently, the greater a technology's combined potential leverage and likelihood of success, the higher its priority should be.

CONCLUSION

Regardless of the evaluation method they use, R&D managers can select projects wisely when their priorities are clearly defined. With a broad perspective on what their organization needs to accomplish, managers have more complete criteria to use in evaluating the technical merits and projected commercial benefits of R&D projects. Establishing and following a technology plan gives R&D managers just such a perspective.

Notes

1. E.D. Weil and R.R. Cangemi, "Linking Long-Range Research to Strategic Planning." *Research Management*, Vol. 26, No. 6, pp 32–38.
2. L.W. Steele, *Innovation in Big Business* (New York: Elsevier, 1975), pp 93–94.
3. J.B. Quinn, "Long-Range Planning of Industrial Research," *Harvard Business Review* (July-August 1961), pp 90–91.
4. Steele, *Innovation in Big Business*, p 104.
5. Fusfeld, "How to Put Technology into Corporate Planning," p 53.
6. Harris, Shaw, and Sommers, "The Strategic Management of Technology," pp 32–33.

CREATING A STRATEGIC TECHNOLOGY PLAN

Adrian Timms and William D. Torregrossa

The benefits of strategic technology planning are well known, and many technology managers are eager to work with a strategic plan. The planning process, however, is not a simple matter. It requires making choices and entering into a process of managed evolution. The authors have faced some of these choices, and the evolutionary steps they have taken are outlined in this chapter.

Why a company decides to adopt technology planning determines how it goes about it. In many cases, the adoption of strategic planning systems is motivated by a crisis; for example, a CEO is unfamiliar with the technological basis of the company's business and hence frustrated because the technology development process is not working. In such cases, organizations are often willing to pay high consulting fees to progress rapidly up the strategic-planning learning curve.

In the authors' case, readings in R&D management led to an interest in technology planning. There was no crisis, just the desire for excellence and the recognition that technology was becoming an increasingly important means to gain competitive advantage. The future role of technology in the relatively low-tech food products industry, it was felt was too important to be left to chance. The authors opted to use internal resources to increase the staff's technology-planning skills.

TAKING THE FIRST STEPS

The approach taken to implementing strategic planning emphasizes home-grown learning, which invariably entails making mistakes and having to live with certain inefficiencies during the early stages. For example, it took a detailed analysis of the technologies required to produce existing projects to drive home the importance of structured planning for decision making and of finding the appropriate level of planning detail. This investment of analytical sweat brought to light things that should have been obvious, including crossovers in technical functions, new opportunities, and wasteful practices.

ADRIAN TIMMS is the associate manager at the Hershey Foods Corp Technical Center in Hershey PA.

WILLIAM D. TORREGROSSA is the director of R&D planning and analysis at the Hershey Foods Corp Technical Center.

ESTABLISHING LEADERSHIP AND SUPPORT

Although the push for strategic technology planning does not have to come from the top, having the support of senior executives is vital. In many cases, the CEO is a driving force in implementing strategic technology planning. In the authors' case, the impetus for planning did not originate at the executive level, but senior-level executive support did exist. Senior management was receptive to the planning efforts, but the burden was on the authors to sell the planning process and enlist the support of all levels of management.

The company's current CEO encourages investment in R&D, which allows R&D management to focus on determining the right mix of R&D activities rather than continually justifying R&D projects. Knowing that support for technology planning is maintained only when executives feel that there is value in the level of investment, the authors instituted systems to evaluate R&D investments in a business manner. As part of this process, financial managers work with the development team, which helps to sell the R&D efforts to senior executives. An additional benefit of these managers' involvement is that they provide a much needed alternative perspective on strategic technology planning, which is particularly critical during the early stages of the process.

Selling Concepts Internally

One manager who reported to an R&D director was given full-time responsibility to determine what was needed in the program, to experiment with alternative approaches, and to implement one that suited the company's needs. Having one person devoted full-time to this job ensured that the planning would take place.

From the outset, a customer-oriented approach to technology planning was taken (the customers being the internal business managers). When any manager is brought into these activities, individual needs and motives must be addressed. Key players were selected to participate in the programs and to champion the system in their areas. These individuals each have their own reasons for backing the program, and their needs are addressed accordingly.

Hurdling Obstacles

Several obstacles were encountered in the form of attitudes and questions. An R&D manager looking for support for such a program should be prepared to address these types of issues. Some typical questions were:

- If we are already a successful company, why do we need strategic technology planning?
- Shouldn't any competent manager know instinctively what an operating unit's technology needs are?
- Isn't technology planning just for the high-tech guys? Our industry is low tech.

One way to address these questions is through seminars and training workshops presented by technology planning consultants—an effective use of outside consultants. To some extent, however, the key players need to be supportive of the program beforehand so that they can deal with objections from their staffs early in the process. A combination of one-on-one meetings with key managers and general awareness-building techniques is helpful here. One such teaching device is the use of the language of strategic technology management in meetings and correspondence.

> **Strategic technology planning should be a routine management activity that is intrinsic to an organization's culture.**

Planning from the Top Down or Bottom Up

The majority of technology-planning practitioners are consultants. Perhaps because of the source of their incomes or because their experience has taught them that it is effective, consultants usually address senior management directly and put the burden on these executives to provide functional decision makers with the motivation and commitment to perform the technology-planning activities. There is no doubt that this approach is expedient, especially in the context of a relatively brief consultancy.

Much of the literature on technology planning says that the process must be top down—that is, that the impetus must come from senior management. This view is valid insofar as it is essential for the motivation for technology planning to be shared at the top. Strategic technology planning, however, should be a routine management activity that is intrinsic to an organization's culture, not a one-time process prompted by senior management. For continuous strategic planning that obtains input from the practicing engineers, the authors aim to strike a balance between top-down and bottom-up development of planning, as well as pay deliberate attention to selling concepts at the engineers' level.

Focusing Resources at the Shop Level

Organizing personnel and activities in groups associated with chosen technology areas not only helps to focus resources on strategic thrusts but facilitates grassroots, bottom-up planning. Individuals with the commitment to lead the group should be encouraged to make the plan happen. Two methods for accomplishing this are discussed in the following sections.

Centers of Excellence. One of the aims of technology planning is to identify key areas for strategic thrust. Some of these are core areas of the company's business, and in such cases there has been no waiting for technology planning to start setting them up. Instead, centers of excellence were used as vehicles for planning and coordinating development related to a specific core technology.

A center of excellence is not a building; it is an organizational construct, a multidisciplinary congregation of people who share the same strategic objectives for that technology. The mission of a center is twofold: to maintain strong links between its technical and strategic business goals and define its long-term technical goals, and to continually pursue world leadership in that technology.

Even in an organization delineated by functional departments (e.g., by academic discipline), a task force approach can be used to bring collective focus to a particular technology. The center is not a task force because it is permanent until such time as that technology is divested; it can even be treated as a budgetary cost center. Because the technologies chosen are of primary importance to the corporation, divestment is unlikely. It is incumbent on each center, however, to be on the lookout for potential substitute technologies and to plan accordingly.

Centers of excellence provide opportunities to bridge departmental, rank, and divisional boundaries among participants. These bridges not only lead to better strategic vision but disseminate technology information and planning techniques to a wider audience. The authors' current stage of evolution involves only technical representatives in the centers, but plans call for them to ultimately include nontechnical business managers.

Through the centers, strategic technology planning can be performed even before the adoption of formal planning processes. The centers will continue to function effectively and to become an increasingly integral part of the overall strategic planning system.

Technology Champions. Technology champions actively take on leadership roles in a technology development group, playing a role analogous to that of a brand manager. A brand manager is charged with developing and deriving maximum benefit from a brand asset; a technology champion does the same for the technology asset. The champion must maintain development impetus, forecast and strive for the future state of the art, promote effective use of the latest technology, and use all available channels to sell management on the need for continued investment in that technology.

MANAGING AN EVOLUTIONARY PROCESS

Implementing a technology-planning program is an evolutionary process, and it should be managed as such. Strategic objectives were drawn up for implementing technology planning as if the planning process were a major development program in itself. This plan should evolve as the technology manager developing it learns more about the plan and process; however, in its early form, it will help the managers prepare for the successive stages. The authors have emphasized the evolutionary nature of planning activities and regard the planning workshops as a laboratory for developing the appropriate planning techniques.

One initial action was to put down on paper what was learned from literature and seminars as part of a grand design for an overall strategic technology planning process that was tailored to the organization. At first, the intent was to implement this process by starting with one small business unit and introducing the process to all business units sequentially; however, on examining the potential culture shock that would ensue, an alternative was chosen—a phased, evolutionary approach. This would permit the development of sound methods built on basic principles. In addition, working with untried techniques straight from the planning department could be avoided.

> Two factors will hinder efforts to institute technology planning: silence, and activities that are inconclusive or not useful.

More experience with the techniques showed that it was appropriate to involve non-R&D managers; if the multidisciplinary team concept is to be realized, their eventual involvement is imperative. During the formative stages, the plans were analyzed and drafted and then reviewed with the R&D staff. After staff members had acquired additional formal technology planning skills, they began to perform some of the planning activities themselves, and R&D management facilitated the overall process.

Structuring the Program

Technologies readily lend themselves to hierarchical structure and thus many levels of detail. For purposes of scanning technology and acquiring intellectual property, detailed distinctions among technologies may be desirable. Fairly quickly, it was found that the level of detail appropriate to strategic allocation of resources can be (to the technologist) embarrassingly simple.

Much the same is true for projects. A laboratory does not need to be unusually large to find itself handling more than 200 projects. If these projects have been independently instigated by a wide variety of customers, the technology manager will probably need to group hundreds of projects into a manageable number of programs. As the strategic planning of R&D becomes more big picture-oriented, the definition of programs is expected to lead to the definition of projects.

From an analytical viewpoint, there is a distinction between technologies and projects as they relate to business goals. A one-to-one relationship among products, technologies, and R&D programs rarely exists; the concept is a simplification frequently used to describe technology-planning techniques. In assessing the needs of these development programs, the technology manager is bound to encounter analytical complexity before learning to differentiate these entities and establish their various interrelationships.

Evaluation techniques should be appropriate to each strategic unit. Technologies can be regarded as capabilities—as a set of strategic assets rather like tools that could be put to use through projects. Most R&D projects apply technological capabilities to producing measurable commercial results rather than to developing the technology itself; other R&D projects are dedicated to furthering technological capabilities.

After projects are divided into those that develop and those that merely apply technology, measures of value, risk, and reward need to be consistent with the types of projects. For example, the concept of return on investment for a developing technology becomes strained because return on investment should apply only to specific applied technology projects; however, individual technological capabilities can be evaluated for their relevance to future business needs and opportunities.

The Triad. The triad is a grouping of business goals, technologies, and R&D projects. In the center of the triad are various planning techniques that emphasize the links among the three strategic entities. The techniques used to evaluate projects and technologies are at opposite ends of a spectrum of technology-planning methods. Technology management consultants typically have specialized expertise associated with different points along this spectrum. For example, some consultants are particularly adept at applying portfolio management techniques to R&D projects.

Design Principles for Technology Planning

At the outset, it was decided that the following ground rules would govern any planning systems that were adopted:

- The planning process and tools must be simple and easy to operate. The logic of the planning system should be straightforward and useful; meaningful decisions should result.
- There must be multifunctional, multilevel corporate involvement in the planning process.
- The role of technology planners must be to catalyze and facilitate the process of planning, not to make strategic decisions.
- All technology planning must be oriented to product attributes deemed significant by the consumer; that is, attributes that drive the consumer's willingness to purchase the product must be the focus of the planning efforts.

The Need for Visible Output

Results of the technology planning process should be reviewed at regular intervals. These results should be shared with the development team, especially if they help to communicate technology planning principles.

While the technology manager is developing planning methods, the internal customers should be involved in and kept abreast of progress made. Two factors will hinder efforts to institute technology planning: silence (which slows momentum) and activities that are inconclusive or not useful. Especially with R&D colleagues, it is necessary to keep technology planning in their minds through frequent involvement in useful and productive activities. A planned evolutionary process will help to accomplish this.

Technology-monitoring activities are useful in this regard. The authors systematically scan patent data bases and copy relevant items to interested parties, provide reports to those charged with gathering competitive intelligence, and make presentations of company and technology profiles in selected areas. There are plans to publish regularly an awareness newsletter for technical and nontechnical personnel to highlight trends and emerging technologies. All these efforts provide necessary input into technology planning and serve as useful, visible evidence of planning activities.

Training in such techniques as group decision making and focus groups is invaluable to technology planning activities.

The Technology-Planning Toolbox

Setting up a technology-planning system is expedited when certain tools and skills are available, particularly a data base, trained facilitators with effective interactive skills, expert systems software, and graphics analysis packages. These tools and skills are discussed in the following sections.

Data Base Accessibility. R&D management will need to draw on a store of historical and planned project data to make sound decisions on products and processes. A data base is the ideal medium for storing and experimenting with different treatments of the data.

In this case, the authors already had an ongoing data base on which the costs of R&D projects were planned and tracked. The data base had been established primarily to justify budgetary resource requirements for that year rather than to perform strategic planning. Nevertheless, it served as a launching pad for quantitative analysis of the current project portfolio. As the capabilities and needs in strategic technology planning grow, so do the systems and structure of the project data base. This interdependence is reinforced by involving the data base administrator in the development of planning systems.

If technology managers do not currently maintain such a system to track project costs in their organizations, developing one to meet both strategic and operational needs will be both a challenge and an opportunity. The challenge is to design the system with enough flexibility to support experimental data use; the opportunity is to make planning easy and effective by creating such a system.

For example, one data base application in strategic technology planning is resource allocation, which entails coding projects for reporting purposes; such data as cost estimates are totaled and formatted to produce a graphic output. Much of this task is iterative and well suited to a data base. It is recommended that someone with data processing experience and skills be included on the technology planning team.

Interactive Skills. Planning is a creative activity that occurs in a workshop-like process, so we trained ourselves in group facilitation. Training in such techniques as group decision making and focus groups was invaluable when applied to technology-planning activities. Without these techniques, it would have been extremely difficult to achieve the desired results.

Consultants are often used for this facilitating role; however, with proper training, in-house facilitators can be developed and used to lead technology planning as well as other problem-solving groups.

Expert Systems Software and Graphics Analysis Packages. In addition to a data base, expert systems software can help a planning team get through evaluation without the need for a full-time domain expert. Such software can assist managers in analyzing their projects, defining objectives, and focusing resources better. For example, an excellent system for project portfolio analysis is the one developed by Strategic Decisions Group (Menlo Park CA). In addition, it is helpful to present information graphically. Group discussion is greatly facilitated by diagrams and visual representations.

EXPLORING THE CONSULTANCY OPTION

In developing a strategic technology-planning process, immense value can be gained from using technology management consultants. The extent to which consultants are used is dictated by the budget resources available and the management team's willingness to learn from its mistakes.

A variety of high-quality consulting companies specialize in technology management. Because of the strategic nature of technology management decisions, these consulting services are priced accordingly. If the six-figure fees that are standard for help in making R&D management decisions constitute a minor part of a department's budget, this route is an excellent way to rapidly gain knowledge of a variety of techniques. If handled correctly, the use of consultants also resolves some of the political issues that arise when multiple functions are involved in the planning activities.

The authors have corresponded with many companies in other industries concerning the implementation of planning systems. If their experience is typical, technology managers should be prepared to engage several consultants before finding a good fit with their own needs. As in any other field, consultants vary in their areas of expertise and style.

When such factors as gaining support and establishing new responsibilities and relationships are added to those of developing the actual planning processes, strategic technology planning becomes a full-time activity. To accomplish these tasks using only outside consultants, it is necessary to engage the consultants on long-term retainers because the selling process is gradual and requires a continued presence.

However, consultants have a role to play even if a manager chooses to proceed with the planning process using internal resources. As more is learned about the distinctions between different technology-planning approaches, the authors have been able to define more specifically the particular hurdle that a consultant must help to overcome. Condensing the projects list to the appropriate level of detail—that is, from many small projects into a few strategic programs—would not have been effective use of a consultant; however, getting an external audit on the portfolio would have been. The external audit would have helped to maintain a realistic view of the company's competitive position and the true value of its mix of R&D programs.

Strategic technology planning should be a routine management activity that is intrinsic to an organization's culture.

CONCLUSION

Getting an organization to realize the benefits of strategic technology planning requires the technology manager to oversee the process of evolving its planning systems. This applies whether the company is implementing such planning where none exists or improving existing technology-planning systems. It is hoped that technology managers can use the points covered in this chapter to help formulate strategies for achieving this goal.

INTEGRATING TECHNOLOGY PLANNING WITH BUSINESS PLANNING

An R&D organization can understand its mission only after it has coordinated its efforts with its company's business plans. This chapter discusses some strategies for achieving this coordination, drawn from the author's extensive experience and reading of industry literature.

Robert Szakonyi

In too many US companies, technology planning and business planning are not well integrated. In a study of long-range R&D and business planning in 55 US companies, Weil and Cangemi identified the primary reasons for this as "the lack of clearly defined corporate objectives and goals or ones that are too general to be of much guidance to research; and . . . lack of effective ways to identify long-range market needs and corporate needs.[1]

Therefore, the first step in integrating technology planning and business planning is to develop business plans. Marketing must develop two-to-three-year strategies for various product lines and planners and business managers must develop a long-range business plan (i.e., a five-to-ten-year plan) for the company.

In developing these product line strategies, marketing groups need to know:

☐ What customers will need in two or three years.
☐ New products that can fulfill those needs.
☐ The ideal cost/performance characteristics of these new products.
☐ The distinctive advantages of these new products over competitors' products at that time.

In developing a long-range business plan, planners and business managers should answer such questions as:

☐ What businesses should the company be in in five to ten years?
☐ How is the company going to seek competitive advantages in each of these businesses?
☐ How fast does the company want to grow?
☐ How much should company growth come from internal development based on R&D?

ROBERT SZAKONYI, PhD, is the director of the Center on Technology Management at IIT Research Institute, Chicago. He has performed consulting work for many companies in various industries and written numerous books and articles on technology management.

Once product line strategies and a long-range business plan are in place, an R&D organization can respond to the business's needs. An R&D organization, however, should take a different approach to serving product line strategies than it does in responding to a long-range business plan.

The organization should focus on how its R&D expenditures serve product line strategies. An R&D organization should specify what is being done for a product line strategy (e.g., sample raw materials); identify the exact business unit(s) that are being served; identify the nature of the R&D organization's responsibility (e.g., provide technical support for purchasing); indicate the objective of the proposed R&D (e.g., to compare the materials currently used with an alternative); specify a period for the R&D expenditure; and indicate a completion date for the R&D effort.[2]

All R&D expenditures for product line strategies (and for company activities in general) should be laid out so its services to company functions can be clearly understood. For example, an R&D organization might indicate in its plans how it will aid marketing in the early trials of a new product, provide technical support for implementing new manufacturing equipment, or conduct tests for evaluating and improving quality.

In coordinating its efforts with a long-range business plan, an R&D organization should consider broad business matters. Some of the needs that can be highlighted in an overall business plan are lowering costs, improving product performance, developing new product features, and creating a new generation of products.

Before considering how to meet these long-term business needs, an R&D organization must analyze its own technological capabilities. Lowell W. Steele has classified technological capabilities into four categories. Using them, an R&D organization can translate its technological capabilities into terms that a business manager can appreciate:

- Applying the state of the art to the company's existing technologies.
- Using the technologies of competitors.
- Extending the state of the art.
- Developing new technology to supplant old ones.[3]

Next, an R&D organization should coordinate its capabilities with the long-range business plan, which usually consists of different strategies for different company businesses. For example, division A might aspire to be the first to market; B might plan to follow the leader; C might emphasize application engineering (which involves using the same parts in very different applications), and D might follow a "me-too" strategy of producing at low cost.

Brian C. Twiss describes another set of strategies: offensive; defensive; interstitial (i.e., attacking a competitor's weakest point); market-creation; and maverick (by which one's innovation reduces the size of the market for everyone).[4] Steele outlines still another system of categorizing the strategies.

LINKING TECHNOLOGY TO BUSINESS GOALS

It can maintain or harvest its business, grow its present business, extend its business, or diversify.

After R&D managers know what its company's business strategies are, they can identify what kinds of technical contributions to make. Moreover, in comparing the company's business strategies with its technological capabilities, R&D managers become aware of the technical contributions their organization cannot make. They might therefore decide to license technology from outside the organization.

> **The first step in integrating technology planning and business planning is to develop business plans.**

Next, R&D managers must consider the objectives of each business strategy (e.g., to lower costs in a business being harvested or to develop new features for products in a business being grown). Then, R&D managers can select projects. In so doing, they should consider more than just coordinating R&D with product line strategies and a long-range business plan. R&D organizations can also contribute to a company by developing a new technology that changes the company's strategic options.

Unfortunately, most US companies currently are poor at managing technology proactively. Competitors can come from within a business's industry; they can be potential entrants, or they can come from other industries (and provide similar or better products based on other materials or systems than the companies in that industry use). Technology can be used to gain competitive advantages over any of these kinds of companies.

Furthermore, technology can affect—and be affected by—much more than the factors that company managers traditionally consider in trying to gain competitive advantage, for example, capital investment and production output. Technology can also affect, and be affected by, social factors, such as globalization and improvements in productivity and personnel training.

To manage technology strategically, business managers must take two steps beyond simple coordination of R&D and business planning. First, they must consider broader business issues, not just specific businesses strategies, when looking for how technology can contribute to a company's businesses. Second, they must look at how technology can influence business strategies and not just at how technology can respond to them.

Management Considerations

The management considerations involved in integrating technology planning and business planning should be examined throughout the integration process: before a company has integrated technology planning and business planning; during integration; and after integration.

Before Integration. A company at this stage is usually in conflict. For example, at a food-processing company the R&D managers disagreed with the company's goals. The R&D managers were developing new businesses based on new technology, and the senior business manager thought the purpose of the company's technology was to cut manufacturing costs. The senior business manager wanted his company to get into new businesses, but through acquisition, not through internal growth based on new technology.

Such differences in goals occur for many reasons. In an electronics company, the conflict was partly due to the fact that the R&D organization was so isolated from the corporate planning group that it could not deal directly with the planners but had to work through a senior business executive. In an energy-products company, problems stemmed from the senior business managers' failure to make business plans because they believed that plans would prevent their company from seizing opportunities. Without any business plans, however, the R&D organization never could lay the technical basis for the new businesses that they desired.

Even if a company does have business plans, its technology planning and business planning still may not be integrated. Sometimes, business plans are too unclear to provide direction. For example, a chemical company's plans dealt with such general business and technological issues that its R&D organization was helped very little in selecting R&D. At another company, the projections in the business plans were so overly optimistic that the R&D organization was uncertain where the company was going.

While Establishing Integration. Given the kinds of problems that have to be overcome to integrate technology and business planning, company managers must recognize that their efforts to improve integration will take a good deal of time. Two major problems that companies usually confront in trying to establish this integration are that many managers do not understand what planning is about and the company lacks data necessary for creating plans.

In light of these problems, A. Wilkinson outlined a three-phase approach to establishing plans: the audit phase when managers take stock of their businesses and technologies; the exploratory phase when they explore and evaluate their options; and the directive phase when they make decisions.[6]

At the beginning of the audit phase, a company must create a team of managers from various functions (e.g., R&D, engineering, marketing, manufacturing, and finance) to make the plans. Initially, team members should not concentrate on their own functions' responsibilities but look at the broad needs of their company (or operating division) and at their company's technological capabilities. They should first focus on concepts. Later, they should consider numbers related to projected sales, and costs. The objectives of this team during the audit phase should be to draft company objectives, prepare the data needed for evaluating options, and analyze potential gaps in their company's businesses.

During the exploratory phase, this team should evaluate the options they identified and begin preparing the plans required to carry them out. Finally, during the directive phase, the team should decide strategies and resources, or it should make strong recommendations to more senior managers who would make these decisions.

Any team of managers created to make these plans should be prepared for not only organizational resistance but also conceptual problems that will make analysis of technological capabilities and long-term business needs difficult. For example, R&D managers and business managers have difficulty analyzing technology as an entity by itself. Managers tend to see technology in the context of many product and process applications, which can complicate a team's efforts to plan. In addition, some business managers reject the idea that there can be gaps between, for example, a company's overall sales goals and actual sales. Such business managers try to explain these gaps by pointing to the long-term economic cycles that also affect a company's sales. By trying to mask weaknesses, business managers can make planning difficult.[7]

Most U.S. companies are poor at managing technology proactively.

Despite the difficulties of initiating planning, a planning team can succeed. One R&D planner's simple plans produced major changes in a company because they highlighted how some long-term business needs were unmet and how important technological capabilities in his company were untapped. He captured the interest of business and R&D managers and linked R&D closely to business planning. Some R&D managers' efforts to get planning implemented in just their R&D organization eventually prompted their business divisions to improve their planning.

After Integration. According to H. Igor Ansoff, companies that have integrated technology planning and business planning have:

- A forecasting and information system.
- Strategic analytical skills.
- The ability to reallocate resources so that new needs are addressed.
- A strategic plan.
- A project management system for guiding the implementation of strategic decisions.[8]

Some companies have successfully integrated technology and business planning. What distinguishes these companies from most others is that their business managers consider planning important in managing the company. For example, a business planner at one company told the president he cared more about whether the president and other business managers asked the right questions than the direction in which they led the company. In other words, he valued whether planning was really taken seriously—which it was—not whether the president was using the best planning techniques. At another company, a senior business manager explained that planning was considered part of a business manager's job. He mentioned that just as business managers in his company are expected to develop annual budgets and to conduct performance evaluations of the people who report to them, they are also expected to plan where their business will be in five to ten years.

Analyzing Technologies and Businesses Together

The tools for analyzing technologies and businesses together can be grouped into two categories: tools for making R&D responsive to business needs and tools for R&D to create new business opportunities.

Although the two categories differ in some ways, it is the mindset of the managers who use them that really differentiates them. Most managers using the first set analyze how various technologies can be used to respond to business needs. Managers using the second set look for creative ways technology can create new business opportunities.

Tools one through four use matrices for analyzing technologies and businesses together. Each matrix contains different questions about a company's technologies and businesses.

Tool 1. An aerospace company developed its own matrix for analyzing various technologies and businesses together. In the matrix rows, the managers listed the company's various technologies. In the columns, the managers listed the possible contracts these technologies might obtain. Company managers asked three questions:

- What are the chances that a contract will actually result in a request for proposals?
- How strong are the technologies in the areas where there might be contracts?
- What are the chances that the company will win a contract if a request for proposals is made?

Tool 2. John M. Harris, Robert W. Shaw, and William P. Sommers' method for evaluating technologies can also be applied to analyzing technology and business together. Their matrix for analyzing a business is comparable to that for analyzing a technology. Whereas their matrix contains two aspects appropriate to a technology—the importance of a technology and a company's relative position on a technology, their matrix for analyzing a business contains two aspects appropriate to a business—the attractiveness of a business and a company's competitive position. As in analyzing a technology, they developed four categories—high attractiveness and high competitive position, high attractiveness and low competitive position, low attractiveness and high competitive position, and low attractiveness and low competitive position.[9]

Company managers can compare a business matrix and a technology matrix. By noting inconsistences between the relative value of a business to a company and the relative value of a technology that serves that business, managers see where resources need to be increased or decreased. For example, if a very attractive business in which the company had a strong competitive position were served by a technology that was not very important and in which the company had a low position, company managers would have to invest in a new technology for this business.

> R&D organizations can develop technologies that change their organizations' strategic options.

Tool 3. Domenic Bitondo and Alan Frohman developed a matrix in which the rows describe various market share strategies and the columns describe various product line needs.[10] By relating an R&D posture to combinations of market share strategy and product line needs, Bitondo and Frohman related technology to a two-dimensional business situation in just one matrix.

To illustrate, the various market share strategies are to yield market share, maintain market share, increase market share, and enter a new market. The various product line needs concern existing products, improved products, and new products. Different R&D postures are appropriate for various combinations of these market share strategies and product line needs. According to Bitondo and Frohman, when a company enters a new market with a new product, it needs to be a technology inventor. On the other hand, when a company increases market share with an improved product or maintains market share with a new product, it only needs to apply technology. Finally, when a company yields or maintains market share with an existing product, it should be a technology avoider.

Company managers should first identify their product line's needs and the market share strategy they will follow. Then, by locating the appropriate R&D posture for this business situation, company managers have a guide to how aggressively to pursue certain technical programs.

Tool 4. Antonio S. Lauglaug developed another matrix for determining how the level of technology required relates to customer demands about product performance. According to Laughlaug, there are two levels of technology required—high and low, and two types of customer demands with respect to product performance—complex and simple. The matrix, therefore, consists of four categories: low level of technology and complex customer demands, low level of technology and simple customer demands, high level of technology and complex customer demands, and high level of technology and simple customer demands.

Company managers can use this matrix to determine the types of technical programs they should pursue for various kinds of businesses. For businesses with complex customer demands and technology requirements, managers should emphasize design styling and marketing. For businesses with simple customer demands and low technology requirements, they should maintain existing technology and emphasize manufacturing. In businesses with complex customer demands and high technology requirements, company managers should concentrate on applications research, design engineering, and systems integration. Finally, in businesses with simple customer demands and high technology requirements, they should focus on process engineering, automation, and artificial intelligence.

The hallmark of the second set of tools for analyzing technology and business together—those tools for looking at how technology can create business opportunities or advantages—is the use of generic terms for identifying technology's potential. Three such tools are:

- *Tool A.* This approach does not examine technologies as they are conventionally thought of. Instead, it examines a unit of analysis called a strategic technical area. It consists of technical skills or disciplines, the area in which these skills or disciplines are applied, the particular product or service in which the skills or disciplines are used, and the specific market need. Graham R. Mitchell cites an example of a strategic technical area that describes integrated circuit processing. In this case, technical skills include lithographic techniques (i.e., photographic X-ray, electron beam) for defining fine geometries on semiconductors, as well as for use in high-temperature solid state chemistry and thin film processing. They are applied to the fabrication of semiconductor integrated circuits, which are used in a wide range of switching and transmission products throughout the telecommunications industry.[12]

    Similar technical skills and expertise are often widely applicable to product or service areas in different parts of a company. By understanding a set of technical skills and expertise, company managers can drive their businesses in new technical directions.

- *Tool B.* Another approach to analyzing how technology can drive businesses is based on Alan R. Fusfeld's method for evaluating technologies.[13] Fusfeld identified seven aspects of product acceptability: functional performance, acquisition cost, ease-of-use characteristics, operating cost, reliability, serviceability, and compatibility. Each involves a question about how a technology affects products.

    Fusfeld also developed a generic concept for evaluating how much a technology affects these products: technology demand elasticity. Technology demand elasticity (which is analogous to economists' price elasticity) measures the responsiveness of market demand to improvements in function, ease of use, and reliability.

- *Tool C.* A third approach to analyzing how technology creates new business opportunities uses Michael E. Porter's analytical framework concerning competitive environment. The generic categories are industrial competitors, potential entrants, substitutes (i.e., products or services from companies outside the industry that could replace a company's products or services), suppliers, and buyers (i.e., customers who could integrate backwards and compete with a company).

    Porter's analytical framework broadens company managers' perspectives on R&D's potential as a competitive weapon. For example, managers might discover ways to use R&D to reduce the dependency of a product's quality on any one supplier's material, or managers might learn how R&D can help differentiate the company's product so that a buyer does not integrate backwards and thus compete with their company.

> **By understanding a set of technical skills and expertise, company managers can drive their businesses in new technical directions.**

DEFINING AN R&D ORGANIZATION'S MISSION

For R&D managers, the aim of coordinating R&D with business planning is to define R&D's mission so R&D can respond to the company's business needs. Another part of its mission involves creating new business opportunities that are not currently recognized in a company's definition of its business needs.

Given the poor coordination between R&D and business planning in most US companies, R&D managers must accept that it takes time to define this mission clearly. An R&D organization can gain a clear definition of its mission only after it has successfully coordinated with the company's business plans.

Notes

1. Weil and Cangemi, "Linking Long-Range Research to Strategic Planning," p 34.
2. A. Wilkinson, "The Relationship between the Planning and Technical Functions. Part 1—The Short to Medium Term," *R&D Management*, vol. 5, no. 2 (1975), pp 135–137.
3. L.W. Steele, *Innovation in Big Business*, pp 83–85.
4. B.C. Twiss, *Managing Technological Innovation*, pp 57–60.
5. Steele, *Innovation in Big Business*, pp 85–90.
6. A. Wilkinson, "An Evolutionary Approach to the Introduction of Corporate Planning in a Company," *R&D Management*, vol. 5, no. 1 (1974), pp 41–47.
7. Twiss, *Managing Technological Innovation*, p 39.
8. H. Igor Ansoff, "Strategic Management of Technology," *The Journal of Business Strategy*, vol. 7, no. 3 (Winter 1987), pp 37–38.
9. J.M. Harris, R.W. Shaw and W.P. Sommers, "The Strategic Management of Technology," Planning Review, January 1983, pp 28–35.
10. D. Bitondo and A. Frohman, "Linking Technological and Business Planning," *Research Management*, vol. 24, no. 6 (November 1981), p 21.
11. A.S. Laughlaug, "A Framework for the Strategic Management of Future Tyre Technology," p 32.
12. G.R. Mitchell, "New Approaches for the Strategic management of Technology," in Mel Horwitch (ed.), *Technology in the Modern Corporation* (Elmsford, N.Y.: Pergamon Press, 1986), p 136.
13. A.R. Fusfeld, "How to Put Technology into Corporate Planning," *Technology Review* (May 1978), pp 51–52.
14. M.E. Porter, "Technology and Competitive Advantage," pp 64–65.

CORPORATE STRATEGIC TECHNOLOGY PLANNING: A CASE STUDY

This chapter describes one company's system for corporate technology planning. This company's experiences demonstrate that R&D managers must devote as much effort to deciding what they will work on as to planning individual projects.

J.R. Champion

This chapter describes a corporate strategic technology planning system that the author designed in the 1980s for a Fortune 500 company. The system improved the company's use of its R&D resources and increased the performance and effectiveness of its R&D organization. Called the corporate technology planning system, its design, installation, and operation was based on three premises.

The first is that to support their companies effectively, an R&D organization must continuously answers to three critical questions:

☐ What are the company's present and foreseeable needs and opportunities that require R&D support?
☐ What is R&D doing and proposing to do to provide this support?
☐ How can R&D communicate the answers to these questions to company managers to enable them to make intelligent R&D investment decisions?

Many industrial R&D departments do not address these questions well. R&D planning in these organizations tends to consist of a helter-skelter agglomeration of requests from their marketing and manufacturing departments. Plans tend to reflect only the R&D support needs of the requesting departments and not those of the entire company. In addition, the needs expressed are frequently inexact or incomplete because they have been formulated by marketing and manufacturing people who are not experts in assessing technology needs.

The second premise is that the only way to develop truly valid answers to the previous three questions is for an R&D organization to design and implement a process of technology planning that does three things: continuously surveys all of the company's technology needs, recommends, at least once a year, a detailed and comprehensive program of R&D activities to address those needs, and communicates its findings and recommendations to company managers in a clearly documented written plan. Only by engaging in such

---

J.R. CHAMPION has retired from Manville Corp., where he served as senior director of corporate research, director of technology planning, and division general manager. He is now working as a consultant.

planning can company management decide where, how, and when to best invest its R&D money.

The third premise is that because R&D is better qualified than any other organization in a company to identify the company's technological needs and opportunities, it must proactively determine those needs and opportunities and recommend what actions should be taken to address them. In other words, designing and operating the corporate technology-planning (CTP) system and producing the corporate technology plan is exclusively R&D's function. This does not mean that R&D should plan without the rest of the company. On the contrary, to plan well, all the elements of the company must plan together. However, that interaction must take place within the context of R&D's responsibility for technology planning. A company that does not require R&D to undertake this responsibility is not getting what it should from its R&D investments, and an R&D organization that does not press its company management to let it do thorough and comprehensive corporate strategic technology-planning is failing in its duty to the company.

It should be clear that the CTP system is not a project planning system, a system that details how a technical project will be conducted. The CTP system identifies the range of technology objectives that the company must undertake to attain its business objectives and the R&D projects (including broad cost and schedule estimates) necessary to attain them. It is in effect a strategic technology-planning system.

THE ORGANIZATIONAL CONTEXT

The company for which the system was designed is a multibillion dollar corporation engaged in a broad range of commodity businesses based on relatively mature technologies. At the time of the system's introduction, these businesses were divided into 48 strategic business units (SBUs) managed by 13 operating divisions.

The company's R&D operations were centralized and consisted of 360 professionals, technicians, and administrative and clerical personnel organized into 10 technical departments reporting to a vice-president of R&D. The 10 departments were subdivided into 27 sections. The departments and sections were technology specific, and each provided technological support to any division or business unit that needed the technology in which the department or section specialized. The areas of support they provided were typical of industrial research organizations (e.g., new product and process development, existing product and process modification, general technology research and development, and technical service support to marketing and manufacturing). Although the company has in recent years changed its mix of businesses and largely decentralized its R&D operations, many of the decentralized R&D departments continue to use the CTP system.

HOW THE PLANNING PROCESS WORKS

The process (see Exhibit 1) is a common-sense approach. It begins with R&D performing three analyses to the company's technological needs and opportunities. These needs and opportunities are compared with what R&D is presently working on. Then, revisions are made to the existing technology program, the costs

> A company that does not require R&D to undertake corporate technology planning is not getting what it should from its R&D investments.

and benefits of the new program are calculated, and the plan is documented. This document becomes the basic reference and agenda for review and negotiation with, and approval by, various levels of management (i.e., R&D, SBU, division, and corporate).

Phase 1: Technology Assessment

This phase includes blocks 1 through 4 of Exhibit 1. The technological analyses that R&D must perform are identified in blocks 1, 2, and 3.

Block 1: Assessing Business Strategies. To perform this activity, each R&D group and subgroup must understand as thoroughly as possible the business strategies of the SBUs and divisions they support and must analyze these strategies for their technological implications. In addition, if the corporate business strategy includes objectives not covered in the division strategies, which is often the case, (e.g., new business areas not yet assigned to existing divisions), R&D must also analyze that strategy to determine what must be accomplished technologically to help the strategy succeed.

These business strategy analyses should be conducted through detailed discussions with the appropriate business managers and their staffs. Ideally, the analyses should be conducted while the SBU, division, and corporate managers are developing their business strategic plans. In the real world, however, such coordination is not always feasible. Some business managers are slow to develop their strategies. Others are reluctant or simply forget to invite their R&D colleagues to their planning sessions. In the CTP system, the R&D managers are responsible for inviting themselves to the strategy-planning sessions of the business units.

The R&D managers of businesses that have incompletely formulated business strategies must make the best assumptions they can about what those strategies are and confront the business managers, saying "Here is how R&D interprets your business strategy. Here are the programs we are working on for you. Are they still addressing your needs? If not, please tell us what has changed."

If the business managers cannot indicate what has changed (as too often happens), the R&D managers must proceed on their best assumptions and advise the business managers of their intentions. Most business managers would understand and consent to this course of action. In addition, the existence of a well-organized and regular technology-planning system requiring business managers to articulate their business strategies clearly is a strong inducement

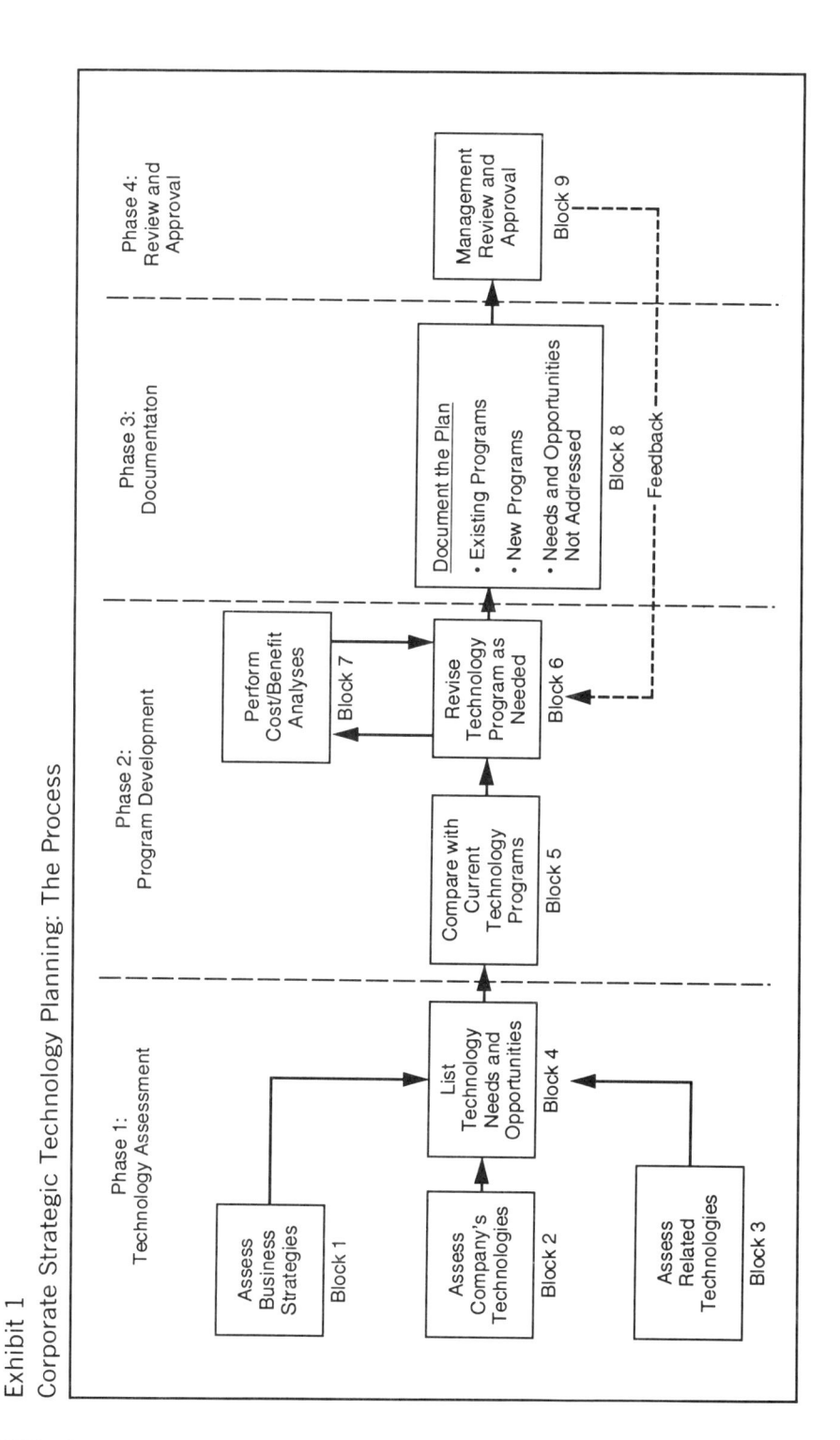

Exhibit 1
Corporate Strategic Technology Planning: The Process

Phase 1: Technology Assessment

- Assess Business Strategies (Block 1)
- Assess Company's Technologies (Block 2)
- Assess Related Technologies (Block 3)
- List Technology Needs and Opportunities (Block 4)

Phase 2: Program Development

- Compare with Current Technology Programs (Block 5)
- Revise Technology Program as Needed (Block 6)
- Perform Cost/Benefit Analyses (Block 7)

Phase 3: Documentation

- Document the Plan (Block 8)
  - Existing Programs
  - New Programs
  - Needs and Opportunities Not Addressed

Phase 4: Review and Approval

- Management Review and Approval (Block 9)
- Feedback

LINKING TECHNOLOGY TO BUSINESS GOALS

to procrastinators to develop and document their business strategies in a timely fashion.

Block 2: Assessing the Company's Technologies.
The information obtained in the assessment of the business strategies is market oriented and tends to be short-term in it focus. It reflects only part, albeit an important part, of the company's technology needs.

> **The R&D functions analysis of business strategies should be done at the same time the SBU, division, and corporate managers are developing their strategies.**

Block 2 addresses a different and usually longer-term set of technological needs and opportunities. It contains a review of the technologies underlying each of the company's businesses: their maturity, their relation to state-of-the-art technology, and how they compare with the competition's technology. This assessment is very important to the company and its businesses and most certainly falls within the scope of R&D's responsibilities and capabilities. It reveals technological needs and opportunities that must be addressed for the company to avoid becoming technologically behind.

Block 3: Assessing Related Technologies.
To complete block 3 of the flow diagram R&D must look beyond the technologies the company is using and must assess related technological developments, usually of a more basic scientific nature, that could indirectly threaten or provide opportunities to the company's businesses. (An example is the potential impact of high-temperature ceramic development on companies that manufacture automobile engines and engine components.) Because industrial R&D practitioners tend to focus only on the company's existing technologies, they must be assigned to examine related technologies. In the company for which the CTP system was designed, the senior scientist had to specify the technologies that the company should be concerned with and ensure that R&D personnel were assigned to monitor the evolution of these technologies.

Block 4: Listing All Technological Needs and Opportunities.
This final step in the technology assessment phase involves documenting all the previously identified needs and opportunities. The reasoning behind their identification must be made explicit, and they must be organized in whatever way is most appropriate for proceeding to the program development phase.

Phase 2: Program Development
This phase comprises blocks 5, 6, and 7 in Exhibit 1.

Block 5: Comparing Identified Needs and Opportunities with the Current Technology Program.
In block 5 the needs and opportunities identified in the technology assessment phase are compared with the slate of projects that R&D is currently working on or planning, to determine what changes, if any, need to be made to the current plan.

Block 6: Revise the Technology Program as Needed. Block 6 merely stipulates that the existing technology program be amended in accordance with the findings in block 5. If R&D has been performing strategic technology planning, most of the changes will consist of updatings (i.e., changes resulting from modifications in business strategies or from new technological insights). If strategic technology planning is being introduced, the R&D organization will probably discover that it is handling things it should not, rather than working on matters it should be addressing. In the company for which the author developed the CTP system, for example, management's first year with the system led it to cancel 20% of the projects that R&D had been working on and to introduce a couple of major programs that should have been started several years earlier.

Block 7: Perform Cost/Benefit Analyses. The cost/benefit analyses indicated in this block are done for every project in the revised list and then are aggregated at the SBU, business department, division, and other appropriate organizational levels. Costs are estimated for R&D expenses and capital expenditures for each year of the project's expected duration, for the capital necessary to implement (i.e., plant and equipment) if the R&D is successful, and for any unusual marketing or sales expenses required to introduce the new product to the marketplace. R&D must work closely with engineering, manufacturing, and marketing to develop these estimates. It must also do some reasonably detailed project planning to forecast projects' durations and costs.

Benefit estimates are stated in two ways: in dollars of additional sales (or sales losses avoided) for the first year of commercialization and four years beyond, and in dollars of gross earnings resulting from these sales. For cost reduction and productivity improvement projects, benefits are stated in dollars of additional gross earnings. In determining these estimates, R&D must work very closely with non-R&D organizations to ensure they agree on them.

Although the company for which the CTP system was developed normally carries individual business results to the level of net earnings, it decided to use gross earnings for estimating the benefits of successful R&D projects because that is the level of earnings on which R&D has the most direct impact. In addition, all estimates of benefits from R&D projects must be made on the assumption that the R&D and commercialization phases of the projects will be 100% successful. This rule has two significant advantages. One, it gives management a feel for a project's potential, an important parameter to consider in deciding whether or not to invest in that project. Second, it avoids the sticky issue of project success probability, about which everyone has a different estimate and which varies as work on a project progresses. When 100% success is clearly assumed in the plan, everyone can make his or her estimate of what the benefits will be. However, though the stated benefit dollars are based on 100% success, the plan document does also indicate whether the project's probability of success is a high, medium, or low, so decision makers have some prediction to work from.

Finally, all cost and benefit projections are clearly stated in this year's dollars (i.e., constant dollars) in the plan. Doing so avoids nonproductive disagreements over estimating the inflation rates of coming years.

This analysis of costs and benefits, though a lot of work, is an absolutely essential part of responsible technology planning for R&D managers. Once the tedious process has been checked thoroughly, however, subsequent analysis of costs and benefits is easier because much of it involves merely updating and because the R&D staff and management will have learned how to perform the process.

Although cost/benefit analyses are a lot of work, they are an absolutely essential part of responsible technology planning.

Phase 3: Program Documentation

This phase is represented by block 8 in Exhibit 1. After preliminary documentation is developed during the program development phase, the plan must be prepared in its final—or nearly final—form. This is the step indicated in block 8.

The plan consists of a nested system of plan documents that contain three types of information and recommendations (including the cost and benefit estimates of each) to the company's business managers:

- Existing programs that R&D proposes to continue and the reasons for so proposing.
- New programs that R&D plans to undertake and why it plans to.
- Technological needs and opportunities that have been identified but are not being addressed and the reasons for their not being addressed.

The last item is important for several reasons and should not be omitted from the plan. If unaddressed technological needs and opportunities are documented, they will not be forgotten. In addition, calling special attention to these issues, which usually have not been addressed because of management decisions, underlines their importance and gives management an opportunity to reconsider them. Every edition of the plan will include these unaddressed items until management decides either to work on or drop them.

When the CTP system was first introduced, R&D managers were reluctant to document their plans sufficiently. This was not acceptable. Clear and complete documentation is a sine qua non of effective planning for the following reasons:

- If a plan is not written down, it is not thought out.
- If it is written down, it is available for future reference. It will not get lost.

- If it is written down clearly, it will be understood by those who will be affected by the plan.
- The plan need be prepared only once. It can be reproduced, and therefore communicated uniformly, any number of times.
- Last but not least, once the plan is written down, preparing it each year is a simple matter of updating it. If it is not written down, it must be recreated anew each time it is articulated.

For these reasons, complete and accurate documentation is necessary in the CTP system. Detailed instructions on plan content and format are provided in a CTP manual that the author prepared as part of the system. In addition to helping in the documentation process, a standardized format is easy to understand, and users can quickly grasp where to look for items of information. An outline of this standardized format is present in a later section of this chapter.

As anticipated, writing down the plan the first time was difficult for the R&D managers in the company for which the system was designed, but they did it. In the subsequent years, the system remained in operation, making the task easier because most of the writing consisted merely of updating.

Phase 4: Management Review and Approval. Block 9 of Exhibit 1 illustrates the final review and approval stage that immediately precedes publication of the plan. The final review should be performed by all managers affected by the plan, including various levels of R&D and operating management and appropriate members of senior management. It can be performed jointly with R&D and operating managers, or the managers can do it independently. If the R&D and operating managers have closely coordinated their efforts (as they were supposed to), the final review will be a quick check of what has been previously agreed on. If the required coordination has not occurred, the final review will be a disaster.

THE PLANNING CYCLE

The CTP system requires that the several technology plans constituting the corporate technology plan be published yearly. Publishing it more frequently is usually not necessary, but publishing it less frequently leads to delays in keeping the company's technology program up to date. It is also highly desirable, but not absolutely necessary, that the technology-planning cycle be timed to coincide with the business strategy planning cycle so business planners and technology planners can work together.

Although technology plans are published once a year, the planning process must be continuous. R&D programs and projects are added, deleted, and modified during the course of the year as the needs of the businesses and the company change. However, unless a technology plan must be redesigned, the changes that take place during the year can usually be included in the next annual corporate technology plan documents.

A NESTED SYSTEM OF PLANS

A company that does not require R&D to undertake corporate technology planning is not getting what it should from its R&D investments.

The corporate technology plan is really a system of unit-level technology plans that are combined to reflect the various operating levels of the company (see Exhibit 2).

The basic building blocks of the system are the business unit technology plans (BUTPs) and the general technology program plans (GTPPs). They are prepared by the appropriate R&D departments. The BUTPs set the technological strategy and program for SBUs. One BUTP is prepared for each SBU in the company. GTPPs cover single projects or groups of projects that involve technology and support more than one SBU or that involve work in areas not related to existing business units.

These plans can be aggregated into plan documents corresponding to various organizational or technological groupings in the company. Exhibit 2 shows aggregations at the R&D department and R&D division levels, at the operating division and operating group levels, and at the corporate level. Fewer or more aggregations are possible, depending on the needs of the company.

In addition to producing BUTPs and GTPPs, each R&D department must prepare its own R&D department plan containing all the BUTPs and GTPPs for which it is responsible. Higher-level plans, such as the R&D division plan and the operating division and group plans, should be compiled by the director of technology planning or the staff of the senior R&D manager. Although the major part of these higher-level plans consist of BUTPs and GTPPs, they should also include summaries and consolidations to help plan recipients better understand the plans' contents.

In the author's company, the CTP system plan covered five years to conform to the planning horizon used for strategic business planning and to facilitate cost and benefit calculations. This conformity between business strategic and technology planning is recommended, whatever the span of years covered in the business strategic plans. It is also very important that the technology plans include all technology programs, regardless of their projected duration. Copies of the completed plans should be sent to all individuals and groups with a direct interest in them.

The BUTP

The BUTPs are the principal components of the CTP system of plans. One BUTP is prepared for each of the company's SBUs. The BUTP brings together, for the business unit it covers, all the information gathered and processed in the technology assessment and program development phases of the planning

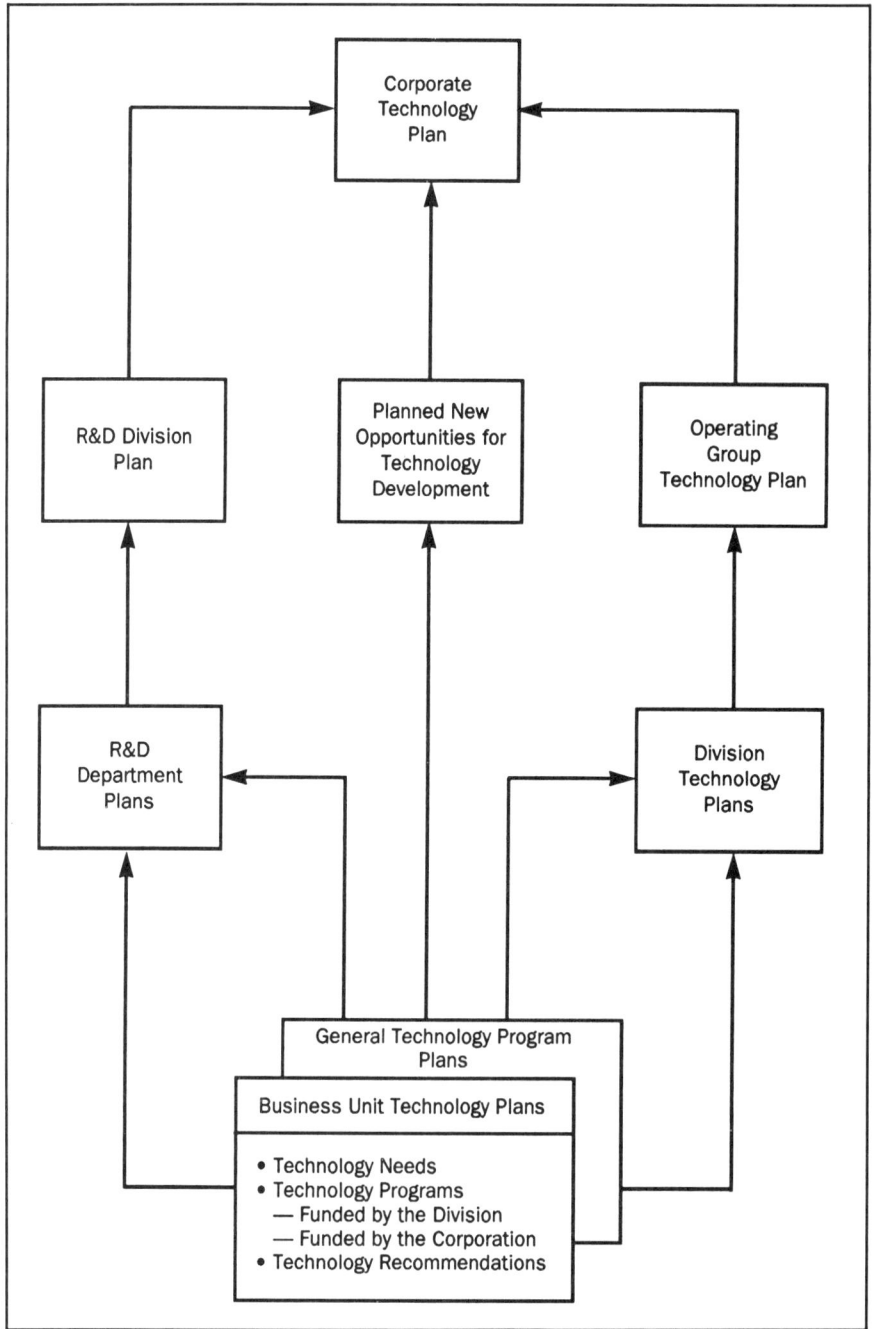

Exhibit 2
The Corporate Technology Planning System of Plans

```
                          ┌──────────────┐
          ┌──────────────→│  Corporate   │←──────────────┐
          │               │  Technology  │               │
          │               │     Plan     │               │
          │               └──────────────┘               │
          │                      ↑                       │
          │                      │                       │
    ┌──────────┐         ┌──────────────┐         ┌──────────────┐
    │   R&D    │         │ Planned New  │         │  Operating   │
    │ Division │         │Opportunities │         │    Group     │
    │   Plan   │         │ for Technology│        │  Technology  │
    │          │         │ Development  │         │     Plan     │
    └──────────┘         └──────────────┘         └──────────────┘
          ↑                      ↑                       ↑
          │                      │                       │
    ┌──────────┐                                   ┌──────────────┐
    │   R&D    │                                   │   Division   │
    │Department│←─────────────────────────────────→│  Technology  │
    │  Plans   │                                   │    Plans     │
    └──────────┘                                   └──────────────┘
          ↑                      ↑                       ↑
          │                      │                       │
          │   ┌──────────────────────────────────────┐   │
          │   │   General Technology Program Plans   │   │
          │   ├──────────────────────────────────────┤   │
          │   │   Business Unit Technology Plans     │   │
          │   │                                      │   │
          └───│  • Technology Needs                  │───┘
              │  • Technology Programs               │
              │    — Funded by the Division          │
              │    — Funded by the Corporation       │
              │  • Technology Recommendations        │
              └──────────────────────────────────────┘
```

process, and it identifies what is to be done and why, by whom, for whom, by when, at what cost, and with what benefits. Because it is such a critical document, its content and rationale are discussed in detail in the following sections. Exhibit 3 outlines the content of the BUTP document.

Section A: Executive Summary. This is a brief description of the plan's content.

Section B: Characteristics of the Business. To inform plan readers and ensure that its R&D preparers understand the businesses they are supporting, these preparers must describe these businesses' salient characteristics in this section of the plan.

The second of these two purposes deserves comment because it is very important. R&D people, oriented primarily to technology, are notoriously ignorant of the businesses their technology supports. This puts them at a significant disadvantage when discussing their R&D programs with business managers. When the CTP system, which requires close and frequent coordination between business and R&D managers, was designed, the R&D staff had to learn about the businesses to describe the key characteristics of those businesses in their technology plans. Their doing so contributed significantly to the effectiveness of the CTP system.

Exhibit 3
Outline of the BUTP

A. Executive summary

B. Characteristics of the business
   1. The nature of the business
   2. The nature of the products
   3. The scope of the business
   4. Major competitors and their share of the market
   5. Patent and licensing status
   6. International aspects

C. Technology assessment
   1. General discussion
   2. The company's current technological strengths
   3. The company's technological weaknesses and threats to it
   4. The development potential of the company's current technology
   5. Achieving and maintaining technological leadership
   6. Technological opportunities

D. Business strategy and required technological support

E. The technology program
   1. Summary of required technological support
   2. Currently approved program versus required support
   3. Estimated benefits and costs
   4. Project lists

The six subsections of this part of the BUTP include the business's commodity or specialty; its major product lines, applications, and markets; its growth over the past five years and projected over the next five years in terms of the company's market share and its sales and gross earnings; its major competitors and their share; its patent and licensing status; and the nature of the company's non-U.S. operations in this business, if any.

Section C: Technology Assessment. This section outlines the business's technological strengths, weaknesses, problems, threats, and opportunities and compares them to those of its competitors. It also compares the business's technology with the state of the art. The purpose of this section is to identify what technological changes the company needs to make to strengthen or optimize its position. This assessment should be conducted for products, processes, and underlying technologies.

The technology assessment section has many subsections. The subsection on the company's current technological strengths identifies the strengths of the technologies underlying the business dealt with in the BUTP and explains how to maintain them. This requirement is included in the plan to guard against the very natural and common tendency of people to take for granted what is working well and thereby ignore possible threats to the company's well-being.

In the subsection on technological weaknesses and threats, R&D planners must identify current weaknesses and potential problems in the technologies supporting the business dealt with in the BUTP and must state what the company should do to solve or avoid them.

In the subsection on the development potential of the company's current technology, planners measure the gap between the current state of the company's technology in the business covered in the BUTP and the maximum potential of that technology. This analysis should be done for products as well as processes. It is important because it indicates whether technological efforts should be devoted to improving the business's current technology or to shifting to a new technology.

The subsection on achieving and maintaining technological leadership answers the following questions about the business for which the BUTP is prepared:

1. Which companies are the technological leaders in the business's industry?
2. What gives each of those companies its leadership position (e.g., product design, process, R&D, technical service, or number of patents)?
3. What technological programs could the company undertake to become a technological leader or to maintain its position as leader?

The purpose of the Technological Opportunities subsection is to guard against overlooking opportunities by focusing too narrowly on problems and threats. In this subsection, R&D planners identify technological options, which, if properly exploited, could give the company an advantage over its competitors or could lead to profitable additions or modifications to the business unit's products or processes.

> The R&D functions analysis of business strategies should be done at the same time the SBU, division, and corporate managers are developing their strategies.

Section D: Business Strategy and Required Technological Support. This section identifies the elements of the business unit strategy that have technological implications and summarizes the technological work needed to support that strategy. The main objective of this section is to record what the business manager and the R&D planner have agreed on to ensure that each functions on the same wavelength.

This section of the BUTP must be written even if the business strategy is not completely defined. Occasions arise in which business managers have not clearly or completely spelled out their business strategy, but they should proceed with their technology plan for the business. Because R&D support cannot wait for a clear business strategy to be developed, CTP system planners are instructed to outline to business managers the assumptions the planners are making about the business strategy and the technological work they are planning to do in support of that strategy. This strategy usually elicits some clarification from the business managers.

Section E: The Technology Program. The purpose of this section is to indicate what will and will not be done about the technological requirements identified in earlier sections of the plan and to provide broad cost/benefit estimates.

In the summary of required technological support subsection, R&D planners must list all major areas of required work. It is not enough to include only those that have been approved.

The subsection that compares the currently approved program with required support is a very important part of the BUTP. It requires the R&D planners to identify what part, if any, of the recommended technology program has not been approved and to state why these parts have not been approved. This process reminds both business managers and R&D planners that the business has important needs for technological support that (probably for very good reasons) are not being addressed. Including this information helps to ensure that it will not be forgotten. These unresolved issues should be repeated in each edition of the CTP until a clear decision is made regarding their disposition.

The estimated benefits and costs subsection summarizes the cost/benefit picture for the total R&D program outlined in the BUTP, including programs proposed but not yet approved. It is the sum of the cost and benefit estimates for each of the projects identified in the next subsection, the project lists. Benefits are stated both in dollars of additional sales and sales losses avoided and in additional gross earnings. The gross earnings are derived from the additional sales and the avoided sales losses, as well as from cost reductions and productivity improvement programs. Benefit estimates should be obtained from and agreed to by the appropriate business managers. Estimates are made for development expenses and capital needs, for technical service, and for the capital required to implement R&D program results. These estimates must be obtained from manufacturing, marketing, and any other non-R&D organization that will be involved in the implementation of the R&D results.

In preparing this subsection of the BUTP, R&D planners must remind plan readers and users that estimates are made in current dollars and that benefits are based on the assumption that all parts of the program, from conceptualization through commercialization, will be 100% successful.

In the final pages of the BUTP come the project lists, a compilation of all the projects that comprise the R&D program supporting the BUTP business unit. For each project listed, this subsection includes: the project name and number, the type of project (e.g., new product, existing product modification, new process, existing process modification), the date the project was approved, its estimated completion date, the estimated cost to complete, its estimated benefits, the estimated capital to implement, the probability of technical success, the technological needs addressed, and whether the project supports overseas as well as domestic business. Exhibit 4 is the project list form used in the author's company.

Tools for the Planner

Clearly this planning system not only involves a considerable amount of work, particularly in the first year of its introduction, but also requires R&D managers and engineers to engage in data collection, analysis, and presentation, which are activities with which most have had little or no experience. To ease these tasks, the author developed some aids and guidelines for the planners to use in the technology assessment and program development phases of the planning process. He also prepared a CTP manual that explains the rationale underlying the CTP process and instructs how to prepare and distribute the written plans. Two of the most important of these aids are a questionnaire that R&D planners use in discussing with business managers the technological requirements of their business strategies and a checklist for the R&D people to use in assessing the state of technologies under review.

MAKING THE SYSTEM WORK

To be effective, a corporate strategic technology planning system must have strong technical and nontechnical leadership. R&D managers and staff should operate according to the following principles:

Exhibit 4
The CTP Project List

19___-19___ CORPORATE TECHNOLOGY PLAN (R&D): _____ PROGRAM

PROJECT NAME	TYPE	Date Approved (month/year)	Estimated Completion Date (month/year)	Estimated Cost to Complete	19___-19___ Benefits			Estimated Capital to Implement	Probable Technical Success (High, Medium, Low)	Need(s) Addressed	I T L
					Sales	Gross Earnings on Sales	Cost Reduction				

STRATEGIC TECHNOLOGY PLANNING

- A conviction that R&D can best support the company by being proactive rather than reactive.
- A recognition that it is essential to see the world through the eyes of the business managers and to talk their language.
- A strong belief in the importance of planning and a recognition of the time and effort necessary to perform it.
- A recognition of R&D's responsibility in getting the business side of the company involved in technology planning.

Business managers must:

- Develop clear business strategies.
- Take the time to interact with R&D.
- Accept R&D as an essential and equal partner in the corporate enterprise.
- Keep R&D from independently developing estimates of business benefits from R&D programs by either developing these estimates themselves or working jointly with R&D to formulate them. This is the best way to arrive at benefit estimates that will be acceptable to both sides.

CONCLUSION

Systems such as the one described in this chapter always work better on paper than in the real world. The CTP system was no exception. Despite considerable success in creating the conditions required to make the system work, the CTP system worked better in some R&D departments and business groups than in others. Nonetheless, all R&D departments, all business groups, and the company as a whole derived significant benefits from working with the system:

- R&D's effectiveness and efficiency improved markedly because it was working on the things the company really needed and because it had a plan against which to measure and adjust its progress.
- Both R&D and the operating groups could manage R&D's support as an integrated set of programs clearly linked to the company's broader business objectives rather than as a heterogeneous collection of individual, loosely related projects.
- R&D rose to the challenge of approaching the company's needs proactively and thus effectively applied its unique talents to the planning and implementation of the company's strategies and operations.
- In their negotiations and deliberations with operating business managers, R&D managers became far more able to state their case clearly and forcefully because they were presenting issues that the planning system had forced them to think through thoroughly.
- The rest of the company, which had tended to denigrate R&D before the introduction of the CTP system, gained a significant understanding of and respect for R&D and began dealing with them as equals.

- As a result, the R&D group's morale and productivity improved.
- Because the corporation made more effective use of its R&D resources, its bottom line improved.

Although cost/benefit analyses are a lot of work, they are an absolutely essential part of responsible technology planning.

ONE CORPORATION'S STRATEGY AND TECHNOLOGY ALIGNMENT: A CASE STUDY

Once the dominant player in the new computer equipment industry, Corporation A lost ground to competitors as the industry matured. This chapter describes how Corporation A used strategic R&D, reorganized, and realigned its products to recover much of its prestige and business in a more demanding and crowded marketplace.

John Peterson

In the early 1990's, Corporation A faced a very different world from the one it had known. Its R&D laboratories had been the primary designer for the computer equipment and business software industries, particularly in the design, development, and manufacture of mainframe computer equipment, applications software, and related technology. It was considered best at optimizing complex computer systems and establishing data communications networks.

Most other industry players had been little more than distribution channels for Corporation A. They sourced most of their equipment and their applications software from Corporation A manufacturing. Then the courts stepped in and changed everything. A consent decree was issued, turning these channels into independent systems providers (ISPs) specializing in computer systems engineering and design, facilities management, custom software, time sharing, and systems integration. After the settlement, each ISP and its affiliates independently made the same business decision because they could no longer afford to depend on a single source for their computer equipment. They introduced computers and software manufactured by others into ISP inventories and customer solutions.

Right after the settlement, Corporation A manufacturing satisfied significant demand for computer equipment and software as many customers established in-house systems groups to manage their data processing solutions. However, after that initial surge in demand, the multiple-sourcing strategies of the ISPs (and other consequences possible of the settlement) left Corporation A manufacturing facing possible major losses in mainframe market shares. This predicament heralded the beginning of a new, fast-paced and continuously changing competitive environment. Corporation A was forced to change.

JOHN PETERSON is manager of technology strategy for a Fortune 500 corporation.

The most immediate result was the need to redesign the mainframe equipment side of the business. Executives had to adjust the strategic objectives and differing cultures of the corporate, manufacturing, and laboratory organizations into new ways of operating, including aligning mainframe computer technology and the strategic objectives of the businesses with the needs of customers and ISPs around the globe. This had to be done in an environment where most of the business remained dependent on the ISPs. The ISPs, in turn, had begun funding computer technology research, architecture, and standards activities through an industry association laboratory, which did not necessarily agree with the directions or the pace of solutions advocated by the corporate organizations.

ORGANIZATIONAL STRUCTURE

Shortly after the settlement, Corporation A was organized into two operating business groups, applications and manufacturing, and a third group that functioned as a corporate resource, the R&D laboratories. The business groups were further divided into major specialty decisions. The applications group included the information services division, which sold mainframe computer, time-sharing, and disaster recovery services; the systems integration division, which sold custom electronic data processing hardware and software systems; and the software division, which wrote and sold generic business applications software. The manufacturing group consisted of the network media division, which manufactured and sold cable and fiber distribution systems; the mainframe computer division, which designed, assembled and sold mainframe computers; the minicomputer division, which designed, assembled, and sold mini-computers, the computer peripherals division, which designed, assembled and sold workstations, terminals, data communications equipment, personal computers and associated products; and the components and electronics systems division, which manufactured microelectronic devices including processors, memories, integrated circuits, and components. The manufacturing group had its own direct sales and marketing teams.

Even before the settlement, business units were associated with a particular product line or family. After the settlement, this "product house" approach was institutionalized. Each product house controlled its products, including channels of distribution and pricing. Each was evaluated on contribution to profit (CTP), which represented the difference between revenue and controllable costs and expenses.

To maximize CTP and reach the largest number of customers, the division's product managers had first tried to solve the needs of different customers with single product solutions. They tried to address not only the markets and customers for which they were responsible but also the bluebirds—unanticipated product applications. (Eventually they learned that although a technology may be adaptable to the needs of multiple customers, a single product rarely fills all the needs of a large heterogeneous customer group.) Sales awards and compensation were tied to the revenues booked through a particular product house.

> Corporation A manufacturing could have set the direction for corporate strategy, but this potential was not widely perceived.

CORPORATION A's MANUFACTURING GROUP

Mainframe computer systems tend to be long-lived. A mainframe can have more than a half-million components and assemblies. Mainframes tend to be major investments with costs, long depreciation cycles, and installed lives of 8 to 10 years. To be successful, Corporation A manufacturing had to look even further into the customers' futures than the other businesses groups did because its development cycle was long and its equipment was expensive and depreciated slowly. Manufacturing needed great market and technical expertise to ensure compatibility and interoperability across and between a customer's multiple data centers.

Corporation A manufacturing could have set the direction for corporate strategy, but this potential was not widely perceived within the corporation. Information services, with its enormous revenue streams, large customer base, and large base of applications for spreading costs over seemed to promise of more efficient and less capital-intensive source of consistent CTP. As a result, applications, not manufacturing, was perceived to be the jewel in Corporation A's crown.

Notwithstanding, the manufacturing product houses generated billions of dollars. In the early 1990s, the mainframe group was the second largest contributor to Corporation A, generating more than $3.5 billion in sales. About 70% of those sales came from external customers, mostly from large multinational corporations. Mainframes's share of corporate revenues was about 23%. It used slightly more than 28% of the capital budget and contributed slightly less than 20% of profits. Its share of cash flow was 30%. Manufacturing also consumed 35 to 40% of the processors, memories, and other components manufactured by one of its own major subdivisions, components, and electronics systems. Manufacturing also funded more than half the basic research work in the corporate labs.

Products

In the early 1990s, the underlying processor and device-level technologies for mainframes were evolving on an 18-month lifecycle. Model 8088 mainframe, in spite of its age, was still a significant strategic factor because of its large customer base. It was also rapidly becoming the primary target of the ISPs replacement and upgrade programs. The 8088 was rapidly being replaced by competitors' mainframes and local area networks. This happened as major customers began to force their suppliers to become full-line providers.

Customers made such large purchases of all types of computing and data networking equipment at one time that major suppliers could not afford *not* to bid on this business and the future replacement opportunities it promised. Corporate structure did not align well with the varying needs of major multinational customers.

Manufacturing's product houses had once had significant market shares in the mainframe world. Rapid growth of distributed architectures and competition value-added resellers and ISPs had significantly reduced any advantage its mainframes may have once provided. Market entry of local area networks and personal computers had trailed that of the competition by more than three years. In the mid-1980s, Corporation A's personal computers had been a major force, but annual budgeting had become a zero-sum game and the applications group, which had very high margins, was too attractive to starve.

On the manufacturing side, commitment of resources, first to the applications group and then to introducing the 8088, caused investments in the PC business to wither. Resources for the PC were repeatedly redirected to develop even more features on the 8088 mainframe. The 8088 mainframe was a masterpiece, best in class, and far superior to competitors' products. It included flexible functionality, advanced big-computer architecture, and modular hardware subsystems that allowed easy upgrading as supporting component and device technology advanced.

By the early 1990s, the 8088 was the best in the world, based on lifecycle cost, quality, reliability, and functionality. A handful of strategic customers were willing to pay a small incremental pricing premium for these attributes. More and more frequently, however, competitors were offering distributed architectures and networked computing solutions, not look-alike mainframes.

Across the industry, development costs for large mainframes were so high that suppliers had been forced to seek overseas sales to hedge against domestic customers' pressure tactics. The industry expected high overseas sales that manufacturers could spread fixed costs and margins over, reducing margin requirements on individual opportunities. Profits not realized directly would be realized indirectly in increased sales. Corporation A manufacturing needed this badly at first because it had developed mainframes and minicomputers independently in different product houses, anticipating federal requirements for separate marketing and sales of computer products and services resulting from the computer industry monopoly inquiry. In addition, the corporation anticipated that the mandates would demand separate manufacturing of mainframes and minis because of the relatively strong shares it had in both markets. Most of its strategic competitors evolved their mainframes from their minis, realizing efficiencies based on common technology and software.

Impact of Technology

Technological evolution contributed heavily to the blurring of many distinctions between computer applications and business practices. Historically, intelligence resided in the system, particularly in the mainframe and some special-purpose peripherals and terminals. Before the settlement, Corporation A had set de facto industry standards. It also managed the timing of new functionality and the introduction of new generations of mainframes, minis, data communications equipment, and standard computer interfaces. This virtual lock, many observers would later argue, was the root cause of the government's inquiry into the computer industry monopoly. After the settlement, however, entry barriers disappeared. New competitors, whose businesses were strongly committed to niche products on which their survival depended, suddenly became industry players.

> Any advantage Corporation A enjoyed based on mainframe attributes or data communications and physical computer interfaces eroded rapidly.

Any advantage Corporation A enjoyed based on mainframe attributes or data communications and physical computer interfaces eroded rapidly. Many factors promoted the erosion, including the establishment of an industry-funded research and development laboratory that soon asserted leadership in setting industry directions and standards; the implementation of ISP and end-customer multiple-supplier sourcing policies; and the distribution of computing processing intelligence to the most efficient, effective, flexible, and functional locations closest to the end user. Network computing allowed smaller computers to be configured to satisfy a customer's business need. In addition, workstations, PCs and other intelligent terminals emerged that provided customers cross-elasticity with the mainframe and minicomputers. (Products that are cross-elastic satisfy customer needs in different ways. Cross-elasticities affect the relationship between the price of a solution and the demand for alternative solutions. Pricing ultimately becomes the key differentiator).

Impact of Technical Sales Tactics

The rules of the sales game changed also. Corporation A manufacturing had long held a major share of the world's largest mainframe computing equipment market (i.e., the U.S.) by marketing the reliability and quality of its technology and support. Corporation A convinced customers' techies that not only was its solution the best technical fit but also the least risky because it included tremendous technical support. Of all the industry players, only Corporation A was willing to throw money at a problem until it was resolved.

New players in the equipment markets were unwilling to try to match Corporation A's technology, its technical expertise, or its support capabilities. New players marketed their products not to engineers but by positioning them with the ISPs and the executives of major prospects. They described their equipment as plug-compatible with the customer's technology and stressed not its technical attributes or lifecycle cost, but its low initial cost. They argued that a low first cost would mean fewer confrontations with bean counters about gold-plated technology. After all, the bean counters did not appreciate the technology, only the costs of the services and their impact on the business. They were also quick to point out that Corporation A was now beginning to charge for postinstallation support. By 1990, due to changes in customer buying behavior, Corporation A's position on the cost curve, and its value proposition, Corporation A was no longer the market leader.

The Problems

Corporation A manufacturing products were not gold plated, but they were carefully designed and engineered. Manufacturing had used custom integrated circuits and processors to ensure unequaled reliability and performance. As a result, the equipment was expensive. Additional burdens were a business value chain and corporate overheads higher than those of strategic competitors. Within Corporate A manufacturing, some managers expected that, in spite of their limited success to date, foreign mainframe competitors would exert extreme price pressures in North American markets. Such potential strategic competitors had deep pockets and would continue to try to buy presence in the US while continuing to recover high margins in their protected home markets. Corporate A manufacturing was trying to expand globally, but lagged foreign strategic competitors in nondomestic installed base, sales and support infrastructures, and in global and regional sales experience and savvy.

Corporate A manufacturing faced a true strategic dilemma. It was under competitive attack in domestic markets while trying to reallocate its assets to strengthen its international and global presence. Additionally, many potential customers perceived that its:

- Costs were high.
- Marketing and sales processed were self-limiting.
- New products were introduced too infrequently.
- Products were overdesigned and too costly.
- Understanding of global markets was immature.
- The core business was too focused on mainframe production when the industry was beginning to evolve to distributed solutions.

The Plan

By late 1990, it was clearly necessary to reposition Corporation A's manufacturing and business units. Executives decided to focus on mainframe computing as the core business improving mainframe efficiencies and pricing. After agonizing over the human costs of the alternatives, executives devised a plan of action with an eye on the current and emerging needs of the strategic customer sets. They again reinforced the product house/business unit concept to gain further control over costs and integrated or eliminated traditional centralized support activities (e.g., market planning) into the business units. The company charged business unit presidents with personally taking more responsibility for controllable costs, forcing major adjustments.

> **Corporation A was under competitive attack in domestic markets while trying to reallocate its assets to strengthen its international and global presence.**

Corporation A integrated laboratories supporting particular business units into the product houses and made them the responsibility of business unit presidents. Corporation A established priorities in a competitive market planning process and targeted opportunities based on their market priority and projected marginal contribution. Only those contributing to predetermined geographic areas were candidates for aggressive pricing action. Priorities for quality, country and ISP-specific customization, and customer risk sharing were established and constantly enforced.

Manufacturing quickly learned that country-and sometimes customer-specific customization of computer and data networking solutions were often necessary and expensive. As a cost-control device, it decided that the only acceptable customization was that necessary to meet the minimum standards for a country or that for which the customer would help pay. Cost reductions received the highest technology priorities, and the corporation scrutinized and generally limited it to internally generated business unit funding. The corporation stretched out and reduced investments in strategic deliverables and in the next generation of mainframes and minis. It established a business growth strategy and assigned responsibility for specific segments to members of the management team. Dominant activities became management of market-specific product and development portfolios, and sales and marketing emphasis shifted from selling product to addressing customer business needs with integrated turnkey solutions. Where business units failed to realize expected efficiencies, Corporation A made additional force adjustments.

1992 and 1993 were record years for manufacturing. A few domestic ISPs continued to gain independence, but the tide appeared to be turning. Although some competitors continued to make inroads with the major multinational customers, overall, Corporation A seemed to be growing its North American equipment markets faster than the industry was. This growth was aided by severe quality problems in the North American market leader's mainframe operating systems.

Corporation A manufacturing achieved significant gains for both major domestic and global computer systems with a shift in bidding tactics. Corporation A manufacturing started responding to big bids with two quotes, the first an exact response to the customer's specifications with pricing comparable to its major competitors, the second addressing the customer's more significant needs. It proposed turnkey solutions that allowed the customer to realize pricing that reflected scale efficiencies well beyond those in the initial bid. Corporation A leveraged competitive advantage from its flexible technology architectures, integrated product solutions (not necessarily based on the mainframe), and increased willingness to share business risks with the customer. Corporation A manufacturing had unilaterally initiated a change in industry-wide sales and marketing practices.

Corporation A's manufacturing business units emerged from the repositioning with better control and understanding of costs. They gained experience and skill at making tough decisions and linking incremental costs to those decisions. They also developed an improved understanding and appreciation of the human costs and contributions to the business. Unfortunately, the experience was double-edged. The business units and their managers emerged with commitment to process and profitability—sometimes at the expense of other divisions. By the end of the early 1990s, this began to emerge as a potential problem. (At the time of this writing, Corporation A manufacturing is restructuring to evolve the business units and divisions to a solution-type structure. Profitability responsibility will be moved out of the product houses to regional marketing and sales executives. The product houses will combine to support specific types of networked computing solutions with turnkey offers, which will be priced by the sales executives. The product/solution teams will be funded by the regional marketing and sales teams.)

Still, emphasis on the four points—customer, quality, cost, and reliability—was never relaxed. At the business unit level, all activities and teams focused first on the needs of the strategic customer set, then on a consistent set of business priorities and objectives.

The Mainframe Computer Business Unit

Corporation A manufacturing included the mainframe business unit, which designed and built two families of large mainframe computers; operating system and applications software; data communications equipment; and (to a lesser extent) workstations, data terminals, personal computers, printers, and other specialized equipment. Mainframe was comprised of three sub-business units: affiliates, which supplied and supported the other Corporation A divisions; international, which marketed to the non-North American markets; and domestic, which focused on the North American countries. Mainframe also included a technology-development division and a few very small activities, including strategic planning, financial results and reporting, and human resources. Internal mainframe marketing activities tended to focus on product marketing and sales support and did not necessarily include broader, longer-term market planning and development.

Mainframe's experience mirrored manufacturing's. Its product management structures dominated. In some sub-business units, the organization focused on customer applications; in others, a traditional portfolio of products were found. Each sub-business unit controlled its products or applications, including the distribution and pricing channels. The technology division worked with emerging technologies to identify breakthrough technological developments that would bring significant cost benefit to the data systems of strategic customers and the ISPs. It also tried out new technologies with strategic customers to define applications and requirements for future products. This experiment provided the sub-business units with working models, not quite product prototypes. The sub-business unit development organizations then manufactured trial models to specific country specifications. Later it introduced them for sale.

> Corporation A integrated laboratories supporting particular business units into the product houses, making them the responsibility of business unit presidents.

Within mainframe, Corporation A evaluated each product and application manager based on contribution to profit (CTP). However, in the early 1990s, the revenue and contribution volumes of the two mainframe generations had been so significant that the rest of the portfolio's cumulative contribution was modest by comparison. To maintain revenue and CTP levels, only the two key products really mattered. Mainframe had been strategically vulnerable. Aggressive pricing from competitors in a few big bid situations would have damaged the unit's profitability and contribution. To make matters worse, this vulnerability was seen in markets in which sales cycles were long and all major competitors (including the deep-pocketed foreign companies) were routinely invited to respond to bids.

The traditional domestic ISP markets, where mainframe had begun to reassert itself, were essentially saturated and relatively slow to grow. Strategic customers sought to reduce their capital expenditures to invest instead in faster growth areas. In every market in which mainframe participated, competitors with deep pockets were desperately trying either to buy market presence or to defend their existing shares. Real sales growth opportunities were in the global markets beyond North America.

Strategy Definition and Ownership

One key contributor to mainframes's successful repositioning was the introduction and implementation of a strategy at the board level. Many board members were technical experts comfortable with addressing issues that could be reduced to binary packages. They were less comfortable in dealing with nebulous and unstructured long-term strategy. The issue was to get board members to take ownership of the "softer strategy stuff."

Fortunately for mainframe, the personal chemistry of the board was not only professional but positive. Members' individual agendas tended to focus on near-term business issues. Agendas were addressable under, not threatened by, the strategy. The strategic planning team also functioned as a part-time staff group. They "owned" the board's meeting agenda. They were also persistent. The chief executive was willing, supportive, and patient. She would not adopt a strategy without a real consensus of board members, who, once engaged, recognized that they had to generate and then enforce the business unit strategies. Doing so involved setting priorities and managing the rules of engagement for the sub-business units.

After some initial hesitation over the time pressures of near-term commitments to the business, board members recognized the importance of the longer-term strategies. They scheduled time in each board meeting to address strategy issues and achieved a consensus on business priorities. In addition, the individual members assumed responsibility for communicating strategies and rationale to the members of their teams (i.e., the sub-business units, the technology development division, and staff). The members of the board personally monitored and enforced strategies implementation and elements within their individual and collective spheres of influence.

Technology

Concurrent with the adoption of strategy as a standing item on the board agenda, Corporation A labs asked the chief scientists in each business unit to develop technology roadmaps to identify technology development requirements. The specifics of the technology roadmaps were undefined, apparently to allow the business units flexibility in approach and identification.

The intention had been to link mainframe to its product, development, and technology investment portfolios. However, the timing, emergence of the mainframe strategy and its elements, ongoing core competency definition activities, and the technology roadmapping requirement all contributed to improve the robust and usefulness of the product. The strategic objectives included the direction of growth, annual growth rates, and target dates for sub-business unit revenues, margins, the related economic value-added (a metric reflecting the cash generated compared to the cash used by the business unit and related to the average weighted cost of capital), and customer satisfaction attributes (the weighted perceptions of independent service providers on the value the business unit brings, based on profit-impact-of-market strategies comparable and satisfaction indices over time).

Looking Back From the Future

In 1992, the technology roadmapping team established scenarios that attempted to define what computing world might look like in a given end year. Because the end year was for internal planning purposes and could be arbitrary, the team picked a 20-year horizon, roughly two historical mainframe product generation lifecycles. The scenarios described the relative competitive advantage that computing could bring to customers; they are illustrated in Exhibit 1.

The team then identified the paths of least resistance leading to the identified end points. They paid particular attention to regional and customer differences. The final description included identification of major factors (i.e., technological, social, political, and cultural) that might influence this evolution. One view of the types of potential national and regional supporting

Exhibit 1
Evolution of Customer Needs

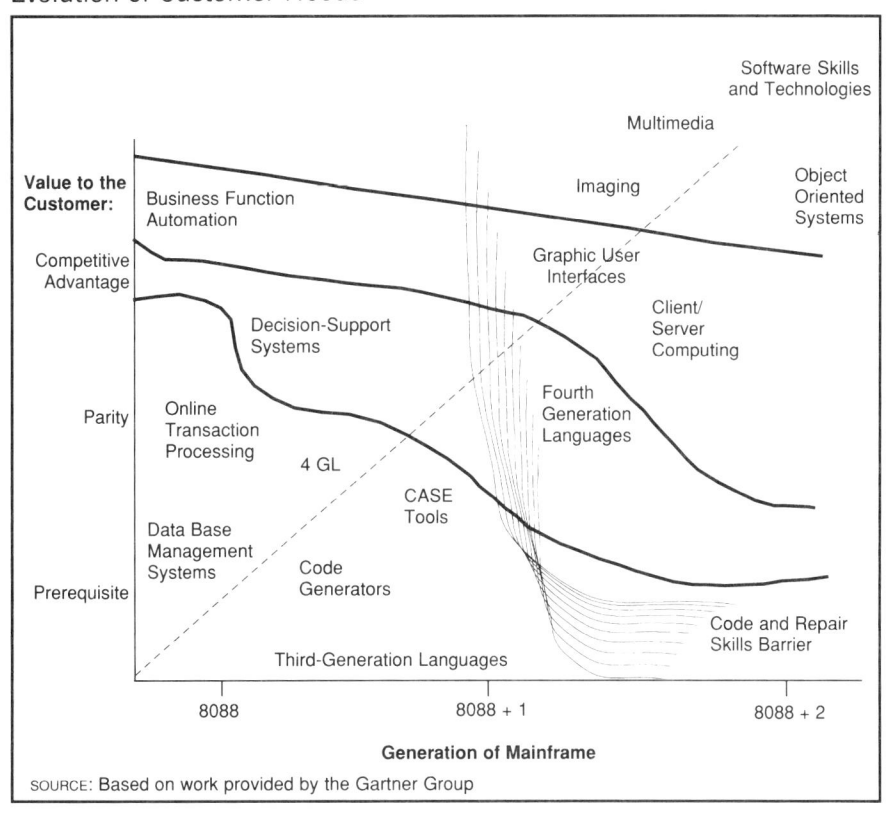

[Chart showing Value to the Customer (Competitive Advantage, Parity, Prerequisite) vs. Generation of Mainframe (8088, 8088 + 1, 8088 + 2), with labeled items: Business Function Automation, Multimedia, Imaging, Software Skills and Technologies, Object Oriented Systems, Graphic User Interfaces, Decision-Support Systems, Client/Server Computing, Online Transaction Processing, Fourth Generation Languages, 4 GL, CASE Tools, Data Base Management Systems, Code Generators, Third-Generation Languages, Code and Repair Skills Barrier]

SOURCE: Based on work provided by the Gartner Group

infrastructures is described in Exhibit 2. It predicts that mainframes will become increasingly important in organizations as the computers are used not only for transaction processing but also for automating business functions. This scenario also suggests that network throughput or bandwidth, not computer size, may be gating factors for computing solutions.

Looking Forward

The team then plotted all major mainframe products and applications forecasts based standard lifecycle curves. Where applicable, it identified and mapped life extensions and replacement products. The team flagged deviations from the standard models in the forecast curves and tried to explain them, based on planning considerations.

Because it takes several years to develop a new mainframe computer, replacement product engineering must often be at work several years before system decline actually starts. The team realized this and identified potential

Exhibit 2
The Evolution of Networked Computing Supporting Infrastructures

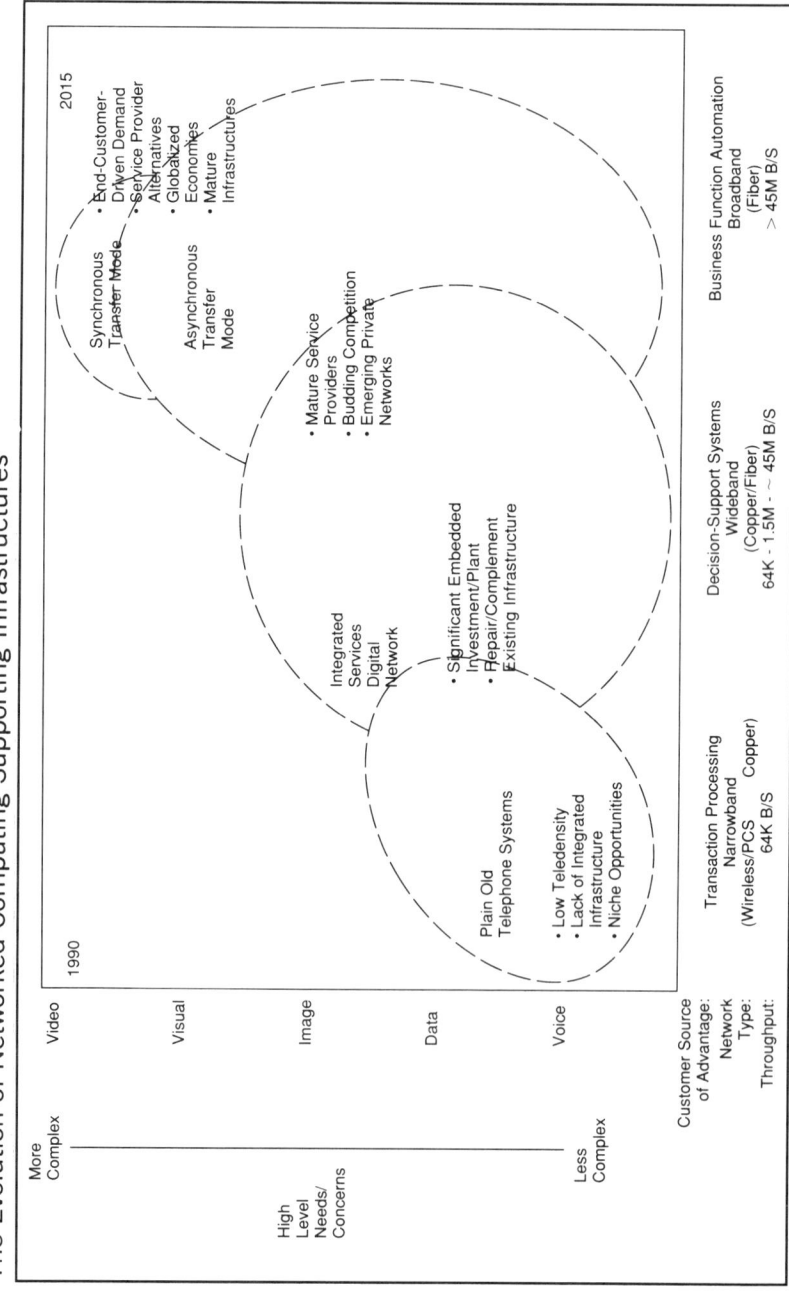

	1990			2015
Video				
Visual				Synchronous Transfer Mode
Image		Integrated Services Digital Network		• End-Customer-Driven Demand • Service Provider Alternatives • Globalized Economies • Mature Infrastructures
Data		• Significant Embedded Investment/Plant • Repair/Complement Existing Infrastructure	Asynchronous Transfer Mode	
Voice	Plain Old Telephone Systems		• Mature Service Providers • Budding Competition • Emerging Private Networks	
	• Low Teledensity • Lack of Integrated Infrastructure • Niche Opportunities			

Customer Source of Advantage:	Transaction Processing	Decision-Support Systems	Business Function Automation
Network Type:	Narrowband (Wireless/PCS Copper)	Wideband (Copper/Fiber)	Broadband (Fiber)
Throughput:	64K B/S	64K - 1.5M - ~ 45M B/S	> 45M B/S

More Complex ←——————————————→ Less Complex

High Level Needs/Concerns

LINKING TECHNOLOGY TO BUSINESS GOALS

technologies to support lifecycle extensions and to support the development of replacement products. Their efforts were taken as indications that Corporation A would allow mainframe to play a proactive role in introducing customer solutions.

> **One key contributor to mainframe's successful repositioning was the introduction and implementation of a strategy at the board level.**

Flagship and other critical emerging products were modeled at a high level. The team identified key common devices and components, and mapped expert forecasts of technology trends and the availability of key common devices and components. This was done on a common time line. It became evident that assumptions differed among product managers and within sub-business units. Because of the complexity of Corporation A's products and applications, product or portfolio managers had difficulty relating product evolution to component and device trends. If taken at face value, the analysis suggested that except in a handful of advanced development projects, current processor and device evolution plans would support planned mainframe product and application portfolios.

The device-level technologies required to support the mainframe data communications equipment and intelligent peripherals is summarized in Exhibit 3. Similar charts were developed for the other mainframe products and applications. Question marks (?) in Exhibit 3 indicate devices and time frames that will not be sufficiently developed to support planned mainframe products. Because it was not necessary to push the envelope of component and device development, investment could be limited to small improvements to existing technologies. This decision could be interpreted as a weak signal that cross-elastic solutions could displace most mainframe computers during the next flagship generation.

Unfortunately, market and technology signals are rarely that clear. Neither the end point nor the evolution scenario dovetailed very well with existing mainframe market and product plans. This did not surprise the team; it was surprised that the "easy" path was well beyond mainframe's plans and assessments. Recognizing that technology development is a strategic business investment, the team flagged some opportunities that helped accelerate some revenues. That, however, did little to address the gap between Corporation A's objectives and the product managers' current product and application plans.

Significant (and Other) Competencies

The gap between Corporation A's objectives and the current product portfolios elicited a knee jerk reaction. Talk quickly turned to "diversification" and "new business opportunities" as executives sought to reduce the risk associated with putting all of mainframe's eggs in one product basket. However, the nature of the business and the risk-averse corporate culture within mainframe discouraged the development of such opportunities. In spite of efforts to create

Exhibit 3
Summary of Selected Device-Level Technologies

Critical Devices/Features	1992 ----> 1995	1996 ----> 1998	1999 ----> 2001	2002 ----> 2005
Feature Size (Microns)	0.5μ	0.35μ	0.25μ	0.15μ?
Gates/Chip	300 K	500 K	2 M	5 M
DRAM b/Chip	16 M	64 M	256 M	IQ
SRAM b/Chip	4 M	16 M	64 M	256 M
Interconnect Levels	3	4-6	5	5-6
Max Power Dissipation	10	16	30	40
Power Supply (Wall)	6 V	3.3 V	2.2 V	2.2 V
I/O Count	500	750	1600	2000
Off Chip Clock (Hz)	50	100	175	250
PWB Layers	7	9	11	13
PWB Line Width	100	60	60	60
uP MIPS	100	250	1000	2000
FPGA Gates	5 K	20 K	50 K?	260 K?
FPGA Clk Speed	60	100	176	260
DSP MIPS/MFLOPS	60	260	>400	>1000?

Illustrative Data

SOURCE: Data Quest & Other Public Domain Data

new product ideas and to facilitate their adoption, most of the new business opportunities that were offered were merely attempts to repackage existing internal tools and subsystems for external sale. Such attempts, though a potential source of additional revenues and profits, could not compare to flagship mainframe computer sales. Nevertheless, based on the

Corporation A manufacturing could have set the direction for corporate strategy, but this potential was not widely perceived.

business interest of the day, funding for new business opportunities was constantly redirected. Even the lower-cost opportunities were abandoned. Projected revenues and contributions did not materialize. Only the new business opportunity review process survived.

To diversify, mainframe had to determine what its competencies were and how to leverage them. A technology value chain established a linear model. It was originally a three-by-three matrix with multiple subelements that plotted basic device-level technologies, (which are microelectronic-type components), systems-level skills and technologies (which represent human value-added), and products and systems that represent what the business units actually sell (see Exhibit 4).

At the recommendation of the strategic planning director, the roadmapping team expanded the technology value-chain model to include the mainframe strategic objectives, the strategy elements, the investment priorities and potential sources of technology. The technology roadmapping team then charted the technologies and skills required to support current and planned products and applications (see Exhibit 5). The team also identified relative industry positions in these areas, comparing mainframe to its strategic competitors.

These competencies were then compared with previous scenarios. Those required for support were flagged and compared to a list of current competencies and capabilities. Most current competencies that did not support the desired evolution were also flagged. They were identified as potential killer competencies, competencies that are so ingrained in a culture that they persist, even though they prevent introduction of new competencies and capabilities. Killer competencies thus promote obsolescence and vulnerability in new products and applications. Such competencies are identified in Exhibit 7 as those to be harvested. If, in the long term, the killer competencies survive, the viability and competitiveness of the business unit may not.

The output of the competency activity yielded a high-level assessment of competencies in three layers. The first layer included the technology value chain; the second layer included the business value chain; and the third layer included mainframe strategic partnering and relationship-building skills (for risk sharing and sourcing). The estimated values over time of the competencies as reflected in current plans were charted against both the core products and applications in the sub-business unit portfolios and the mandated growth rates. The result

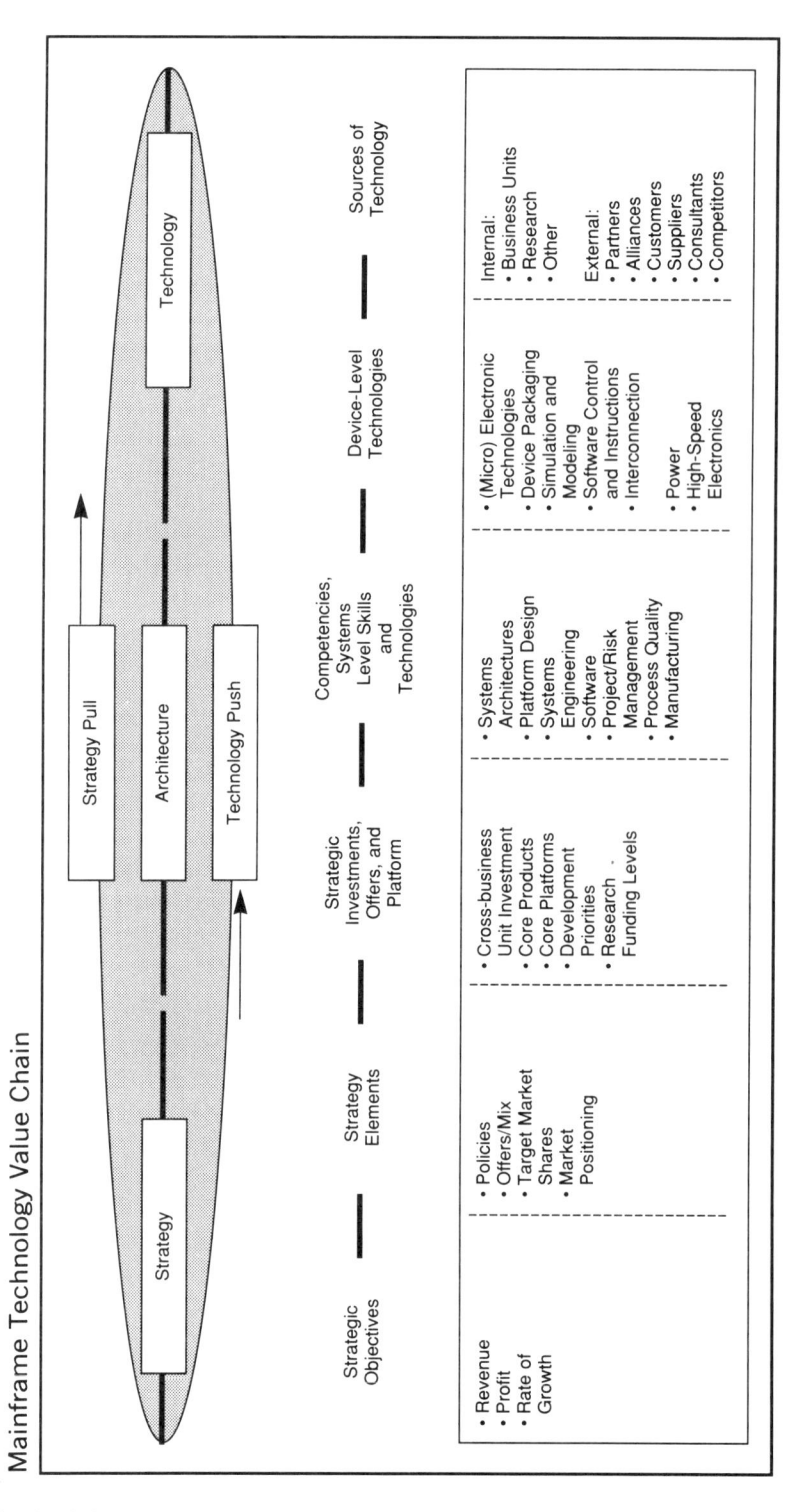

Exhibit 4
Mainframe Technology Value Chain

Strategy	Architecture	Technology
Strategy Pull →		← Technology Push

Strategic Objectives	Strategy Elements	Strategic Investments, Offers, and Platform	Competencies, Systems Level Skills and Technologies	Device-Level Technologies	Sources of Technology
• Revenue • Profit • Rate of Growth	• Policies • Offers/Mix • Target Market Shares • Market Positioning	• Cross-business Unit Investment • Core Products • Core Platforms • Development Priorities • Research Funding Levels	• Systems Architectures • Platform Design • Systems Engineering • Software • Project/Risk Management • Process Quality • Manufacturing	• (Micro) Electronic Technologies • Device Packaging • Simulation and Modeling • Software Control and Instructions • Interconnection • Power • High-Speed Electronics	Internal: • Business Units • Research • Other External: • Partners • Alliances • Customers • Suppliers • Consultants • Competitors

LINKING TECHNOLOGY TO BUSINESS GOALS

Exhibit 5
Required Competencies and Advantage

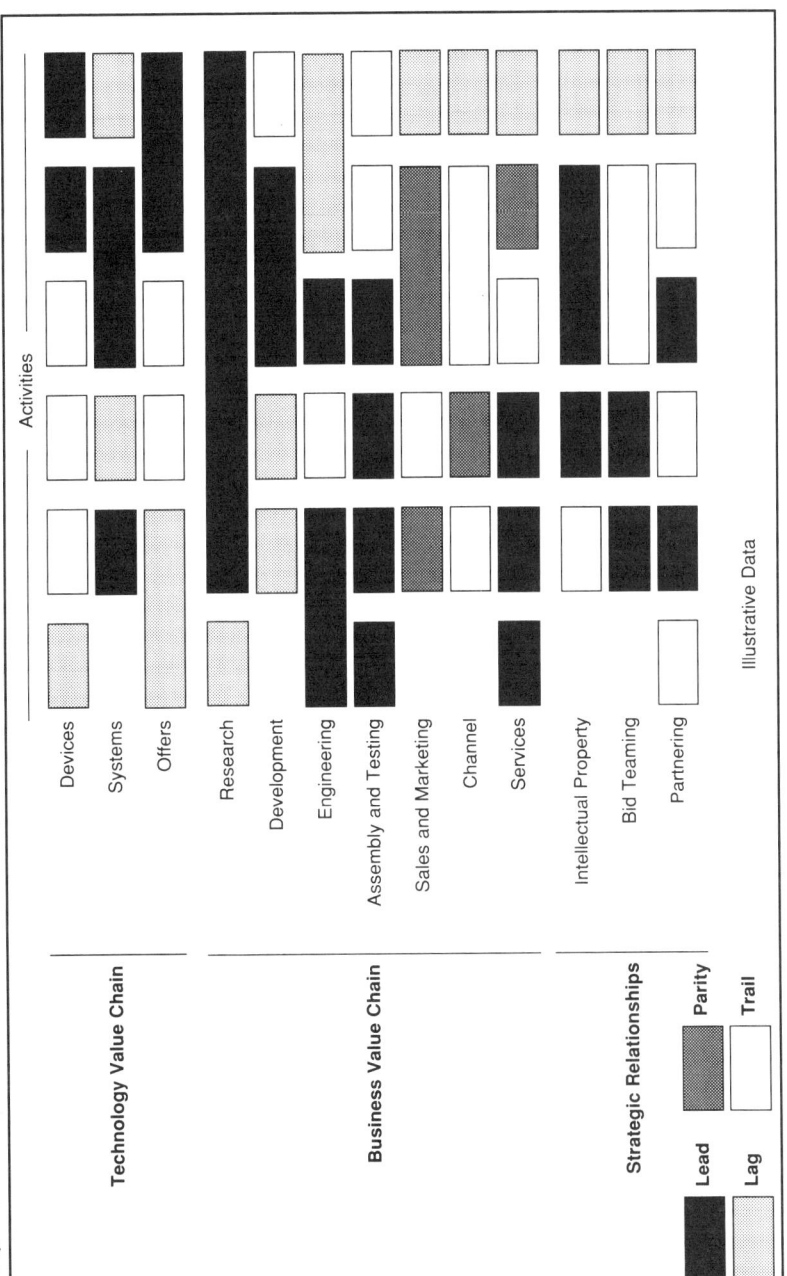

Activities

Technology Value Chain
- Devices
- Systems
- Offers

Business Value Chain
- Research
- Development
- Engineering
- Assembly and Testing
- Sales and Marketing
- Channel
- Services

Strategic Relationships
- Intellectual Property
- Bid Teaming
- Partnering

Lead | Parity
Lag | Trail

Illustrative Data

ONE CORPORATION'S STRATEGY AND TECHNOLOGY

was a first cut in the portfolio of mainframe unit competencies. As illustrated in Exhibit 6, the portfolio can be mapped on a simple two by two matrix that includes recommendations for managing the competencies.

The technology roadmapping team recognized that the approach was both unproved and well beyond the comfort level of some members of the mainframe management team. After considering possible damage to the credibility of the recommendations, the team chose to repackage the data in terms of nurturing and growing projects and capabilities. As a back-up, the team charted the competencies and activities by type of innovation and impact on the customer and mainframe's business value chain. Most of the technology investment was in incremental, conservative, low-risk areas, but the portfolio did include a few higher-risk investments (see Exhibit 7). Thanks to the chief scientist's efforts, most of the surviving projects were in areas that would change some imbedded obsolescent competencies and help grow the business.

The surviving investments could change the way computers and customer solutions were designed and implemented. In addition, the competency portfolio, although conservative and relatively low risk (and perhaps underfunded), did provide capabilities consistent with the original projected scenario. (In general, multiple scenarios should be constructed, and environmental, market, and customer signals should be tracked so they can be modified and regenerated as necessary). If managed successfully, those competencies would carry the business well beyond the lifecycle of current computing solutions and products. This was important because the roadmapping team had concluded that the next generation of mainframe computer, although undefined, was already obsolete. This conclusion was one of many, based on weak trends in networked computing capabilities and seen as a threat in the mainframe 1993 strategic plans. The board acknowledged and acted on this information.

SUMMARY

The world in which Corporation A manufacturing competes has changed drastically in the last decade. In response to changes in the customer, corporate, business, competitive and regulatory environments, the mainframe unit has refocused its core businesses on its strategic customers. That focus, adopted at the business unit board level, accommodates both the pull of business strategy and the push of technology and innovation.

The board established priorities and investment amounts. The plans were communicated personally by individual board members to their teams. At the operational level, functional teams aligned their product and applications portfolios with a set of strategic objectives and elements, and investment priorities management explored then dismissed internally generated new business opportunities as engines for diversification. By aligning its technologies,

Exhibit 6
Competencies as a Portfolio Concept Model

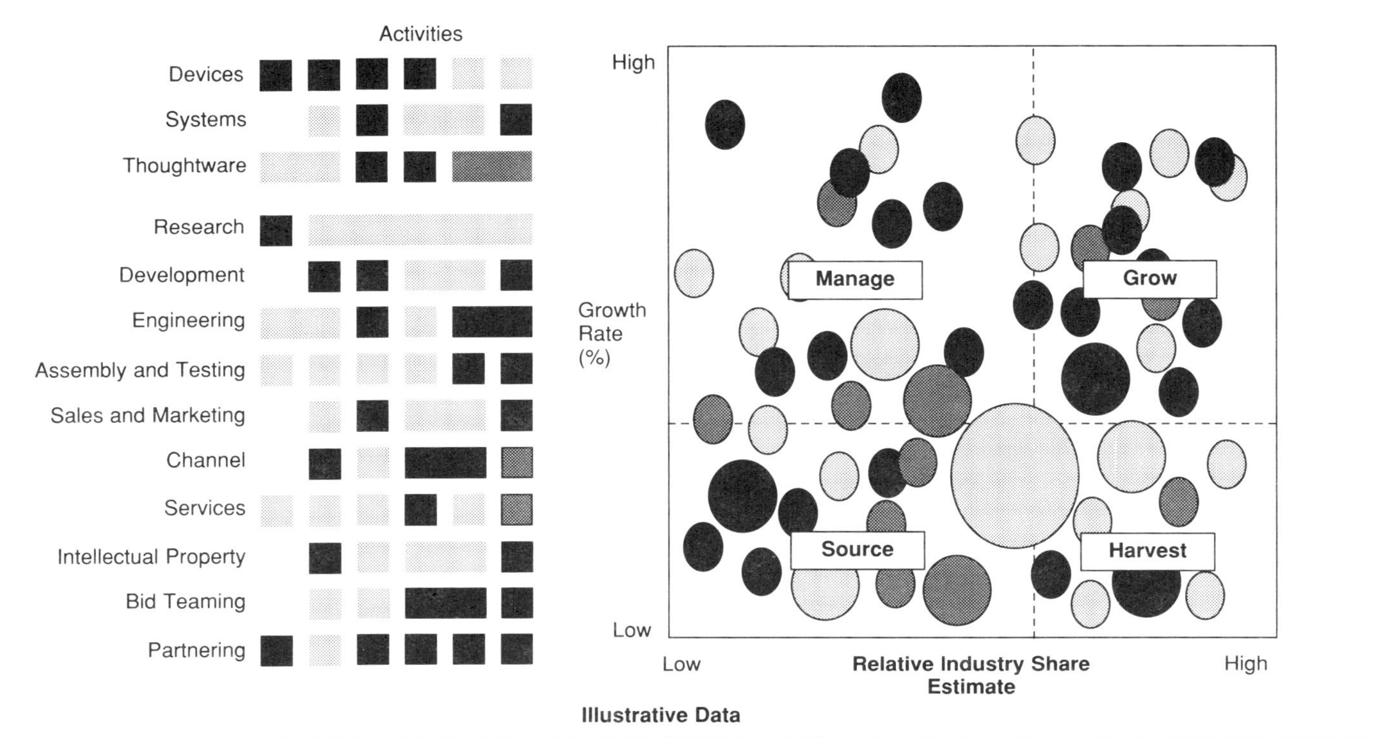

ONE CORPORATION'S STRATEGY AND TECHNOLOGY

Exhibit 7
Low Risk - Slow-Growth Evolutionary Portfolios: Current Mainframe Unit Investments

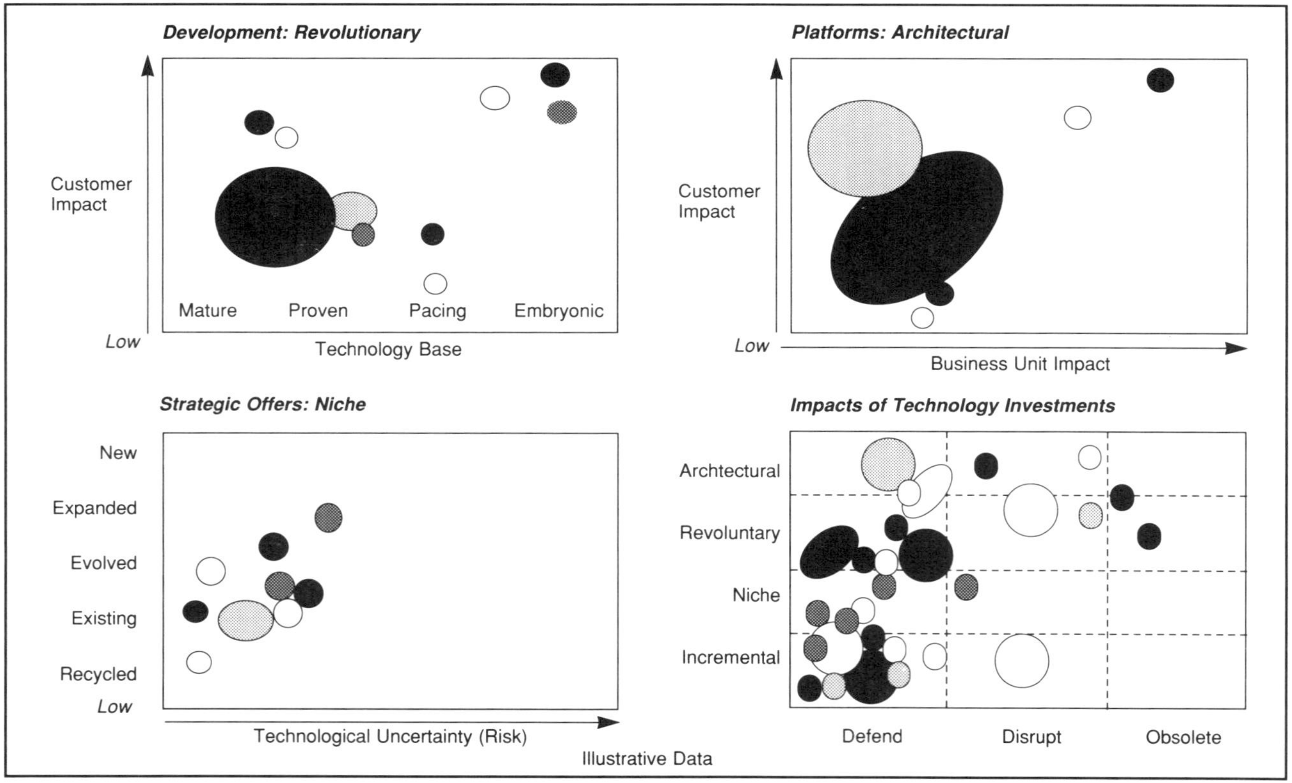

LINKING TECHNOLOGY TO BUSINESS GOALS

competencies, and objectives, mainframe offered its customers integrated computing solutions. By shifting its sales focus and tactics from pushing product to providing its customers unique competitive advantages, the team achieved significant sales. As the structure of the manufacturing group evolves to accommodate teams that could achieve global computing solutions in the future, the strategy and technology roadmapping processes were institutionalized and expanded.

Any advantage Corporation A enjoyed based on mainframe attributes or data communications and physical computer interfaces eroded rapidly.

CORE TECHNOLOGICAL COMPETENCIES

William J. Walsh

The author's company produces more than 200 different families of electromechanical and electronic controls for the automotive and appliance industries. This chapter shows how the company used core technological competencies to meet the twin productivity and quality challenge.

The controls business is an interesting and challenging one. The author's company, for example, produces more than 200 families of electromechanical and electronic controls for the automotive and appliance industries. The product families are in turn classified under about 30 major product-line groups. Typical product-line groups include interior switches, climate controls, engine sensors, and actuators for vehicles. On the appliance side, product-line groups include many timers, water valves, and cooking and refrigeration controls. Like most businesses today that compete in a global arena, the author's company had a requirement to continuously improve both quality and productivity. To do so consistently on a product-by-product basis, however, has repeatedly shown itself to be impractical from the standpoints of both time and labor power. A more efficient way was clearly needed.

As part of a large corporation, the author's division did have one advantage in meeting the twin productivity and quality challenge: it has three well-staffed and well-equipped corporate centers engaged in multiple engineering and manufacturing technologies. Moreover, the company recognized that a number of its key suppliers also had technological resources that the company might more effectively use than had been the practice in the past. Early on, the company decided that a major element of its productivity and quality improvement strategy would be to integrate these corporate and supplier technological resources with the divisional R&D and manufacturing engineering capabilities by means of a comprehensive set of written action plans.

Now the question became how the company should develop these action plans in a way that would cover the great diversity of products it supplied to the marketplace. Over the years, through specialized task forces, the company had fixed or upgraded a large number of products individually. It now wanted a different approach. It has been said, give a hungry man a fish and you will feed him for a day, but give him a fishing pole and you will feed him for life. The company wanted its action plans to be fishing poles—that is, generic steps it could take that would simultaneously upgrade the productivity and the quality of a majority of its products.

WILLIAM J. WALSH is director of engineering at Eaton Corp. in Carol Stream IL.

The company had long recognized that underlying its more than 200 product families were about a dozen technical building blocks that were combined in many ways to form new products. A few examples of these are DC motor-driven gear-train actuators, pneumatic motors, and electric solenoids. Although these building blocks are a useful concept in themselves, each part of the engineering community product development, manufacturing engineering, supplier support, or corporate research typically approached them as the proverbial blind men feeling different parts of an elephant. What the company wanted, on the other hand, was a common focus on some broadly applicable engineering and manufacturing issues that could benefit from the viewpoints of these various sets of eyes.

The thought process in this regard was aided greatly by a timely *Harvard Business Review* article entitled "The Core Competence of the Corporation".[1] In this article, two professor-authors analyze the outstanding success of many Japanese companies, such as Honda Motor Company Ltd. and Nippon Electric Company Ltd. (NEC). They attribute this success to a competitive advantage that derives from exploiting specific and highly developed core competencies. In the case of Honda, this core competency is in engines; for NEC, it is large-scale chip making. In the authors' words, "The real sources of advantage are to be found in management's ability to consolidate corporate wide technologies and production skills into competencies that empower individual businesses to adapt to quickly changing opportunities."

Building the company's action plans around selected core competencies seemed to be a useful approach. For its purposes, however, the author's company took a slightly different focus on core technological competencies. These core technological competencies were defined to be those established and relatively unique areas of engineering expertise and manufacturing capabilities that underlie the division's products. Implicit in this definition was the total teaming of people skills, process capabilities, and materials sourcing that leads to a superior product. Ultimately by defining seven core technological competencies, the company achieved this desired focus.

SELECTING CORE TECHNOLOGICAL COMPETENCIES

At first glance, there is a natural tendency to see a great number of engineering and manufacturing technologies as being core to a business. And if left unconstrained, it is easy to come up with a bedsheet-sized listing of diverse design and processing skills. The inevitable consequence is a totally unmanageable plan for dealing with them. What is needed, therefore, is some fairly stringent selection criteria that will winnow the final choice down to a handful of key technologies for which meaningful action plans can be developed. The author's company, with annual domestic sales in the range of $400 million, adopted the following three criteria in the initial screening of its core competencies:

- The technology must be used in a multiplicity of the company's product lines with aggregate annual sales of at least $25 million.
- The technological competency must add significant customer perceived value to the end product.
- The level of technological competence as currently practiced by the company must effectively exclude all but a small number of potential competitors in a given product-market segment.

Underlying the company's more than 200 product families were about a dozen technical building blocks that were combined in many ways to form new products.

Using these criteria and with a good deal of trial and error to see whether the shoe fit, about 25 candidate technologies were eventually reduced to seven core competencies. The whole process was akin to a standard Pareto analysis. Considerable group effort on the part of multidepartmental teams was expended in reaching agreement about the defining scope of a given core competency. For example, in the case of high-current blade-switch operators, the operating current range was qualified. Similarly, the electronics core was more narrowly defined as extended environment electronics to reflect the severe operating environment encountered by many of the automotive and appliance products.

SAMPLE DEFINITIONS OF SELECTED CORE COMPETENCIES

Engineered Plastics. This involves the ability to design and cost-effectively manufacture in high volumes closely toleranced thermoplastic and thermoset components and assemblies. This competency is used in the vast majority of the division's products.

High-Current Blade-Switch Operators. This involves the ability to design and cost-effectively manufacture mechanically precise and highly reliable electrical switches with an operating range of 100 milliamperes to 50 amperes. Major applications are in automotive convenience switches and appliance program timers.

Molded Elastomeric Components. This involves the ability to design and assemble into products precisely controlled elastomeric elements with specialized environmental properties in the form of diaphragms, seals, inserts, and flow controls. Major applications are in water valves, pressure switches, gas controls, and automotive cruise controls.

Electromagnetic Operators. This involves the ability to design and cost-effectively manufacture AC and DC electromagnetic structures and environmentally protected coils to the highest standards of reliability. Typical product applications are water valve operators and automotive solenoids.

Extended Environment Electronics. This involves the ability to design and cost-effectively manufacture electronic assemblies that interface with mechanical devices and are required to function in the extended temperature, electrical noise, shock, and time environments that are characteristic of the automotive and appliance markets. Typical applications are cooking range and dishwasher controls and automotive climate controls.

Power Thermal Elements. This involves the ability to design and manufacture vapor-driven devices that result in a precise mechanical deflection as a response to a sensed temperature. Major applications are in thermostatic expansion valves, air conditioners, refrigeration, and thermostat controls.

Bimetal Operators. This involves the ability to design and manufacture precisely calibratible, electrically driven, heat-sensitive bimetal motors. Major applications are gas-oven valves and infinite range burner controls.

DEALING WITH NONCORE COMPETENCIES

In screening for an operation's core technological competencies, it will be apparent that there are a number of technologies that are important for the success of the business yet do not meet the established criteria for a core. The author has chosen to classify these technologies into two types: supportive and niche. Supportive technologies are those particular broad capabilities that fail the screening test for proprietariness. In the operating division being discussed here, such engineering and manufacturing capabilities as software engineering, small mechanisms design, metal stamping, and screw machining fall into this category. The company could scarcely produce a single product without these capabilities, yet they do not currently give the company the unique competitive advantages associated with a core technical competency.

Niche technologies, on the other hand, meet the proprietary and uniqueness requirements but currently lack the required sales base to be designated as core. Examples for the subject company are piezoelectric sensors and strain gauge load cells. It is expected that with time and effort some supportive and niche technologies will be brought up to core status. In fact, this must happen if a company is to remain competitive because all core technologies have a finite life span.

METHODOLOGY OF EVALUATING SELECTED CORE TECHNOLOGICAL COMPETENCIES

The goal of a core technological competency study is to develop a meaningful and efficient action plan to enhance each identified engineering and manufacturing technology using all the technical resources at an operation's disposal. Once the initial selection of the cores has been made, it is useful to take four additional steps in preparation for the action plan write-ups. These prior steps are:

1. Rank the selected cores in terms of the relative product dollar sales they support for the entire operation.
2. Rank the selected cores in terms of their relative importance to the functioning of each of the operation's products.
3. Make an estimate for each core of its current engineering and manufacturing competency relative to the known state of the art for the industry.
4. Estimate for each core its remaining life expectancy in light of other known competing technologies.

Estimating the dollars of product sales that each core helps to support is a rather straightforward though time-consuming step. It is one, however, that is useful in attracting management's attention and financial support of the whole core technological competency process, particularly for those cores that individually underlie a high percentage of the company's products. Exhibit 1 summarizes these results graphically.

Estimating the relative importance of each core technology to the organization's product lines is a little more complex because of its subjective nature.

Exhibit 1
Core Technological Competencies

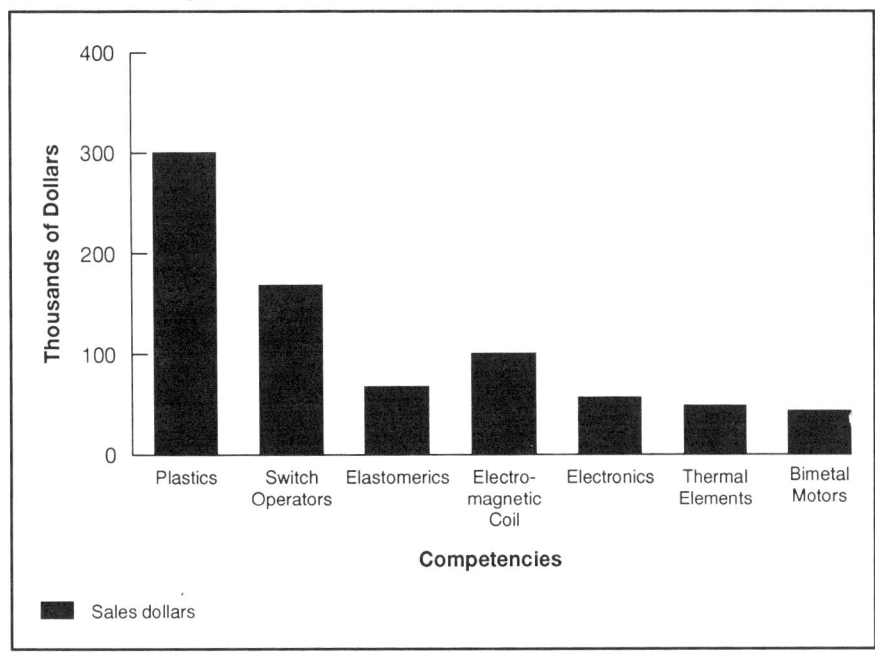

The method that the author employed was to list all the division's major product-line groups and then rate from 1 to 5 the relative impact that each of the seven selected core technologies exerted on those product groups. A rating of 1 was considered to have a minor impact on the product-line group, whereas the maximum 5 rating was taken to mean "of critical importance." Exhibits 2 and 3 show the ratings made for the automotive and appliance product-line groups respectively. These numeric evaluations were made for each product line by small groups of knowledgeable individuals representing manufacturing and engineering. Once these evaluations had been made, the numeric ratings were summed up vertically. These vertical summations were taken to be a measure of the relative importance of each core technological competency on the entire operation. The results of this summation are shown in Exhibit 4.

It will be noted that there is a spread of more than 6 to 1 in the importance ratings. Although the highest rating given to injection molding came as no surprise because of its near universal application in the company's product lines, some other results ran counter to intuition. In particular was the impact that molded elastomer components exerted on the overall product portfolio. It was comparable with electronics technology, which had long received a far greater share of resources and managerial attention. As in the case of relating core competency to dollar sales, first ranking core technological competencies by their relative importance is a useful management yardstick in the development of the action plans. Sooner or later the inevitable issues of costs and priorities have to be addressed relative to action plans, and these two measures supply a good deal of the needed perspective. Besides summing the importance ratings vertically in Exhibit 4, it is also interesting to do a horizontal summation for each product line. Relatively high values of 11 and greater indicate that a given product line makes high use of the division's established cores. In the vast majority of these cases, the results show high market shares and relatively higher profitabilities for these particular products, thus supporting the core competency theory.

MAKING STATE-OF-THE-ART AND LIFE EXPECTANCY ESTIMATES OF EACH CORE TECHNOLOGICAL COMPETENCY

If a core technological competency study of the author's company had been made 20 years ago, one of the selected cores would undoubtedly have been wax power element. It was a technological building block of many of the company's products, and the company had brought the application skills of the technology to a very high level. These elements make use of the large piston force available as a result of the expansion of a wax material as it changes from a solid to a liquid state. Properly formulated, the transition can be made to take place at any desired temperature over very wide operating limits. This technology is still generally used in engine thermostats, but its many other former applications, such as in vehicle emission control actuators,

Exhibit 2
Automotive and Specialty Products Core Competency Rating

Product Area	Engineered Plastics	Extended Environment Electronics	Electro-magnetic Coils	Molded Rubber Components	Blade Switch Operators	Power Thermal Elements	Bimetal Operators	Product Summation
LBS	3	1	5	2				11
DRCV								0
KNS	3							3
LPCO	3			2	2			7
HTV	4			4				8
TXV						5		5
ATC	4	4		1			2	11
CRCL	4		4	4				12
WETS	3							3
CONVSW	5	2			5			12
AUXSW	5				5			10
VTD	3	4		4	4			15
SPI	1	5						6
Competence Summation 0338m	38	16	9	17	16	5	2	

Notes:
1 Not very important
3 Important
5 Critically important

CORE TECHNOLOGICAL COMPETENCIES

Exhibit 3
Appliance Core Competency Rating

Product Area	Engineered Plastics	Extended Environment Electronics	Electro-magnetic Coils	Molded Rubber Components	Blade Switch Operators	Power Thermal Elements	Bimetal Operators	Product Summation
AWV	5		5	5				15
APS	3			4	5			12
APT	5		2		5			12
ADT	5		2		5			12
SYMTR	1		4					5
RNGT	4				5	5		14
WBT	4				5	5		14
INFSW	4				4		5	13
ACC	4				5	5		14
AGR				5				5
AGV				3			5	8
CYX	3				2	5		10
SID	3	4	5					12
TOD	3	5						8
AEC		5						5
PS			5					5
Competence Summation	44	14	23	17	36	20	10	

Notes:
1 Not very important
3 Important
5 Critically important

Exhibit 4
Core Technological Competencies

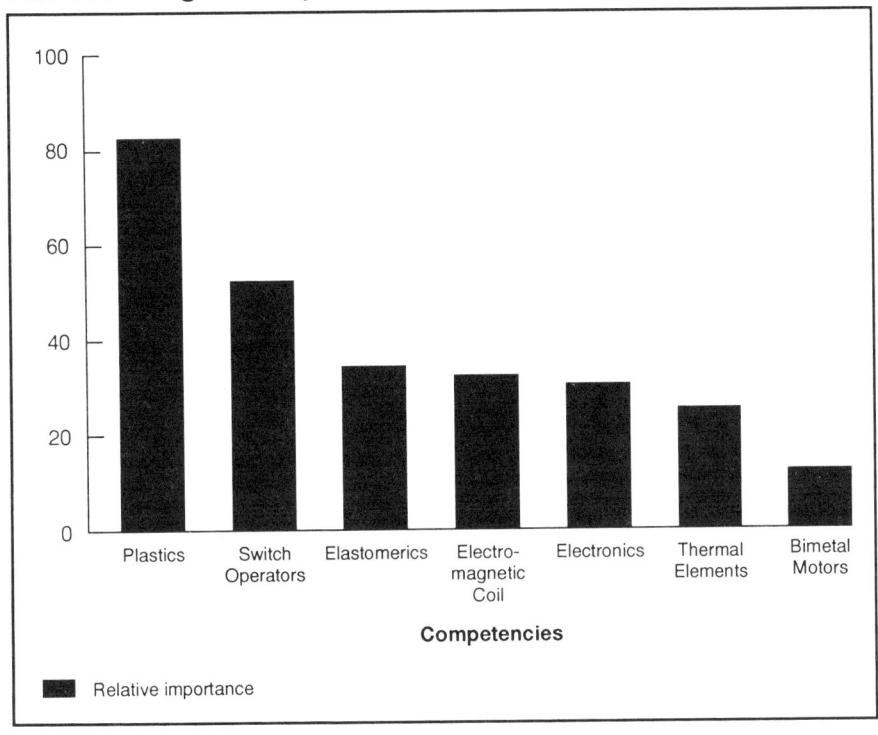

```
100

 80

 60

 40

 20

  0
      Plastics  Switch    Elastomerics  Electro-   Electronics  Thermal   Bimetal
                Operators               magnetic                Elements  Motors
                                        Coil
                                   Competencies

      ■ Relative importance
```

heat motors, and ice-maker controls, have fallen into general disuse. Solid state thermistors coupled with the widespread use of electronics technology have all but displaced this technology.

All core technological competencies are vulnerable to this kind of displacement and as often as not form some totally unrelated technology. But technological displacement is only a part of the danger in today's globally competitive world. The existence of many geographically dispersed competitors with continuous efforts to significantly improve a core's technological limits through new processing techniques all but guarantees an ever-higher state of the art. Thus a company must continuously evaluate its core technological competencies against the known, global SOA.

The author has found that the 3 x 3 matrix shown in Exhibit 5 can give a good overview of a company's core technological competencies. Each core technological competency is placed in one of nine boxes depending on its estimated engineering and manufacturing strength relative to the known state of the art as well as on its remaining life expectancy. Life expectancy is defined here as the point at which there is approximately a 50% decline from mature use levels. Again, making these estimates is best done by a jury that can evaluate

Exhibit 5
A Hypothetical Map of Six Core Technologies

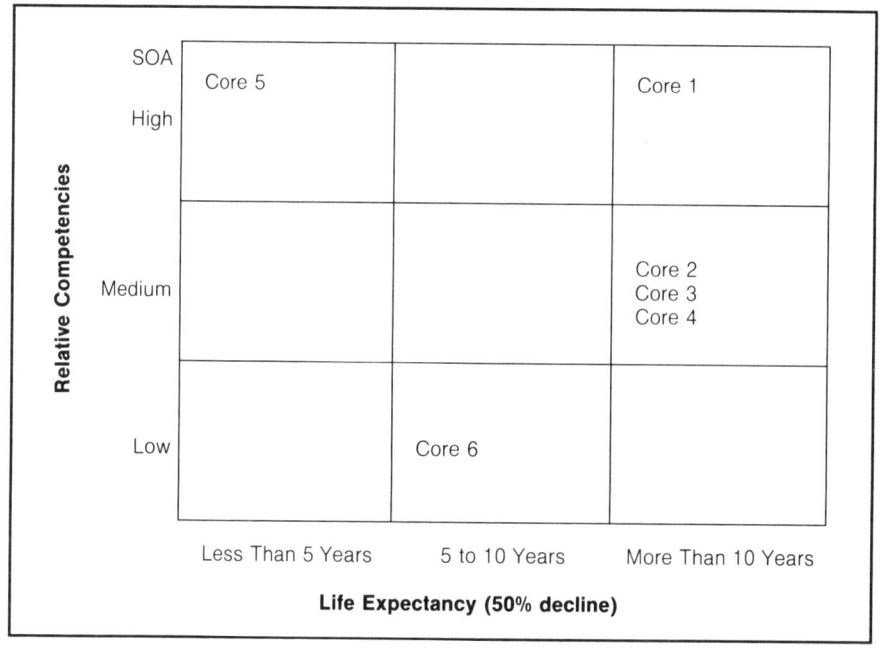

	Less Than 5 Years	5 to 10 Years	More Than 10 Years
SOA / High	Core 5		Core 1
Medium			Core 2, Core 3, Core 4
Low		Core 6	

Relative Competencies (vertical axis)
Life Expectancy (50% decline) (horizontal axis)

a given core technological competency from the perspective of several different disciplines. At first, there may be some divergence of opinion as to how a core technological competency should be ranked, but a consensus will often be reached after some clarifying discussions.

For the controls industry, the life expectancy of certain core technological competencies (e.g., electrical solenoids and thermoplastics) spans the foreseeable future. The author has therefore put outer time limits in the range of 20 years. Needless to say, for someone working in the fast-moving computer or similar industries, outer life expectancy limits of several years would be more appropriate.

Exhibit 5 shows a hypothetical example in which a group of six core technological competencies might fall for a typical company with a rather moderate pace of life expectancy. In this case, it was 10 years. It will be noted that for this company, core 1 is near to the SOA and has a maximum life expectancy. Here is an example of a strong competitive advantage not to be low. Cores 2, 3, and 4 also have long life expectancies, but the indication is that the engineering or manufacturing competence is not up to the SOA. Perhaps some relatively modest action program efforts can improve these three cores. Cores 5 and 6, on the other hand, present management with immediate major challenges. Although the capability of core 5 is world class, it appears that at least as far as the company's product lines are concerned it has reached the terminal stage of its life span. Many electromechanical component suppliers find themselves in this particular box as a result of the electronic revolution. Their very survival depends on their ability to make the major financial and cultural investments required to quickly assimilate this pervasive technology. In the case of core competency 6, remaining life is not the issue, but the capability relative to the SOA is dangerously low. Product lines with slipping profitability, poor manufacturing efficiency, and field quality are symptomatic of the root problems associated with this core. In this instance, the company must quickly marshal its internal and supplier engineering and manufacturing resources in a unified action plan to rehabilitate the core.

> **Building the company's actions plans around selected core competencies seemed to be a useful approach.**

The relative capability and life expectancy matrix, such as the one shown in Exhibit 5, has the advantage of providing management with a snapshot of the condition of the basic technological and manufacturing infrastructure of the entire business. It should be recognized that these core capabilities often reside undocumented in the hands of a few employees, in the output of a few critical machining processes, or in the input of a few special suppliers. It is not surprising then that core competencies, if left to their own, have a way of quietly eroding over time until some breaking point is reached.

DEVELOPMENT OF THE ACTION PLANS

The end objective of a core technological competency analysis is a set of written action plans that integrates the internal and external resources of the company toward significantly improving the capability or extending the life of each selected core. The goal is a seamless leveraging of technical resources of the corporate advanced engineering facilities and supplier capabilities with those of the operating division. The author's company wrote its particular action plans with a five-year time frame in mind but with the clear understanding that these were living documents that would be updated at least yearly. This time span, tied in with corporate strategic-planning horizons, gave the needed perspective for the long-range program planning of the corporate centers and was consistent with the required major capital investment planning time frames. Obviously different businesses will require appropriate adjustments to the selected time span.

The action plan format, as shown in Exhibit 6, consists of a one- or two-page document for each identified core and addresses the following seven areas.

Current Status. This is a brief paragraph summarizing the current capability of the core relative to the SOA and its probable life expectancy. Strengths and weaknesses as perceived both internally and by major customers should be addressed here.

Exhibit 6
Core Competency Analysis Sheet

```
Technology:                           Type:        Rank:

Current Status:

Key Products Where Used:

Competitive Threats:

Recommended Strategy:

Key Divisional Programs:

Key Corporate Programs:

Key Supplier Programs:
```

Key Products Where Used. In this section, the major product lines supported by the core and their aggregate impact on sales dollars are listed.

Competitive Threats. Here there should be a frank assessment of both the engineering and manufacturing aspects of this competency relative to emerging alternatives. Similarly, the status of this competency as currently practiced by key competitors should also be addressed.

Recommended Strategy. Based on the Current Status and Competitive Threats sections, the overall strategy to be pursued for this core should be stated. This might involve transitioning to a new base technology, major replacement of existing processing equipment with advanced machinery, or entering a licensing agreement.

Key Divisional Programs. Once an overall strategy has been written, those specific engineering and manufacturing programs to be carried out at the operational level should be established. This should always include anticipated results, estimated cost, and expected completion dates.

Key Corporate Programs. When the company has access to a corporate engineering center, those program activities should be documented that can be directed toward enhancing the operating division's core competency. Generally, these will be longer-term but more basic activities and may be applicable to more than one operating division.

Key Supplier Programs. The core technological competency approach makes clear the potential importance of a supplier's engineering and manufacturing contributions. As a vital part of the team, these should be fundamental contributions that go well beyond the usual day-to-day serving of the account. In this section, those significant improvement programs that can be leveraged from supplier expertise should be documented in terms of results, investments, and completion dates.

> **Core technological competencies were defined as those established and relatively unique areas of engineering expertise and manufacturing capabilities that underlie the divisions's products.**

SUMMARY

The core technology competency approach provides a number of advantages; among these are:

- It provides the basic landmarks in the mapping out and interlinking of long-range developmental strategies for divisional, corporate, and supplier engineering resources.
- It provides an important and strategic three-dimensional view of a company's business beyond the traditional thinking of products and markets; that is, it deals with the technological infrastructure of the entire business.
- It provides the common meeting ground between product engineering and manufacturing processes.
- It often results in rapid generic improvements in productivity and quality.

Although not specifically discussed, the development of core technological competencies yields additional benefits to companies. In particular, core technological competencies:

- Allow companies to quickly bring to market with minimal risks products that are designed using these proven building blocks. The need for additional capital equipment is often minimized.
- Are often reservoirs of proprietary advantage in the form of patents and trade secrets.
- Can be an ever-advancing line of competitive advantage as the company continues to perfect the cost-effectiveness, utility, and quality level of a particular competency.

Note

1. C.K. Prahalad and G. Hamel, "The Core Competence of the Corporation," *Harvard Business Review* (May–June 1990).

CONDUCTING A CORPORATE R&D RETRENCHMENT

Mark T. Hehnen,
Norman E. Johnson, and
Edward L. Soule

For the past 10 years, Weyerhaeuser Co rapidly expanded and diversified its business. Then, in quick succession, it was forced to downsize, refocus on core areas, and retrench, in agreement with corporate strategy. This sudden change in corporate policy required realigning corporate R&D and technology commercialization quickly and cost-effectively.

Since April 1989, Weyerhaeuser Co has been changing its direction. To increase shareholder value, it is currently seeking to regain the leadership position in the forest products industries. During the previous five years, the company had shifted its focus from developing new products to seeking manufacturing excellence in the production of high-volume, commodity goods. Although the rationale for this shift is familiar to industries throughout the country, an understanding of how the switch affected Weyerhaeuser's corporate research and technology commercialization efforts, which had been strongly supported from 1985 to early 1989, may prove valuable.

A less-than-desirable performance during the first half of the 1980s prompted Weyerhaeuser to decentralize in 1985. It became a quasi holding company, shifting attention toward customers and the marketplace and away from large, raw-material resources. Soon afterward, research, development, and engineering were realigned to satisfy this new corporate structure. Whereas the focus of research had been on process, it was now on developing new products to meet customer needs.

What was once a centralized research, development, and engineering organization was split into four components; three were to serve the three main sectors of the newly reorganized company, and the fourth was formed as a holding-company research, engineering, and commercial development unit. That unit can be characterized as follows:

☐ Its mission was to help Weyerhaeuser gain strategic and competitive advantage through the use of technology.
☐ Its goal was to develop, within three years, many new business and product opportunities in all states of commercialization for various Weyerhaeuser businesses.

MARK T. HEHNEN is director of technology commercialization at Weyerhaeuser Co, Tacoma WA.

NORMAN E. JOHNSON is vice-president of corporate research, engineering, and technology commercialization at Weyerhaeuser.

EDWARD L. SOULE is vice-president of corporate research and engineering at Weyerhaeuser.

- Its job was to:
  - Provide technology support services to the operating companies.
  - Keep abreast of key technologies that could threaten the company.
  - Evaluate, manage, and develop technologies that would offer business opportunities and competitive advantage.
  - Develop commercial opportunities from technologies that had a strategic fit with the company.

The commercial development function of the unit was responsible for:

- Developing new products and processes beyond the scope of other corporate units engaged in new product development.
- Staying abreast of key technologies that could threaten or benefit the corporation and ensuring that technological information was included in corporate strategies.
- Maximizing the income from developed technologies through selective sales and licensing agreements.

From the outset, the mission of the corporate group was to create commercial and strategic advantage through technology. To accomplish this, the company decided to leverage existing and developing technologies in several core areas:

- Strategic biology—To support the company's plant-growing activities.
- Materials science—To improve the price/performance ratio of materials, particularly composites or laminates that contain wood and other raw materials.
- Environmental sciences—To better use the technology and skills the R&D staff had developed over many years.
- Process technology—To further the development of sensors.

Lacking an express corporate strategy, R&D management selected these four areas because they represented significant commercial opportunities supported by demographic, economic, and technology trends as well as technologies important to Weyerhaeuser. In addition, these were areas in which the R&D staff had acquired strong skills during the last decade. As a result of this effort, many new materials, products, and business opportunites were developed; in fact, some major commercialization efforts are currently under way.

The primary strategy was to build around a technology base that the R&D group had largely mastered. In addition, the areas R&D management chose to pursue were supported by significant market trends. It was important not to venture too far from familiar markets, and when Weyerhaeuser did, it tried to establish marketing alliances with companies that had strengths in those markets.

R&D management's philosophy was to think big but start small, and its approach was to make existing businesses more successful and to give them first choice to implement a new technology. In addition, R&D management set up boards of advisers to help improve its business decisions; R&D management also established quarterly meetings with the president and chief executive officer to verify their support and obtain their advice.

> **Weyerhaeuser's aim was to make existing businesses more successful and give them the first choice to implement a new technology.**

A CHANGE IN DIRECTION

By the end of 1988, however, the bloom was off the rose for new business and product development. Wall Street clearly told Weyerhaeuser that shareholder value must be increased. Corporate finances indicated that the company was involved in too many areas and that it needed to focus on those areas that had once made Weyerhaeuser the leader in its industry. At the shareholders meeting in April 1989, senior management announced the beginning of a major refocusing effort. The financial situation was inarguable; in part, the culprit was diversification, which stretched scarce managerial resources, talent, and capital over too many businesses.

Weyerhaeuser consequently decided to cut back its diversification efforts and focus on core strengths. For example, a major diversified business group that had led the company into such fields as nursery products, skin care, diapers, and salmon ranching was discontinued; other such businesses were sold or closed. In 1987, there were five new business development groups spread throughout the company; by 1989, the only commercial development group left was corporate R&D.

Weyerhaeuser was not the only company forced to undertake such retrenchment. Over the years, Weyerhaeuser had expanded and diversified as rapidly as such companies as Campbell Soup, Exxon, and Kodak; now these companies were also reassessing their new businesses.

This turn of events strongly suggested that R&D should reexamine its commercialization program. Ideally, this should be done in the light of a new corporate strategy, because corporate R&D and technology commercialization management had constantly stressed the need to match technology strategies to corporate strategies. However, the new corporate strategy was not to be determined until the results of ongoing, business-by-business strategic reviews were in. In the meantime, the corporate R&D and technology commercialization function tried to understand its current situation and what its options might be.

A RATIONALE FOR TECHNOLOGY COMMERCIALIZATION

As R&D management examined its situation and considered the lessons learned, it reflected on how the departmental rationale and implementation approach had evolved over the preceding five years. Certain core technologies are critically important to the future of Weyerhaeuser, whatever form the company may assume. R&D management determined which technologies fulfilled this role by assessing which ones were sources of competitive or strategic advantage for the business units. Every core business activity has essential technologies that are important for gaining competitive advantage. For example, if growing trees is fundamental to a company, various forestry technologies are essential because they maintain strategic positions in pulp making, paper making, or sawmilling.

In addition, some technologies cut across business lines; they form core competencies.[1] In 1985, when R&D management identified four potentially key technologies, the order of priority was materials science, plant biotechnology, environmental sciences and equipment, and process technologies. Although Weyerhaeuser still considers these areas essential, the order of priorities has changed with the emerging corporate strategy—environmental sciences and process technologies now lead the list.

Applying Core Technologies to Support Strategic Excellent Positions

Strategic excellent positions are key business areas that are developed by the allocation of resources.[2] Appropriate resources must therefore be committed to those core technologies that support strong strategic positions across the corporation. Once strategic excellent position has been developed, it can be maintained only through consistent, long-term corporate support through the ups and downs of business cycles. Developing strategic excellent positions is a medium- to long-term activity that can be easily disrupted when the length of the cycle between reorganizations is shorter than the time required for commercializing a new product.

At this juncture, Weyerhaeuser faced the critical strategic choice of whether or not to invest at the corporate level in developing core competence technology assets to support its strategic excellent positions. To make the correct decision, several questions had to be addressed:

- Should the company's individual businesses be responsible for their own technologies?
- Should Weyerhaeuser stay among the leaders in the industry in supporting new technology at a time when competitors are retrenching to increase shareholder value?
- How can the company use its technological base to greater advantage?

Underlying these questions is the principle that if a company decides to invest in an R&D program, it must commit resources to seek maximum return on that investment. It must find the people and create the organization to support this effort. Maximizing return means not only seeking the benefit that accrues to the main businesses from having strong core technology competence but seeking value through all legitimate avenues.

> Some technologies cut across business lines, forming core competencies—these were essential to the business.

THE CURRENT APPROACH TO TECHNOLOGY COMMERCIALIZATION

In Weyerhaeuser's experience, creating new products or businesses that fit easily into its existing business units was limiting and difficult for the R&D group to accomplish because of natural impediments in the internal system. Consequently, R&D management now considers all the following commercialization options (none of which are mutually exclusive) as early in the technology commercialization process as possible:

- A new product for an existing business.
- A new Weyerhaeuser business.
- A joint venture.
- A spin-off venture.
- Licensing or cross-licensing.
- Trading technology for strategic position.

To maximize R&D resources, studies were conducted to determine how other companies commercialize technologies. Through conferences, visits to exemplary companies, conversations with those who had experience hiring consultants, and training courses, much was learned about the options available. The commercialization approach Weyerhaeuser R&D finally adopted emphasizes corporate strategy and exploring all options before making an investment.

R&D now seeks to align research programs with business and corporate strategies. If these strategies are not written down, R&D management must push, tug, and cajole senior management to explain them; R&D management then writes down these strategies and tests and retests them.

In addition, R&D management follows and assesses important demographic, social, economic, political, and technical trends that can benefit or threaten the company. Technologies that support the primary business and corporate strategies are pursued relentlessly, and an aggressive technology commercialization effort helps uncover opportunities outside the company.

Weyerhaeuser R&D now evaluates and sponsors commercialization projects by establishing rigorous business plans, seeking early market involvement, and developing cross-functional teams that include members from the marketing, engineering, research, manufacturing, and finance divisions. It is always easier to start a project than to stop one; this approach tends to discourage premature project termination. Every successful project has its dark days, and many managers understandably are reluctant to kill projects in their early, difficult stages. The commercialization process must be sufficiently rigorous at each stage to keep projects from going on forever yet not so tough as to stop innovation.

R&D management now aggressively seeks to extract value from promising ideas; the organizational environment is designed to be conducive to innovation and commercialization, so these ideas are heartily encouraged. The importance of a separate organizational home for projects not yet ready to be picked up by operational units is emphasized. These temporary homes act as greenhouses to nurture young projects until permanent organizational homes can be found.

Ideally, once they are developed, a product or an emerging business can be sold to a buyer—at the right time. The best time to approach a prospective buyer is after the product's value is apparent but before its commercial value is threatened by outside forces. Weyerhaeuser has tended toward gaining the early involvement of the businesses in order to gain their support sooner.

SEIZING OPPORTUNITY

The typical corporation has immense resources to prevent errors of commission but almost never has the equivalent resources to prevent errors of omission. A technology commercialization process such as Weyerhaeuser's is at least one attempt to gain shareholder value by not missing new opportunities.

Notes

1. C.K. Prahalad and G. Hamel, "The Core Competence of the Corporation," *Harvard Business Review* (May–June 1990), pp 79–91.
2. Strategic excellent positions are defined and discussed by C. Pumpin in *The Essence of Corporate Strategy* (Brookfield VT: Gower, 1987).

SECTION 2

MANAGEMENT OF R&D

This section details the steps that R&D managers can take to make the most of their departments' and their companies' R&D efforts.

According to Robert Szakonyi, an important element of creating and implementing successful R&D projects is usually overlooked in a project's creation and implementation: the need for a solid infrastructure within the R&D organization. He describes how to build this infrastructure and then how to maintain it in the first two chapters in this section.

John J. Moran describes how R&D managers can gather and use customer input in the structuring and selection of R&D projects.

Stephen Hellman discusses R&D strategy and coordination in multinational settings, describing how a US-based consumer products companies increased productivity and output by implementing some basic management strategies. In the following chapter, Harold G. Wakeley looks at three aspects of R&D productivity: a short history of methods used to measure and improve R&D productivity, organizational issues that influence R&D productivity, and the development of R&D productivity measures.

Stephen Hellman describes how R&D cost accounting (i.e., charge-backs) in a multiclient consumer produts R&D group helped to align its company's mix of R&D projects with its business goals.

Teaching and motivating personnel is an important part of any manager's job, but it is particularly challenging for R&D managers. In separate chapters, Timothy J. Kriewall and Stephen J. Fraenkel offer suggestions for motivating and rewarding R&D personnel. Ron K. Bhada discusses how to teach management to engineers.

Finally, Harris M. Burte discusses management in the technology base, that area of expertise that falls between truly basic knowledge-driven research and short-range development or product introduction. He provides several short case studies to illustrate the importance of this area.

DEVELOPING AN R&D STRATEGY AND STRENGTHENING R&D ADMINISTRATION

> In the creating and carrying out of R&D projects, one important element of the projects' success is often overlooked: the need for a solid infrastructure within the R&D organization. By developing an R&D strategy and strengthening the R&D administration, an effective environment can be established.

Robert Szakonyi

Most R&D organizations focus almost completely on selecting, planning, and managing projects and then transferring them to business operations. However, in addition to this efforts, a solid infrastructure is needed—one that increases the probability of a project's success. Two key elements of the infrastructure of an R&D organization are an R&D strategy and R&D administration.

An R&D strategy is important because it helps an R&D organization select projects in terms of a broader perspective rather than just piecemeal. When an R&D organization has a strategy, it will more likely select projects that match its technical strengths, the capabilities of its company, and the demands of the marketplace.

R&D administration encompasses many things—from purchasing instruments for the laboratory to handling grievances with the R&D organization, from determining the pay scales of R&D people to designing the work areas of a laboratory. This chapter discusses three areas of R&D administration: strengthening the technical skills in house; deciding whether to contract out for technical skills; and determining the number of people in various groups in the R&D organization. The better these areas of R&D administration are handled, the better projects will be planned, managed, and transferred to business operations.

DEVELOPING AN R&D STRATEGY

To develop an R&D strategy, an R&D organization first must ensure that certain conditions are in place:

☐ A perception that an R&D strategy can solve a problem.

ROBERT SZAKONY, PhD, is the director of the Center on Technology Management at IIT Research Institute, Chicago. He has consulted for many companies in various industries and has written two books and more than 40 articles on technology management.

- A planning staff within a large R&D organization—or, in a small R&D organization, a strong commitment by the R&D line managers to devote enough effort to doing R&D strategic planning.
- A way of linking R&D strategic planning to R&D operations.
- A method to do strategic marketing.
- The active support of senior business managers.
- One or more previous efforts to develop an R&D strategy.
- A series of concrete efforts that produce tangible results on their own but also allow R&D people to improve at R&D strategic planning.

The Perception That an R&D Strategy Can Solve a Problem

Although the idea of developing an R&D strategy may be accepted, usually the development of an R&D strategy will not come about *unless* an R&D strategy is perceived as solving a problem. That is, the effort that goes into developing an R&D strategy—and the conflicts that may occur about how R&D resources should be allocated—are so great that unless an R&D organization needs to have an R&D strategy in order to solve a problem it usually will not make the effort or face the conflicts.

For example, the head of an R&D organization in a chemical company championed the need to develop an R&D strategy because he was anxious about his R&D organization's ability to deliver the new businesses that senior business managers expected from the R&D organization. When this R&D organization's company had sales of $100 million, this R&D manager was always confident that his R&D organization could develop one new business worth $40 million each year. After the company reached $1 billion in sales, however, this R&D organization was expected to develop four new businesses, each worth $40 million every year. Although this R&D manager was confident that his R&D organization could develop one or two of these new businesses each year, he was not confident that it could develop four of them a year. In addition, at this time the R&D organization was also under pressure to be more productive in its use of R&D resources. For these reasons, therefore, this R&D manager perceived developing an R&D strategy as a way of solving his problems related to what the R&D organization was expected to produce.

The R&D managers of a food-processing company recognized the need to develop an R&D strategy for other reasons. These R&D managers were forced to consolidate within one laboratory all of the R&D on coffee that previously was carried out in several laboratories. To accomplish this task, they recognized that they had to have an R&D strategy to help establish priorities and coordinate all of the R&D being done on coffee. Later, when they were forced to consolidate the R&D being done on beverages, these R&D managers perceived the need to develop an R&D strategy related to R&D on beverages.

In sum, while acknowledging the value of an R&D strategy, unless R&D managers perceive that an R&D strategy will solve a problem, an R&D strategy usually will not be developed.

The Creation of a Planning Staff

Although line managers within a large R&D organization can develop an R&D strategy, in practice, if there is not a planning staff that facilitates the development of an R&D strategy, there usually will not be one.

For example, if a natural resources company, R&D managers have recognized the need for an R&D strategy for quite a while. However, because no one was appointed to facilitate the development of an R&D strategy, this R&D organization still does not have an R&D strategy.

In contrast, in a pulp and paper company there has been an R&D planning group for eight years. Although this company now has an R&D strategy, it was not easy to develop. The R&D planners ran across two types of problems: analytical and organizational.

The analytical problems surfaced when the R&D planners first attempted to facilitate the development of an R&D strategy. They found that there was no accepted methodology for developing an R&D strategy. Although other R&D managers, consultants, and academicians all had their opinions on what an R&D strategy should be like and how it should be formulated, almost none of then had a complete picture of the process. In addition, the opinions of these various R&D managers, consultants, and academicians often were in conflict with each other or were irreconcilable. Thus, the R&D planners in this company had to develop their own methodology.

The R&D planners also found that members of the R&D staff were not interested in developing an R&D strategy. Because of this, the R&D planners had to spend 70 percent of their time during the first few years persuading the R&D staff to do R&D strategic planning and then educating them with regard to how an R&D strategy can be developed.

The experiences of these R&D planners are typical. Someone usually has to develop the methodology to be used in doing R&D strategic planning. Someone also has to be the day-to-day champion of R&D strategic planning— or it will not get done. Theoretically, R&D managers in a large R&D organization can handle these responsibilities. In practice, R&D line managers in a large R&D organization normally have so many other responsibilities that they neglect R&D strategic planning. Thus, an R&D planning group usually proves to be necessary to accomplishing R&D strategic planning.

In a small R&D organization, on the other hand, R&D line managers are the only ones who can develop an R&D strategy because staff positions usually do not exist. Thus, to get R&D strategic planning done, the R&D line managers in a small R&D organization must add the responsibilities of an R&D planning

Unless R&D managers perceive that an R&D strategy will solve a problem, an R&D strategy usually will not be developed.

group to their normal responsibilities. These R&D managers usually will not be able to devote much time to developing a planning methodology. However, because they have responsibility for managing the R&D groups, if the R&D managers in a small R&D organization do develop an R&D strategy, they should have less difficulty in getting this strategy accepted and implemented.

The Linking of R&D Strategic Planning to R&D Operations

To get an R&D staff to develop an R&D strategy, R&D planners (or R&D line managers) must find ways to relate R&D strategic planning to R&D operations. This connection between R&D strategic planning and R&D operations has two aspects.

First, R&D staff members must be able to see that their interests can be served through developing an R&D strategy. Thus, R&D planners must find a mechanism that involves R&D strategic planning while serving the interests of the R&D staff.

For example, R&D planners at a candy company established a cross-disciplinary forum involving a variety of R&D people. One of the purposes of this forum was to get these R&D people to talk with each other. Although they all carried out R&D on chocolate, they seldom interacted. This cross-disciplinary forum turned out to be a useful mechanism not only for improving communication, but also for getting R&D strategic planning accepted. Once involved in this cross-disciplinary forum, these R&D people became interested in coordinating their technical activities. However, they lacked a common language to describe their technical activities and an analytical framework through which they could evaluate their various technical activities. With the help of the R&D planners, they learned how to use the tools of R&D strategic planning, which helped them in both areas. Thus, through this mechanism R&D people not only found a way to coordinate their technical work, but also in the process learned how to do R&D strategic planning.

Second, for the R&D strategy to be meaningful, the R&D projects that are actually selected and carried out have to be linked to the strategy. In other words, an R&D strategy is not worth much if it does not affect which R&D projects are selected and carried out.

An R&D organization in a household products company addressed this problem by viewing the development of an R&D strategy as involving two phases. During the first phase, the senior R&D managers defined the overall direction of the R&D strategy. During the second phase, middle-level R&D managers defined the specifics of the R&D strategy through the projects they selected and carried out.

Getting Strategic Marketing Done

Although it is important for an R&D organization to link its strategic planning to its operations, this is not enough. For an R&D organization's strategic plans to pay off, those strategic plans must be clearly linked to customers' future needs. Therefore, besides doing R&D strategic planning, an R&D organization must also ensure that strategic marketing is done.

> **R&D staff members must be able to see that their interests can be served through developing an R&D strategy.**

Following are the approaches that three R&D organizations are taking to get their strategic marketing done.

First, in a chemical company, an R&D planner also serves informally as the strategic marketing planner. The R&D planner gained these responsibilities—at least informally—because (1) no one in marketing was doing the job and (2) the R&D planner had the ability to deal with marketing questions. This R&D planner spends 40 percent of his time dealing with marketing questions (e.g., what will the trends be like in three to five years, and how may the public think about environmental issues in the future).

Second, in a dairy products company, a marketing person who is knowledgeable about technology was assigned to the R&D organization to do strategic marketing as part of investigating opportunities for new business development. This marketing person leads a group made of four technical people at the company's headquarters and six other technical people who work in various regions of the country. An example of one of this group's goals is to help the company go from selling food products that are commodities to selling specialty products in other areas, such as selling fats to the cosmetic industry. Good strategic marketing is one of the keys to helping this company make such a transition.

Third, at an appliance company, market research people are invited to serve on R&D project teams. The advantage of this approach is that many individual market research and R&D people communicate with each other. The disadvantage is that because each project team only focuses on a specific area, strategic marketing is done in a piecemeal way. The challenge in making this approach work is getting the market research and R&D people to actually develop a close collaboration. One way in which this company tries to make this happen is through conducting very critical reviews of project plans. Not only senior R&D managers, but also senior business managers, ask hard questions related to the market implications of the R&D being carried out.

The Active Support of Senior Business Managers

For an R&D organization to get its R&D strategy integrated with business plans, the active support of senior business managers is required. Those R&D organizations that have been able to gain the active support of senior business managers for their R&D strategy were able to do this through a variety of ways.

The R&D organization in a chemical company received active support because the chief executive officer is one of the major proponents of R&D in the company. As opposed to most of the other senior business managers in this company, this chief executive officer previously managed the division of the company that sells to industrial customers. Because the customers of this division are more knowledgeable about what R&D can contribute (e.g., many of these industrial customers want their suppliers to do more R&D), this CEO realizes how valuable R&D can be. Thus, he not only has supported, but encouraged the R&D organization to do strategic planning.

At an aerospace company, the R&D organization was able to gain the support of senior business managers because it deals with strategic business executives rather than the managers of the individual businesses. These strategic executives achieved their positions through a reorganization. Their new management positions were created to ensure that strategic issues that normally lie outside the interests of the individual managers were addressed effectively. Therefore, one of the strategic business executives' main responsibilities is to approve the individual business managers' business plans, particularly with regard to how these business managers plan to utilize new technology to grow their businesses. Having senior business managers with this kind of assignment has made it much easier for the R&D organization to gain the support of senior business managers.

At an electronics company, a key factor that allowed the R&D organization to first gain and then maintain the support of senior business managers was continuity in leadership among senior business managers. Continuity in leadership was important because it took a few years for the senior business managers to appreciate the benefits of having an R&D strategy. By learning year after year what the R&D organization was trying to accomplish with an R&D strategy, these senior business managers were able to appreciate the benefits. The R&D organization, in turn, could build on the progress that it had made with these senior business managers in previous years.

One or More Previous Efforts to Develop an R&D Strategy

An R&D organization's first efforts at developing an R&D strategy normally will not yield the desired results. Unavoidably there is a learning process through which an R&D organization must go. This learning process may take from three to five years.

For example, in looking back at his R&D organization's first efforts at developing an R&D strategy, the head of R&D in a chemical company called these first efforts "naive." He said that during the first year of R&D strategic planning the R&D organization just categorized the obvious. In the second year, the R&D organization began understanding R&D strategic planning better and consequently gained new insights. For example, it recognized for the first time the importance of new technology in improving the company's manufacturing processes. After five years of doing R&D strategic planning, the R&D organization in this company has more or less institutionalized the process of R&D strategic planning. Now one of this R&D manager's greatest challenges is maintaining the vitality of the process so that the R&D people continue to learn new things by doing R&D strategic planning.

A Series of Efforts That Produce Results

One way in which an R&D organization can maintain the vitality of the R&D strategic planning process is through carrying out a series of concrete efforts that produce tangible results on their own, while allowing R&D people to get better at strategic planning. The experiences of a candy company's R&D organization illustrate what a series of such efforts might be like.

> For an R&D organization to get its R&D strategy integrated with business plans, the active support of senior business managers is required.

The first effort consisted of the establishment of a cross-disciplinary forum involving R&D people who looked at different technical aspects of chocolate. This effort was successful because it allowed the R&D people to coordinate their activities. It also introduced them to the tools of R&D strategic planning.

A second effort involved a benchmarking study in which the R&D people evaluated the strengths and weaknesses of 36 of their company's technologies in relation to the strengths and weaknesses of six competitors' technologies. The R&D people gained two things through this effort. First, they gained a much better understanding of how their company stood in relation to competitors. Second, they learned how to analyze their technologies. For example, they learned how to think more precisely about how their technical work could improve the performance of the company's products (e.g., by improving the flavor of cocoa beans).

A third effort that the R&D organization is considering involves using the techniques of portfolio management to evaluate the potential and payoffs of various technologies. Because the effort is not yet underway, the exact benefits that would result are not currently known.

In conclusion, the value of carrying out a series of concrete efforts is that together they serve as stepping stones to improve R&D strategic planning. Moreover, because they produce tangible results along the way, they also help elicit support for the R&D strategic planning processes.

EXAMPLES OF STRATEGIC GOALS OF R&D ORGANIZATIONS

Following are descriptions of some of the strategic goals of R&D organizations. The R&D organization of a household cleaning products company, for example, identified infection control as one of its major strategic goals because it realized that 80 percent of its products involved infection control in some way. The value for this organization of focusing on infection control is that by emphasizing the prevention of infection, rather than just the removal of dirt and stains, this R&D organization's company can better differentiate its products from the competitors'.

To make this strategy viable, the R&D organization had to demonstrate to the marketing department and to senior business managers the value of emphasizing infection control. As part of its mission, this R&D organization educates the general public about infection control through a variety of seminars. It also sponsors many clinical studies that investigate the conditions underlying infection control.

One of the factors that led this organization to focus on infection control was having a head of R&D who is a biochemist rather than a chemist. For more than 15 years, the heads of R&D in this company had been chemists. Partly because of this, they had focused more on typical areas of new product development in a consumer product industry. Having a head of R&D who viewed the issues related to new product development differently laid the basis for developing new strategic goals in R&D.

The R&D organization of a consumer product company, which produces toothpaste, has identified one of its strategic goals as cavity prevention. Having such a strategic goal means much more than simply improving toothpaste in order to cut down on tooth decay. It also means researching how cavities could be prevented without the use of fluoride, in case governmental regulations ever force the company to change what it puts in toothpaste. Researching how to prevent cavities also involves considering how less familiar kinds of tooth cavities can be prevented. For example, one challenge that this R&D organization is addressing now has to do with preventing tooth decay in the roots of teeth, which normal brushing of teeth cannot prevent.

R&D organizations of food processing companies also have identified strategic goals. For example, an R&D organization in a soup company is focusing on new ways of preserving foods. Because many of the company's products are preserved through chemical means (which result in there being too many fats and oils), this R&D organization has set a goal of helping the company produce food that is preserved through other ways.

An R&D organization of a breakfast cereal company has as one of its strategic goals to improve the packaging of breakfast cereals. Rather than be bound by the traditional packaging of a box with a liner, this R&D organization is looking for ways to package breakfast foods in a simpler way.

Finally, an R&D organization that supports company businesses related to the retail sale of cooked food has the strategic goal of remaking the businesses by developing a system of precooked food for the retail outlets. A new system of precooked food would improve the quality and variety of the company's products. In addition, it would allow the staff in the retail stores to be free from some of the manufacturing aspects of cooking food and instead to focus on serving the consumer.

THE LIMITED PROGRESS OF MOST R&D ORGANIZATIONS IN DEVELOPING AN R&D STRATEGY

Most R&D organizations do have an R&D strategy—and have not established the preconditions required for developing an R&D strategy. For example, most R&D organizations have not perceived an R&D strategy as a way to solve a problem. Many large R&D organizations do not have an R&D planning group, and the R&D managers of most small R&D

organizations do not do R&D strategic planning. Also, few R&D organizations have found a way to get strategic marketing done. Finally, in most companies, efforts to do R&D strategic planning do not have the active support of senior business managers.

> **The primary challenge facing most R&D organizations has to do with establishing the preconditions required for developing an R&D strategy.**

In addition, a close look at the strategic goals of those R&D organizations that do have them, reveals that many of these goals were arrived at intuitively, not analytically. That is, many of the strategic goals that R&D organizations have are intuitively obvious. For example, most of the strategic goals that food processing R&D organizations have formulated are similar and quite predictable: to ward out fundamental threats to the company's businesses, to improve food safety, and to develop salt-free and low-fat foods.

Most R&D organizations have not conducted benchmarking studies aimed at comparing their technologies to companies' technologies. Most R&D organizations do not carry out technology forecasting studies aimed at understanding what technological advances may be occurring in the next 5 to 10 years and what those technological advances may mean.

Even those R&D organizations that have made significant progress in developing an R&D strategy admit that they still have much to do to improve the strategic planning process. For example, the R&D organization of an electronics company has an R&D strategy, but the strategy is not accepted by the managers of the company's businesses. The R&D organization of a chemical company also has an R&D strategy, but it only sets priorities within product categories, not across product categories. Because much of the value of an R&D strategy lies in setting priorities across product categories, this R&D organization still has much work to do in getting R&D strategic planning accepted. The R&D organization of an appliance company also has developed an R&D strategy, but it has not found a way to integrate its R&D strategy with the marketing and manufacturing strategies in the company. In addition, parts of the R&D strategy in this company can be altered at the whim of anyone who at a later date becomes involved in the planning process—without necessarily informing anyone else about the changes in the plans.

In sum, some R&D organizations have made significant progress in developing an R&D strategy. Most R&D organizations, however, have barely started. The primary challenqe facinq most R&D organizations, therefore, has to do with establishing the preconditions required for developing an R&D strategy. After doing this, they will be able to meaningfully address questions concerning what the R&D priorities should be and how R&D resources should be allocated.

STRENGTHENING R&D ADMINISTRATION

One way in which R&D organizations can strengthen technical skills in house is by recruiting technical people from other industries. For example, the R&D

organization of a food processing company gained a great deal by bringing in an engineer from the petrochemical industry. Because this engineer had a different technical perspective on the handling of oils, he was able to develop technical solutions that engineers who had spent their career in food processing could not. At a toy company, an engineer with an extremely varied career in many types of industries, including the machine tool and aerospace industries, has provided valuable help in identifying technologies that can be utilized from other industries. Finally, at a tobacco company, a medical doctor from the health care product industry has provided a totally new perspective on how research might alleviate some of the health problems related to smoking.

When R&D organizations do recruit young scientists or engineers, one good practice that they could follow is to recruit them as interns while they are in graduate school. For example, an R&D organization in an electronics company recruits most of its young scientists and engineers in this way. By being able to evaluate in depth the technical skills and work habits of these scientists and engineers before making a commitment, the R&D organization finds that it makes better choices. The organization hires about one-half of these interns permanently.

Deciding about Contracting for Technical Skills Outside

Because R&D organizations increasingly find that they do not have all of the technical skills that they need in house, many of them are contracting out for the necessary technical skills. R&D organizations follow a variety of strategies with regard to what technical skills they maintain in house and what technical skills they receive from outside.

In deciding what strategy to follow R&D organizations can consider the following two questions: In which technical areas should the company be able to use skills from any source? and how much should the company control the technical skills that it needs?

Although there are a variety of possible ways in which R&D organizations could answer these questions, the actual strategies that R&D organizations formulate fall into roughly four types.

- *Strategy 1.* The company should have complete control over skills in all technical areas.
- *Strategy 2.* The company should have some control over skills in all technical areas.
- *Strategy 3.* The company should have complete control over skills in the most important technical areas.
- *Strategy 4.* The company should have a little control over skills in all technical areas.

Strategy 1: Be Self-sufficient. The basis of this strategy is the belief that all technical knowledge about the company's products and processes is vitally important and must be kept within in order to maintain the company's competitiveness. For example, one consumer product R&D organization rarely contracts out for tests concerning safety because competitors may learn important information based on the kinds of tests run by the outside contractors. However, R&D organizations following this strategy do contract out on occasion in areas in which they lack any expertise (e.g., in biotechnology).

> Questions about the proper ratio between groups can only be answered in terms of a particular context.

Strategy 2: Use Strategic Alliances. Under this strategy, R&D organizations work as partners with the outside contractor, rather than have the contractor do all of it. The organizations often perform some of the technical work themselves, in order to deal on an equal basis with the contractor. Many R&D organizations have developed strategic alliances with their suppliers. For instance, many food processing R&D organizations have developed strategic alliances with ingredient manufacturers. One chemical R&D organization has 42 different strategic alliances with outside organizations. Its overall aim in utilizing so many strategic alliances is to bring many new technologies into the company in order to improve its speed and flexibility in developing new products. A health care products company has used strategic alliances with universities and small companies for the purpose of developing technologies that are central to this company's businesses, but which this company's R&D organization does not have the technical skills to develop.

Strategy 3: Maintain the Technical Skills at the Highest Level in House Only in Areas That Are Directly Related to Product Development. Companies that follow this strategy between knowledge related to product development, which they consider proprietary, and other technical knowledge, which they consider nonproprietary. Although each R&D organization has its own definition of what is nonproprietary, some areas of technical knowledge that R&D organizations often consider nonproprietary are packaging, manufacturing equipment, toxicology, and safety. Many R&D organizations contract out for analytical studies, especially studies that involve the use of expensive research instruments.

Strategy 4: Contract Out Much of the Work. There are some R&D organizations that follow a strategy of contracting out a great deal of their work. For example, one R&D organization, which consists of 25 technical people, has been contracting out much of its work for over 20 years. Another R&D organization, which has 20 technical people, contracts with university professors to do most of its research. As might be expected, many of the R&D organizations that follow this strategy do not pursue an aggressive policy of new product development.

The following analogy illustrates and compares these various strategies. Assume a homeowner needs to build and paint an outside deck. The homeowner might have four choices on how to get the job done: the homeowner could build and paint the deck; the homeowner could build and paint the deck with the help of a contractor for specialized building tasks; the homeowner could build the deck, and have a contractor paint it; and a contractor could build and paint the deck.

All four ways of getting the deck built and painted could work very well. The choice depends on the handiness of the homeowner and the amount of time the homeowner has. Homeowners who are extremely handy and have enough time to build and paint a deck should build and paint the deck themselves: Strategy 1. Homeowners who are only fairly handy and have very little time should have the deck built and painted by a contractor: Strategy 4. Thus, there is not any best way to have a deck built and painted.

This is similar with regard to doing the technical work in house or contracting for it outside. There is not a right answer for all R&D organizations because it depends on an R&D organization's circumstances and values.

Determining the Number of People in Various Groups in the R&D Organization

Two questions are helpful in determining the number of people in various groups in the R&D organization. The first question concerns the right ratio between different groups or personnel within an R&D organization (e.g., between technical support groups and other R&D groups, between technicians and R&D professionals, and between administrative support personnel and R&D personnel). The second question concerns the optimum number of personnel that an R&D manager should supervise.

Right Ratio of Groups or Personnel. Questions about the proper ratio between technical support groups and other R&D groups, between technicians and R&D professionals, and between administration support personnel and R&D personnel can only be answered in terms of a particular context. There is not one universal answer.

The right-ratio questions are similar to asking "What percentage of their salary should people save? Some individuals will argue for saving 5%, others for saving 10%, and still others for saving 15% of their salary. Who is right? The answer is "No one". All individuals need to make their own decisions based on their personal circumstances.

The same holds for ratios related to various technical groups or personnel. There is no right answer for all R&D organizations.

However, what complicates matters related to determining the right ratio between technical groups or personnel is that most R&D managers do not have sufficient knowledge to adequately compare the tradeoffs involved in one ratio with the tradeoffs involved in another ratio (e.g., how does a ratio of one analytical support person for *four* line technical people compare with a ratio of one analytical support person for six line technical people). In reality, R&D managers make decisions about these matters based mostly on their gut feel.

Unless R&D managers perceive that an R&D strategy will solve a problem, an R&D strategy usually will not be developed.

On the whole, R&D managers do not understand well how technical support personnel, technicians, and administrative support personnel work. During their career, R&D managers normally have worked primarily in product development or exploratory R&D, not as technicians or technical or administrative support personnel. Therefore, they are inclined to err on the side of ensuring that there are enough R&D professionals in product development or in exploratory R&D rather than erring on the side of having enough technical support personnel, technicians, or administration support personnel. Their rationale is that it is easier to contract out for technical support personnel, technicians, and administrative support personnel, who usually have less specialized skills, than it is to contract out for R&D professionals. However, R&D managers should consider that they, on the whole, do not understand the tradeoffs between various technical groups because they do not understand how those technical groups or personnel work, therefore, how can they know if they have the correct number of R&D professionals in product development or exploratory R&D to maintain the proper ratio?

It might be mentioned that R&D managers make decisions about technical support personnel, technicians, and administration support personnel that, on the whole, are not unlike the decisions that senior business managers make about R&D people. Because senior business managers usually do not understand well how R&D people work, they often make decisions about R&D budgets or R&D personnel that are based either on certain arbitrary ratios (e.g., spending on R&D should be a certain percent of sales) or on arbitrary figures (e.g., the maximum number of people in an R&D organization should be such-and-such). Similarly, because R&D managers usually do not understand well how technical support personnel, technicians, and administrative support personnel work, they also often rely on arbitrary ratios or figures when making decisions about the operations of these groups or personnel.

Optimum Number of Personnel Whom an R&D Manager Should Supervise. In this area, R&D managers' knowledge is also inadequate and they often make decisions based on their gut feeling. For example, an R&D manager at one chemical company believes that R&D managers should not supervise more than six technical people because, if the staff were larger, the manager would not be able to know the substance of the staff's work adequately. At another chemical company, an R&D manager believes that R&D managers can supervise up to 20 to 30 technical people.

There is no right answer for all R&D organizations concerning the optimum number of people an R&D manager should supervise. Right answers in this area must be determined, however, not only for an specific R&D organization by taking into account its situation, but also for each R&D manager within that R&D organization by taking into account his or her particular situation.

CONCLUSION

To increase the probability of success of R&D projects, R&D managers should improve the infrastructure of the R&D organization. Specifically, they need (1) to develop an R&D strategy, and (2) to strengthen R&D administration. By doing this, they will also improve how projects are selected, planned and managed, and transferred to business operations—in turn, increasing the probability of success of the R&D projects

R&D MANAGEMENT: A BALANCING ACT

Manek R. Dustoor

In this era of corporate *rightsizing* (a polite term for downsizing), where does research and development fit? Many senior managers fail to consider the following, simple question: What are our expectations of R&D? An unwillingness to address this question early in the strategic planning process forces many corporations to re-examine their R&D needs, often after considerable wasted effort and the dislocation of human resources that downsizing inevitably entails.

How should a company view its need for R&D? The need is not static and must be re-evaluated at reasonable intervals, to ensure that R&D continues to offer a tangible contribution to the company's profitability, the ultimate measure of any function's business value.

All established R&D organizations, as well as those in their embryonic stages, must accurately and compellingly rationalize the need for their continued efforts, in terms of their potential contribution to their customers. That need is easily rationalized when R&D represents a critical element for survival in an industry—for example, pharmaceuticals or computer software—for which innovation and being first to market distinguish the healthy survivors from the also-rans. Paradoxically, rationalization of this need seems to be most challenging for those companies that first grew on the fruits of R&D successes and subsequently failed to nurture both a spirit of innovation and the capability to capitalize on the results of successful discoveries.

R&D RESOUCES: CENTRALIZED OR DISTRIBUTED?

Deciding whether to concentrate or distribute R&D resources is the subject of a long-standing debate. Opinions on this matter swung to one extreme during the 1980s, the heyday of leveraged buyouts. During that period, the acquiring entity often sought quick returns and organizational savings that made easy prey of large, centralized R&D departments that had more historical significance than relevance to the fast pace of business envisioned by the acquiring entity.

Manek R. Dustoor, PhD, is Director for Technologies & Concepts, the R&D arm of Haworth, Inc., a manufacturer of office furniture systems.

The financial holding companies that typically made the buyouts had little understanding of the R&D needs of the acquired companies, other than what could be represented in simplistic, financial terms. Without careful consideration of the potential long-term losses from disbanding the centralized R&D service, the acquiring company viewed the savings from disinvestment in R&D as an immediate return to offset the costs of the acquisition.

This desire to demonstrate a lean mentality by virtually eliminating the centralized R&D function remained in favor until a company's bank of ideas ran dry and company management recognized that restoring the R&D function was necessary for regaining competitiveness in mature markets. Senior management and R&D managers then had to determine how to rejuvenate R&D and where it fit within the organization: as a centralized entity or distributed under one or more business units.

Centralized R&D

A concentration of R&D personnel working to shape the company's future seems to make perfect sense. However, this simple structure fails to place proper emphasis on offering direction for downstream implementation of the results of the R&D effort.

Simply creating the R&D function without giving thought to the process of capitalizing on the ideas generated by R&D virtually guarantees the eventual scrutiny of the R&D need. Some organizations—those large enough to justify the expense—formalize the transition from technically proven concept to full-scale production, by testing concepts in scaled-down manufacturing facilities, within the R&D complex. These real-world test drives offer a measure of comfort to the manufacturing risk takers, allowing them to evaluate the merit of new concepts that are presented in a recognizable context.

With centralization, all business units share the costs of R&D. A business unit with high growth potential may merit a disproportionate amount of the R&D resources until it reaches a level of business self-sufficiency, after which the allocation can become more equitable. This capability to pool R&D funding also allows for the hiring of well-paid technical specialists.

Distributed R&D

The centralized approach has its counterpart in the distributed one, in which R&D is independently positioned under one or more business units, often administratively reporting to marketing or manufacturing. Although the distributed approach brings a degree of real-world relevance to the R&D effort, it also runs the risk of concentrating R&D efforts on those projects with immediate returns, even when this may endanger the long-term health of the business unit.

Each business unit's current profitability constrains the outlay that unit can afford for R&D. The fortunes of the various R&D groups rise and fall with the fortunes of their respective business units, and the inevitable swings of business cycles have a negative effect on efforts to foster the proper environment for creative product development. With the distributed approach, business units may find it difficult to justify the expense of well-paid, full-time technical specialists; instead, part-time, external consultants may have to suffice.

Making a Choice

In determining which avenue to pursue, centralized or distributed R&D, management must balance numerous factors and expectations. In a start-up business, R&D is often indistinguishably meshed with other groups, such as marketing, sales, manufacturing, and management. As the business grows, so does the complexity of managing the R&D function, until the company reaches a crossroads: Should R&D continue as is or should it be legitimized as a separate department that is corporate in scope and not dedicated to any one business unit?

To ensure the ongoing well-being and effectiveness of the R&D department, management performs numerous balancing acts involving such issues as organizational structures and reporting channels, project selection, recruitment, budgeting, performance measurement, and motivation.

Compared to other functions, R&D often has an equitable position in start-up businesses that recognize the survival implications of developing a niche entry product. As these companies mature, however, management places increasing financial importance on manufacturing, marketing, and sales, while relegating the R&D function to a maintenance role rather than the aggressive entrepreneurial one that dominated the early years of the business. This natural imbalance reflects the changes within the growing ranks of corporate executives, most of whom are hired based on the bottom-line performance they demonstrated in previous positions. This imbalance also signals the need for realistically reassessing the goals for R&D and restoring a level of equity within the business.

A Case History

An example from the automotive industry illustrates the potential pitfalls that companies encounter in the balancing act between centralized and distributed R&D. This case history involves the evolution of a centralized R&D laboratory for a large corporation with a dominant international presence in the manufacture of bearings, as well as the manner in which the R&D lab eventually outlived its usefulness.

R&D was responsible for the development of a sleeve bearing for automotive crankshafts. This component serves as the interface between the reciprocating piston of an engine and the rotational motion imparted to the crankshaft and eventually to the driven wheels, through the transmission. This rather innocuous part gets almost no attention unless an automobile engine seizes because of a failed bearing.

To develop this bearing, R&D had to balance several, traditionally conflicting, material requirements. For example, the development required a soft, weak material to offer lubricity and thus avoid scoring the crankshaft. However, the material also had to be strong enough to take the phenomenal loads exerted by the explosion of the compressed gases on each power stroke of the piston. Moreover, these bearings were expected to survive throughout the useful life of the vehicle. Senior management took considerable interest in the early design and development of these bearings.

Development efforts during the 1960s and the 1970s produced increasing levels of sophistication in these bearings; the life cycles of these products eventually exceeded those of most other engine components. The rapid business success of these unique, patented bearings allowed the company to dominate its market and justified the formation of a centralized R&D laboratory, which was expected to duplicate the successes of past development efforts. Considerable expense was incurred, but without the urgency of survival implications, the R&D laboratory produced only incremental enhancements to the original development.

With the growing financial success of the Bearings Division, the company brought in a caretaker management that did not share the original management's commitment to development. R&D personnel grew increasingly alienated by the lack of support for their mission, and R&D management's preservation instinct fostered a slew of projects that were optimistically sold to the funding divisions. As the objectives for these projects grew more extreme, the true probabilities of success were rarely voiced and the financial burden continued to mount.

Increased competition and the expiration of patent protection transformed the once-unique product into a commodity item faced with intense price pressures. Failing to appreciate the benefits of well-executed R&D, management pushed for another development miracle, setting increasingly unrealistic time frames for execution. Development efforts were further hindered by the long-standing alienation between the business division and corporate R&D. With losses mounting, management initiated a search for a friendly buyer and R&D became a target for cost cutting that was aimed at slowing the financial hemorrhaging. R&D resources eventually dipped below critical mass, and following the sale of the business division, this relatively large R&D facility, with a worldwide reputation of accomplishments, was soon permanently shuttered.

In retrospect, what should have been done differently? Expectations for the product's life cycle should have been more realistic and the corporation should have been less dependent on the Bearings Division for its long-term growth. This high sales (though not necessarily high profit) division should not have dominated corporate decision making to the extent that it did. Each business unit should have defined realistic expectations for its funded research, and executive bonuses should have been tied to successful transitions of products from R&D to manufacturing. Company management should have taken a more holistic view on the use of R&D resources for fostering new business opportunities within the corporation's range of competency, though not necessarily restricted to automotive bearings. The short-term, risk-averse management that became entrenched in the Bearings Division was not suited to accepting the business risk necessary for carving out new opportunities. Unfortunately, this scenario has been played out all too often in what were once considered highly successful businesses that other companies hoped to emulate.

> A centralized structure often fails to place proper emphasis on providing direction for downstream implementation of the results of the R&D effort.

THE BALANCING GAME

R&D management must have the political skills necessary for fostering close, productive relationships with other functions (e.g., sales, manufacturing, distribution) as well as balancing the needs of top performers. Senior management, accustomed to the short-term performance measures of these other functions, must be constantly updated on R&D accomplishments, which should remain directed toward long-term deliverables.

The Project Mix

R&D managers often face a dilemma in organizations that have not reached the level of maturity necessary for recognizing the unique character and needs of the R&D function, in human terms and in the resources required for success. In a typical cycle, senior management raises the bar to unattainable heights and then—rather than lower the bar to reasonable levels—cites R&D's failure to deliver as the reason for cutting resources. To avoid such situations, R&D managers must establish targets that, although not necessarily technically challenging, offer the needed publicity and recognition. This can be accomplished with a mix of about 25% to 40% near-term goals that are in support of production or marketing, with the balance being long-term strategic goals. With this approach, R&D runs the risk of being too successful, in which case R&D may be forced to continuously dedicate a portion of its development effort to near-term support. However, this is a more manageable problem than one in which other departments become vocal antagonists of R&D.

A related issue involves the choice of long-term projects and deliverables. Playing it too safe ultimately fosters mediocrity in a function for which risk taking is an acceptable *modus operandi*. For example, volunteering to focus R&D resources on finding incremental cost reductions in established products or processes, although offering immediate, tangible financial returns, may cast the R&D group as being unable to conceive and deliver the next leap forward for the company.

However, the other extreme of taking on challenges for which R&D has little expertise is equally foolhardy, because this plays into the hands of those who cite a failure to achieve goals as the basis for retrenchment in the R&D ranks. When asked for input on the choice of long-term projects and deliverables, the R&D manager needs to research the resources required for success and present these requirements in the proposal to senior management. Volunteering to storm the bastille, although noble, may be life threatening for the R&D function. A proven record of developments in line with the company's strategic goals is the only insurance for sustaining an R&D function.

R&D managers routinely face this balancing act: accepting reasonable challenges, but not impossible missions that are set up by other managers. Many senior managers expect the R&D manager to deliver the final solution: a low-cost, backward- and forward-compatible design, with demonstrated manufacturability and a one- to two-year return on investment. The likelihood of success would increase if a cross-functional team from various disciplines (finance, manufacturing, quality, marketing) were given a joint charter to meet those goals. More often than not, however, team involvement is limited until R&D has resolved nearly every issue, often with only a single researcher fully dedicated to the project. This is a formula for failure, not success.

Another increasingly widespread means for performing this balancing act is to involve risk-sharing partners from complementary businesses that can exercise a degree of control on their niche markets. Although many companies use this approach, partnering should not degenerate into a risk-avoidance tactic, with each partner offering the other the dubious honor of taking the first step across the threshold into the unknown. Well-executed partnerships have levels of risk and opportunity proportionate to the capabilities (i.e., resources, sales, market share, growth potential) of the players.

Winning Friends

R&D management can gain valuable allies by demystifying the work of R&D personnel, who may be viewed as privileged because they seemingly do not have to contend with the daily crises faced by professionals from other departments. Cross-training stints can build the needed bridges between professionals from R&D and other departments.

Another simple way to foster a broader appreciation of the unique challenges facing the R&D function is to limit the geographic isolation of the department. When deciding where to locate the R&D function, management must strike a balance between creating a functional, results-oriented environment and offering a degree of protection from critics whose input would be damaging during the ideation phase of projects. Although individuals within the R&D department might justifiably see the value of a pristine research campus within a natural setting that nurtures creative thought, management must weigh the benefits of this setting against the negative feelings that might arise among those excluded from such privileged spaces. If R&D productivity can survive this inescapable jealousy, the company can opt for this idyllic setting.

> As a company matures, management often relegates R&D to a maintenance role rather than the entrepreneurial one that dominated the early years of the business.

Recruiting

Management must also find the right balance in terms of the hiring opportunities within an R&D organization. For small departments of five to 10 individuals, the R&D manager should make the extra effort to recruit the most experienced, talented candidates that the budget can support, because much will be expected of the few. In addition to the required technical competence in their respective research areas, these candidates must be able to provide sales support, troubleshoot problems for clients, interact with customers, negotiate supplier contracts, and assist manufacturing with processing problems. Willing participation in these other tasks limits the vulnerability of a fledgling R&D group and offers the necessary exposure for developing research goals that better match the evolving needs of a growing business.

In larger R&D groups—those with more than 25 to 30 people—recruitment efforts should reflect a different balance of skills and experience. For such groups, R&D managers should seek a wider spectrum of talent and not load the bases with superstars. This strategy allows the most experienced people to remain involved with interdepartmental support functions, as well as oversee and guide junior staff in their research projects. It also offers a career ladder for entry-level or less-experienced personnel, allowing those people to avoid the disillusionment of stuck careers that sometimes occurs when most of the staff is considered senior.

In mature organizations where R&D has demonstrated its value and has grown to a significant size (approximately 100 or more people), the support functions also become legitimized and several of the previously mentioned issues are no longer applicable. The politics of larger groups then becomes unavoidable, and R&D management must draw on a different set of skills to protect the legitimate interests of the group. For these R&D departments, and for moderate-sized ones, recruiting efforts should focus on candidates from within the company ranks, to increase the likelihood of fostering internal allies. However, care must be taken to minimize the in-breeding that can stifles open, creative thinking.

Job descriptions are an often overlooked aspect of these recruitment efforts, especially for medium-sized corporations in which relatively few personnel are engaged in R&D, compared with all of the other functions. These descriptions must accurately specify development's role and the company's expectations for R&D.

Human Resources and Compensation departments often succumb to the lowest-common-denominator syndrome, developing generic descriptions of perceived development roles, regardless of whether those positions are in product engineering, manufacturing, or marketing. Such job descriptions fail to recognize the unique nature of R&D and the special attributes of its personnel. Consequently, individuals are either promoted into R&D roles or recruited externally based on superficial definitions of required skills—for example, "a proven track record of delivery on time and within budget." Although this is a laudable accomplishment for almost any position, it fails to recognize the other character traits essential to a discovery function (as opposed to one that focuses on implementation).

Communication

Another political balancing act involves the appropriate degree of communication with other groups within the company. R&D management needs to clearly understand and identify with the strategic goals of the corporation and then exercise the necessary judgment to execute its mission. Although it makes sense to share progress on R&D projects with those who can appreciate the technical nature of the accomplishments or disappointments, many managers have neither the time nor the inclination to understand the workings of the development process. For them, progress reports must be offered on a need-to-know basis, without ever intentionally misleading them.

A project in the early phases of development—idea generation and concept feasibility assessment, for example—requires the protective leadership of managers who understand the intricacies of the mission and can offer a level of protection until the project reaches a stage at which public scrutiny no longer jeopardizes an intelligent decision on the future of the project. The embryonic stages of development are not the right time for enforcing ritualistic, traditional management tracking devices, measures of success, or performance indicators. Instead, R&D managers must nurture, challenge, humor, and, to a degree, sustain the strategic focus that should justify the activity.

THE REPORTING CHANNEL—HOW TO INFLUENCE IT

The fragile nature of R&D rests in its fundamental charter, which involves the nurturing of new ideas and demonstrating the feasibility of concepts that may still have business risk associated with them. Because the ultimate gauge of success rests with market acceptance, which may be several months to a few years away, an element of trust must be present among senior managers, who might otherwise find it easy to question the returns on the company's R&D investment. In this environment, the reporting channel can have a significant effect on a company's commitment to R&D.

The Sympathetic Executive

Surprisingly few corporations elevate R&D to an equal partner status with the other functions, such as manufacturing, marketing, sales, finance, human resources, and quality. This is especially true of those companies in the middle ground (sales of more than about $50 million and less than $1 billion).

Before suggesting an alternative structure, however, R&D management must dispassionately consider the

> Each business unit must define realistic expectations for its funded research, and executive bonuses should be tied to successful transitions of products from R&D to manufacturing.

pros and cons of being administratively slotted under another function. The challenge is to identify the impediments to achieving stated objectives and then pursue the structure that makes the most sense at the time. For example, if the senior executive in charge of sales understands the nature of R&D, that individual can forge alliances with his or her counterparts to gain support for development and even shoulder the budget burden by burying R&D under more acceptable department headings. If this officer holds a strong position in the executive ranks, the R&D manager should stick with that reporting channel until the company implements a more effective organizational structure.

Because the R&D function is especially vulnerable to budget cuts, R&D management must secure the strongest possible ally among the executive ranks; a weak representation is tantamount to a slow death. In time, as goals are achieved and corporate confidence in the role of R&D grows beyond the critical stage, R&D management should push for equal status and senior management representation. Sound judgment is essential for determining when to make this push; poor timing could push R&D down the organizational ladder.

Converting the Skeptic

What if the members of the reporting channel have little appreciation of the process for effective R&D? This can be very frustrating for the R&D manager and the R&D group that is not effectively sheltered from unsympathetic decisions.

The following suggestions can help the R&D manager who must delicately maneuver through the process of re-educating a senior manager about the benefits of effective R&D:

- Diplomatically generate support within peripheral operations that can influence this executive's appreciation of R&D's role.
- Garner support from peers by offering to help meet their short-term needs, and thus trade help from the R&D group for publicity that brings credit to the group and, by association, to the senior manager.
- Avoid any actions that might bolster the senior manager's negative impression of R&D.

- Solicit affidavits from the customers of R&D; nothing has a greater influence on executive opinion than satisfied customers (and potential sales).
- Develop channels of communication with key clients who not only place great emphasis on R&D, but also do not hesitate to inform senior management about the link they perceive between R&D and future sales.
- Introduce senior managers to the ideas ot well-regarded consultants, futurists, and strategic thinkers; this motivates strategic thinking and encourages senior managers to view the company's R&D resources as a means for achieving business objectives.

The R&D manager who successfully converts a senior manager from a skeptic to a supporter creates the strongest possible champion. The stubbornness that limited receptive thinking is now channeled toward supporting the R&D role.

The difficulty of this re-education mission is often compounded by another aspect of the current downsizing and management delayering era, in which decision making gets increasingly concentrated in fewer individuals with wider responsibilities. In this environment, senior managers have little time for speculative thinking about the risks and benefits of development efforts that may be four to five years removed from their contribution to company profits.

The Negative Approach

The negative approach—which casts the R&D champion as an impartial player—can be an effective tactic for gaining senior management support of the R&D function. Rather than promote all the benefits to be gained from support of the R&D function, the manager dwells on the losses the company would suffer through inaction.

Ferreting out facts that executives readily accept takes some effort, but the payoffs are generally worth the effort. Whe R&D manager needs to assess information on competitors, using such sources as annual reports, sales literature, award write-ups, design journals, patent claims, technical publications, and key customers. From that information, the R&D manager can develop profiles of competitors' winning strategies that threaten the company's future business. An effective presentation to senior management benchmarks those corporations perceived as winners, highlights their R&D approach, and compares it with the company's R&D investment.

INCENTIVES

The company's senior managers need to recognize that R&D performance cannot always be measured in terms of revenue-producing deliverables; R&D success includes both individual and team accomplishments. Being left out of the traditional reward system (in which employees enjoy financial rewards for a combination of individual and corporate accomplishment) fosters a sense of discriminatory isolation in R&D personnel. Such systems create the perception that members of the marketing, executive, and manufacturing management ranks are enjoying the rewards R&D's creative accomplishments.

Tradition-bound executives with a hunger for bottom-line measures of accomplishment shy away from the so-called soft side of motivating and rewarding R&D personnel. A company often incurs significant expenses in such activities as distributing memos that laud the efforts of the marketing department, conducting photo shoots of the top sales team, and publishing write-ups about the highest productivity manufacturing cell. The publicity and the expense from these activities can exacerbate the sense of isolation felt by R&D professionals.

Well-executed partnerships have levels of risk and opportunity proportionate to the capabilities of both players by focusing recruiting efforts on candidates from within the company ranks, R&D management can increase the likelihood of fostering internal allies.

Clearly, not all recognition must come via the paycheck (although that still ranks high). Other possible avenues for recognizing the efforts of R&D staff include the following:

- An increased budget allocation for discretionary expenses.
- Greater signature authority.
- An awards dinner.
- Points allocated toward a research sabbatical.
- Freedom to attend more seminars or continuing education courses.
- Approval for an expensive research tool.
- An hour with the company president.
- A write-up in the company newsletter.

THE BUDGETING GAME

The support of those managers who control the company's financial decision making can be tremendously beneficial to the R&D budgeting process. Although most R&D managers would be delighted with a budget allocation that is close to their request, the funding for R&D is determined primarily on the basis of historical expenditures and not on the anticipated value of the risk commitment.

Because the time frame for returns on R&D investments always seems to extend beyond the budgeted periods, the company's financial decision makers must accept a risk that often defies the ritualistic financial bracketing that business managers and senior managers routinely employ. The R&D manager must engender recognition of the proposed programs and present them in recognizable terms that prompt a reasoned response from senior management. Ambivalence on the part of senior management puts R&D in the tenuous position of gaining only superficial support that may be withdrawn at any turn of the company's business fortunes.

Although few R&D managers cherish budget planning time, the exercise becomes less burdensome if it represents the culmination of the R&D manager's efforts throughout the fiscal year, rather than a painful flurry of activity over the course of a few weeks. In brief, the negotiations that surround the budgeting process should proceed smoothly if the R&D manager operates in a selling mode throughout the year, employing the tactics discussed in this chapter.

INTANGIBLE MEASURES OF SUCCESS

The most obvious measure of long-term success is evident when manufacturing and marketing embrace the efforts of R&D, resulting in products or services that are market winners. The intangible indicators of a successful R&D function include reduced isolation from other functions, more requests to "check with R&D," an enhanced sensitivity to R&D's resource needs, a willingness on the part of senior managers to listen to the R&D manager, formal incorporation of R&D goals under corporate strategic plans, increased frequency of spontaneous interaction between R&D management and senior management, a facilitative attitude on the part of senior management, and a generally healthy environment of inquiry and feedback. The R&D manager who successfully fosters most of these traits can then focus on the joy of purposeful discovery.

SETTING R&D PRIORITIES: A CUSTOMER-DRIVEN PROCESS

This chapter describes a process R&D managers can use to gather and use customer input to structure and select R&D projects.

John J. Moran

Two issues occupy the minds of many R&D managers: getting the greatest return on R&D investment through setting appropriate R&D spending priorities and listening to the voice of the customer to increase market awareness in R&D organizations. R&D managers often complain that their departments try to cover the waterfront instead of setting priorities, that they have no systematic way of setting priorities in the laboratory, and that their staffs are not as close to customers as they think they are.

At the same time, R&D managers are operating in an increasingly difficult environment. Foreign competition and quality requirements are increasing, new product success rates are too low, and companies are pressuring R&D departments to adopt a multifunctional or team perspective. All this is asked of R&D managers in an era of downsizing. Today's R&D managers must work smarter; working harder is simply not an option. The only way to survive is to change, to aggressively look for new tools, new approaches, to improve the management of R&D processes.

This chapter suggests such an approach, a new process that uses customer input to improve the R&D decision process for allocating resources and setting priorities. The process combines three elements:

- Quantitative market research—not for conclusive measurements but for the framework it provides for structuring customer input.
- Qualitative market research—to provide an in-depth, second-level understanding of the customer.
- Quality function deployment (QFD)—to relate customer input to specific R&D decisions or priorities under consideration.

The following sections show how these tools can be used to help set priorities in new product development projects, an area in which much improvement is needed.

JOHN MORAN is a certified management consultant and a principal with J. J. Moran & Associates, a marketing/quality consulting firm located in Santa Monica, CA.

R&D AND NEW PRODUCT DEVELOPMENT

When customers buy a product, they are purchasing a bundle of attributes that more closely matches their needs, applications, and preferences than competitive offerings do. Unfortunately, it is often difficult to define this bundle of attributes and to factor these attributes into the new product development (NPD) process. The following elements compose a process that combines existing market research and quality tools:

Structured Input > In-Depth Discussion > Implementation

Each of these elements is an existing tool. In the example this chapter explores, the tools are:

Perceptual Mapping > Focus Groups > Quality Function Deployment (QFD)

In this process, perceptual mapping and focus groups are used to probe and understand customer requirements. The results are then fed into a QFD matrix, where the impact on design decisions (in this example, alternative projects) is determined.

The following are some sample product improvement projects. They address three broad product areas:

- *Product performance* - Speed, quality, storage capacity, and reliability features.
- *Product customer support* - Customer training, product documentation, and telephone support.
- *Product services* - On-site maintenance, supplies, billing, installation, and start-up.

It is interesting to note that these attributes are often the same elements that many firms use to measure customer satisfaction. Therefore, many organizations may already collect similar metrics. In the present example, the R&D member of the multifunctional product team asked for market input to help the team make informed choices among several product improvement projects under consideration. The R&D member felt that the team always chose the technical, product-performance option because that alternative was familiar, not because it would better serve the customer. Marketing agreed to run a series of focus groups to address this concern, and the customer would be allowed a vote on project priorities.

Process Elements

The first step was to look for structural input. A "map" was used to capture and structure the preferences of focus group participants before discussion started. This step allowed respondents to gather their thoughts before they could be swayed or drowned out by more aggressive respondents. Several structured or quantitative approaches could have been used at this point.

Exhibit 1 shows the perceptual map used to gather respondents' ratings of importance and satisfaction for three bundles of product attributes. At the beginning of the focus session, the product attributes (i.e., product services, product customer services, and product performance) were defined, and respondents were asked to plot the satisfaction and the importance of each attribute for their particular application or environment.

Group responses should be plotted to present both the consensus opinion and those reactions that lie outside the consensus. Discussion can then focus on the motivations, needs, and applications behind these choices. It is critical

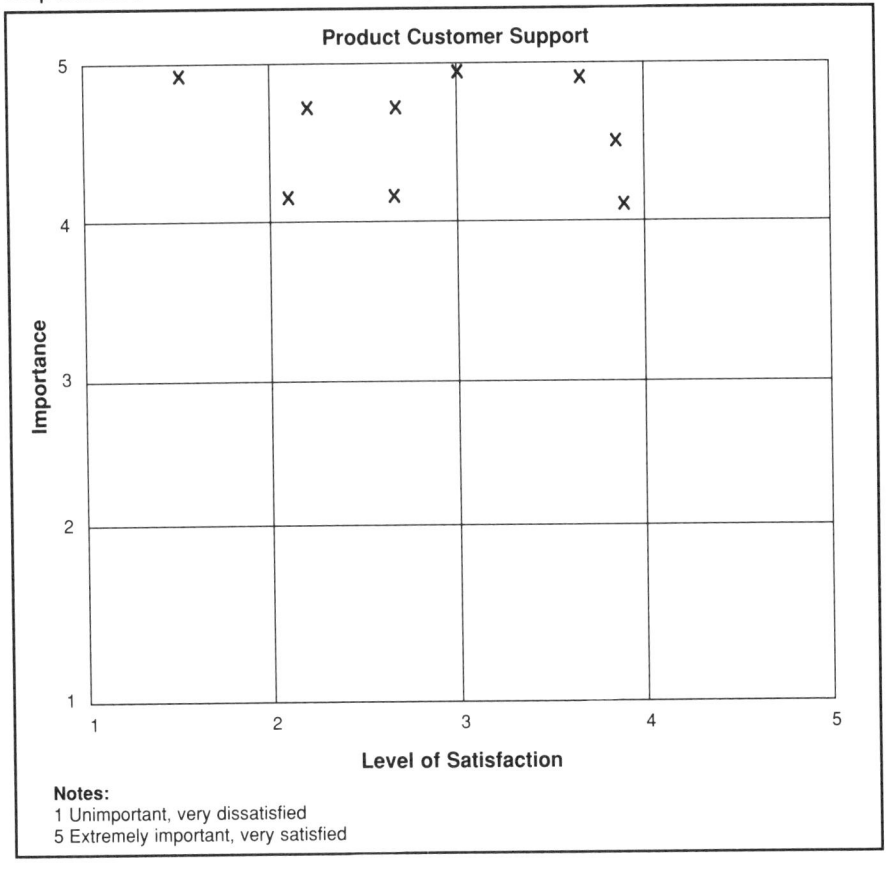

Exhibit 1
Importance and Satisfaction Rating

Product Customer Support

[Scatter plot with Importance (y-axis, 1-5) vs Level of Satisfaction (x-axis, 1-5)]

Notes:
1 Unimportant, very dissatisfied
5 Extremely important, very satisfied

Exhibit 1 *Cont.*
Importance and Satisfaction Rating

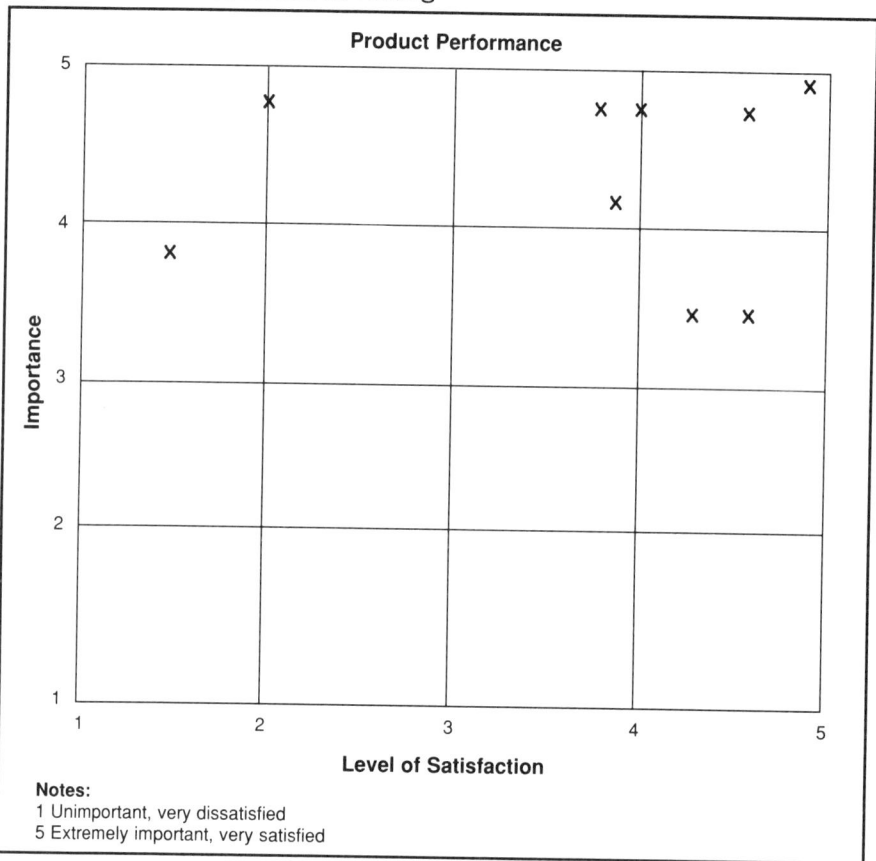

Product Performance

[Scatter plot with x-axis "Level of Satisfaction" (1 to 5) and y-axis "Importance" (1 to 5). Data points plotted as X marks at approximately: (2, 4.8), (3.2, 4.8), (3.5, 4.8), (4.2, 4.8), (5, 4.8), (3.5, 4.2), (1.5, 3.8), (4, 3.5), (4.3, 3.5)]

Notes:
1 Unimportant, very dissatisfied
5 Extremely important, very satisfied

to understand why a customer finds a product attribute important, why it provides satisfaction, and focus groups are an excellent medium for understanding this.

Exhibit 1 shows the group's results. Each individual's response is represented by an *x*. The results show:

- Product customer support was seen as highly important, but respondents varied widely in their levels of satisfaction.
- Product services were less important, but group's satisfaction level was both higher and more consistent in this area.
- Two members of the group appeared to have distinctly different levels of satisfaction with regard to product performance.

Exhibit 1 *Cont.*
Importance and Satisfaction Rating

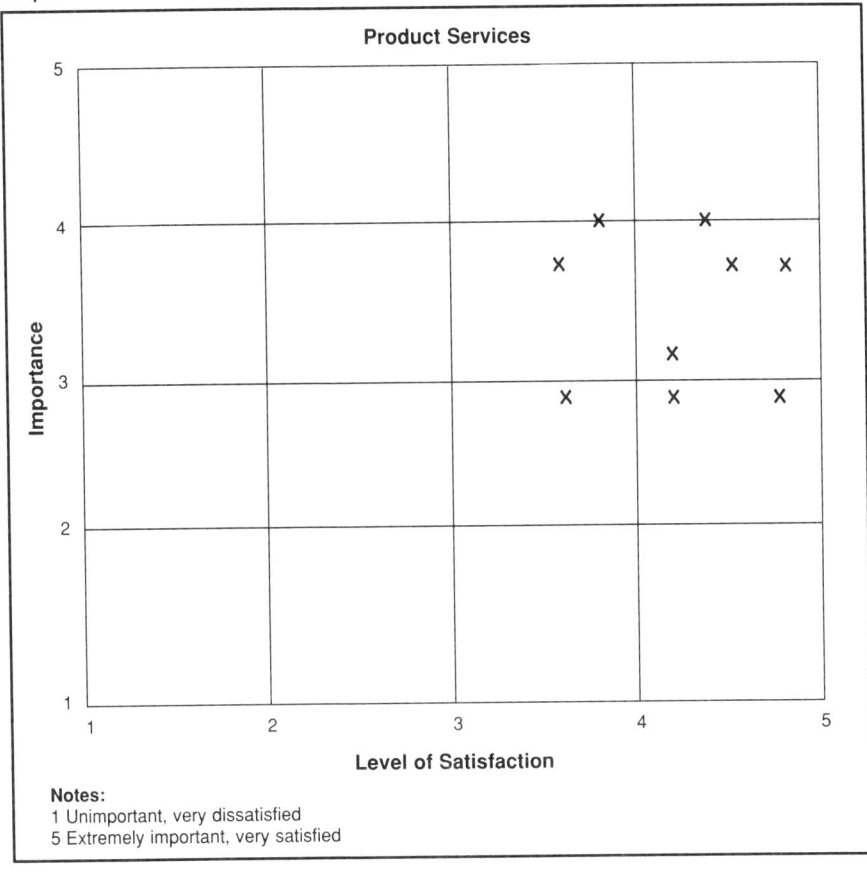

Product Services

[Scatter plot with Importance (y-axis, 1-5) vs. Level of Satisfaction (x-axis, 1-5). Data points plotted at approximately: (3.5, 4.1), (4.5, 4.1), (3.5, 3.7), (4.5, 3.7), (5, 3.7), (4, 3.2), (3.5, 3), (4, 3), (5, 3).]

Notes:
1 Unimportant, very dissatisfied
5 Extremely important, very satisfied

This visual feedback is an important dimension that adds considerable value to a group discussion.

The product team used the importance and satisfaction ratings to create a QFD matrix that emphasized customer input in setting project priorities. Exhibit 2 shows a simplified QFD matrix in which the plotted results of the focus group have been related to specific alternative projects under consideration. Even though the focus group results are not conclusive, they provide an important means to understand better the data that has been collected. Tying these preliminary results to specific alternatives can be extremely valuable. Although a complete description of QFD is beyond the scope of this chapter, a brief description of Exhibit 2 follows.

Exhibit 2
Implementing the Voice of the Customer

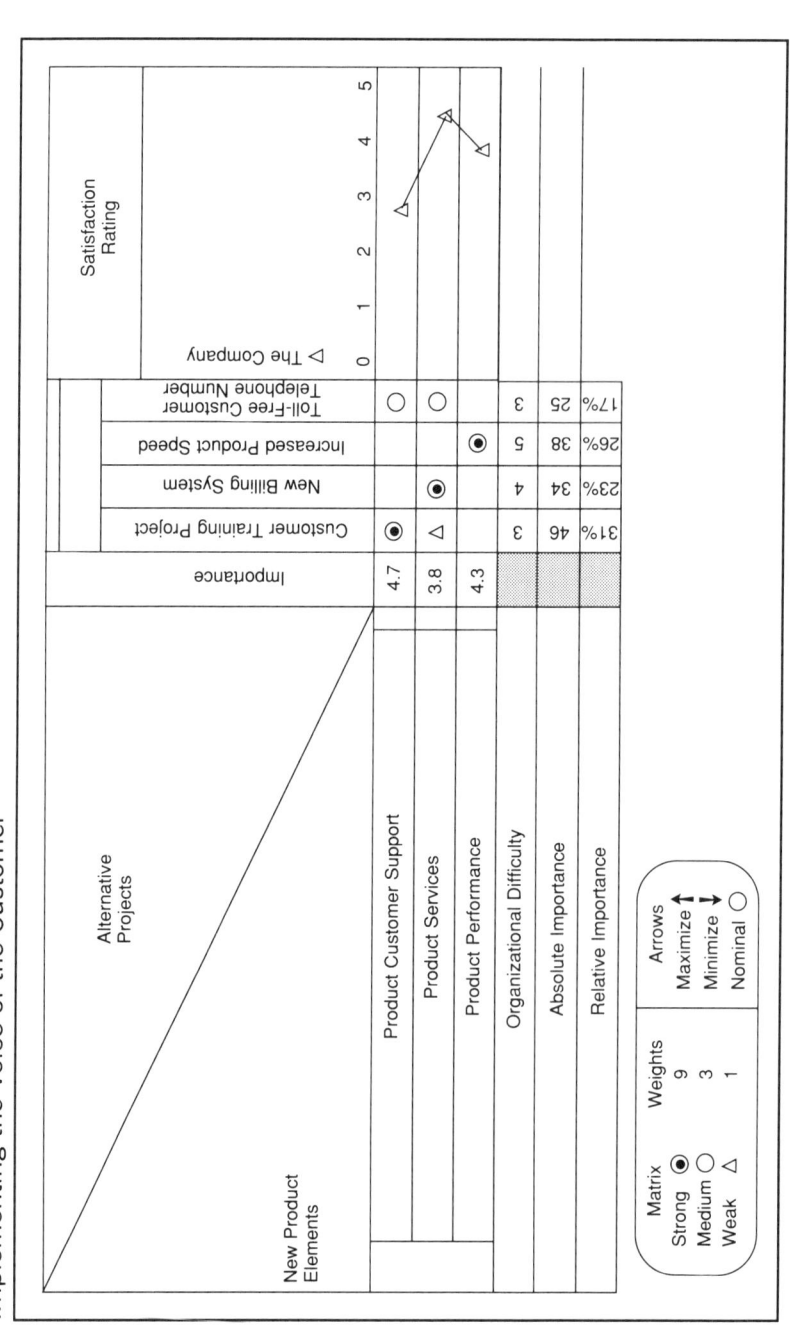

New Product Elements	Importance	Customer Training Project	New Billing System	Increased Product Speed	Toll-Free Customer Telephone Number
Product Customer Support	4.7	●			○
Product Services	3.8	△	●		○
Product Performance	4.3			●	
Organizational Difficulty		3	4	5	3
Absolute Importance		46	34	38	25
Relative Importance		31%	23%	26%	17%

Satisfaction Rating: △ The Company

Matrix		Weights		Arrows	
Strong	●	9		Maximize	← →
Medium	○	3		Minimize	→ ←
Weak	△	1		Nominal	○

MANAGEMENT OF R&D

Process Results

Exhibit 2 is the output of the team's QFD exercise, sometimes called the house of quality. The importance and satisfaction scores for each of the new product elements represent the average scores presented in Exhibit 1 (i.e., the averages of the nine respondents or data points plotted on the various axes).

> When customers buy a product, they purchase a bundle of attributes that more closely match their needs than competitive offerings do.

Against these new product elements are arrayed four alternative projects originating in R&D and other departments. For each project, the team comes to consensus on two pieces of information:

- The impact that project will have on each corresponding new product element. Strong = 9; medium = 3; and weak = 1 or none.
- The organizational difficulty in implementing that project (1 = easy, 5 = very difficult).

At this point, the absolute and relative importance (shown at the bottom of the chart) for each is calculated by summarizing the product of the importance factor and the matrix weight for each new product element. For example:

$$\text{Customer Training Project Absolute Importance} = (4.7 \times 9 = 42.3) + (3.8 \times 1 = 3.8) = 46.1$$

Exhibit 2 captures the satisfaction and importance scores and, more significantly, relates them to the four alternative projects under consideration. Thus, customer input helps set project priority. Exhibit 2 shows the customer training project as a clear first choice for the following reasons:

- Its relative importance was the highest.
- It had strong impact on the most important, least satisfied new product element—product customer support.
- It was relatively easy to implement (the organizational difficulty score was 3).

BENEFITS

This just-described process has been used for a wide range of assignments, including new product development, investigations into customer satisfaction, and project design. With new product development, a conjoint analysis front-end can be useful to measure the relative attractiveness of specific product design attributes. For new products that rely on customer satisfaction and service, the perceptual-mapping front-end approach previously described works best.

This process provides a way to use customer input to rank projects according to priority. It facilitates the use of customer input as a major consideration in making the trade-offs that are necessary in managing new product and R&D efforts.

Another advantage is the process's simple, modular approach. Built around focus groups, the process can be implemented in stages by using a mapping approach to improve the focus group discussion or by using a simple QFD matrix to decide how the information developed in the focus group will be used. The team can also use the process to better understand decisions that might otherwise require more expensive research into statistically representative populations or different market segments.

Designers using QFD can not examine a design element in isolation. An assessment of customer reactions, competitive impacts, and the consequences on other product design elements are necessary.

The holistic QFD approach facilitates a team-based decision process, whereas deterministic models are better suited to the functionally isolated approaches of the past. Even advocates of deterministic models soon realize the true value of the QFD model is not the accuracy of its results but the discipline of the decision process. QFD is only a process; it does not really produce a result or answer. As such, it highlights continuous improvement as a critical strategy in achieving lasting, profitable results. This emphasis on improving the process can be a powerful force to facilitate the creative, risk-embracing, failure-accepting culture needed for continuous improvement.

THE QFD MODEL AS A DECISION MODEL

QFD differs from many business models in that its major value is holistic rather than deterministic. The purpose of other models is to determine an answer. For example, the major output of the project schedule model is the completion date and determination of the critical path. The financial model calculates both peak cash flow requirements and return on investment. The market model generates units sold, revenue, and market share. QFD is different. It does not calculate an optimum design. The only calculation it involves is the absolute importance of design elements in meeting customer requirements.

The QFD exercise primarily consists of recording symbols into a matrix to agreed-on team conclusions. It is really a large, sophisticated, graphic checklist! QFD's holistic, interactive approach in documenting team consensus is its value.

CONTINUOUS IMPROVEMENT

The customer-driven process described in this chapter can be continuously improved in several ways. The elements in the process can be improved (e.g., by substituting conjoint analysis for perceptual mapping) or by extending the process to include other process elements that would improve the decision or priority objective.

Another approach or tool called organizational mapping can be easily designed as a fourth element that plugs into the QFD matrix to help define the team and measure the communication needed to address a specific project or task. Organizational mapping is a graphic tool that defines a team or other organization as a series of connected nodes not unlike a local area network. Organizational mapping measures the frequency and importance of the communication links between individuals (or nodes) on specific tasks or projects. The resulting map provides insight into team interaction, structure, and communication. It shows how team members actually interact and provides an organizational communication metric that can facilitate improvement in team design and performance.

It is critical to understand why a customer finds a product attribute important, why it gives satisfaction.

SUMMARY

Getting closer to the customer and setting R&D priorities are complementary requirements. Providing the R&D manager with a process that uses customer input to set priorities and evaluate trade-offs is an important step to help increase the return on R&D spending.

R&D STRATEGY AND COORDINATION IN MULTINATIONAL SETTINGS

Stephen Hellman

Large multinational corporations frequently operate a variety of R&D laboratories in diverse locations worldwide. This chapter discusses how a US-based consumer products company seized the opportunity to increase productivity and output by implementing some basic management strategies.

Large multinational corporations frequently operate a variety of R&D laboratories in diverse locations worldwide. In many cases, such a network of laboratories has not been deliberately planned and managed as part of a worldwide strategy. Rather, it may be the result of the acquisition of affiliate companies with established R&D sites, unofficially established labs that grew from a dissatisfaction with the support provided by geographically (and often culturally) distant central labs, or local quality assurance or technical service facilities that have broadened the scope of their charters to become involved in product and process development, or even basic research. In cases in which worldwide R&D has grown in this unplanned fashion, significant opportunity exists to increase productivity and output by implementing some basic but helpful management strategies.

ONE CASE HISTORY: CONSUMER PRODUCTS R&D

A particularly good example of the opportunities that exist to increase worldwide R&D productivity involves a consumer products company in the personal products industry. This US-based company maintained a large laboratory near its headquarters operation, with a broad mission that included basic research; product, process, and package research; sensory evaluation; analytical methods development for quality assurance; and regulatory intelligence and submissions activities. The laboratory developed products for both US and international markets; these products represented slightly more than 2% of its division's sales on a worldwide basis.

Interaction between this main lab and overseas labs had been quite minimal. Virtually all contact resulted from one-of-a-kind initiatives that required collaboration with one or two specific labs at affiliates. For example, if a decision was reached to replace a production line with specialized capabilities at a location in France, the main US lab and the French lab would be represented on a task force (along with manufacturing representatives, of course) charged with identifying options, developing a recommendation, and managing

---

STEPHEN HELLMAN is a director of technology planning at Clairol Corp. in Stamford CT.

implementation. Such an activity would generate interaction between the labs involved, but it would be of a sporadic nature, occur only in a narrow area of endeavor, and generally involve the same laboratories time after time. Other more remote labs were hardly ever involved, and interaction with them was nonexistent.

Compounding this lack of interaction were certain organizational barriers. Although R&D at the headquarters location enjoyed close and frequent interaction with marketing and manufacturing management on a worldwide basis, the labs at the overseas locations did not. Most often, such labs reported to manufacturing, and incentives among their management led to minimizing expenditures in any way possible, particularly for an R&D group that did not represent the main goal of the organization. Travel to other R&D labs was discouraged for its attendant costs and time demands, and providing technical support to fire-fighting at a manufacturing facility received higher priority than R&D activities.

Many of the overseas R&D facilities were very small. Many labs had fewer than 10 scientists—frequently, fewer than 5. The small scale of such labs raised issues of critical mass, succession planning, and corporate vulnerability when retirement or unexpected separation occurred.

GLOBALIZED R&D MANAGEMENT STRATEGY

To address realities such as these, several steps were taken to help remove barriers to collaboration between laboratories worldwide:

- Identifying all labs worldwide along with a point of contact and language options.
- Arranging site visits by R&D management to build rapport and develop a capabilities profile for each lab.
- Holding an annual worldwide R&D meeting to promote technology transfer and build rapport.
- Developing a worldwide R&D quarterly report describing each lab's activities.
- Creating a long-range worldwide R&D strategy, including definition of roles for each lab.

Each of these topics is discussed in some detail in the following sections.

Identifying Worldwide Laboratory Sites

For a small company, identifying R&D sites may be a trivial exercise. In larger companies, however, this is not necessarily an easy task. Conflicting information is common, arising from a highly flexible definition of what constitutes a real R&D site. Particularly when routine communication has not occurred, rumor and outdated impressions are often the only available starting points.

Sometimes a two-person plant engineering department is regarded as an R&D group, perhaps because five years ago while adapting an old production line to handle a new product introduction the group members inadvertently created a new product that was quite successful. Occasionally, a plant quality assurance department may get creative and develop prototypes of new products (usually without required stability testing, sensory testing, or other standard testing). In such instances, further investigation shows that R&D does not in fact occur on any substantial level and that a consistent supply of R&D resources is not available for high-profile projects when the need arises.

If no reasonable leads exist for determining where overseas R&D is located, the human resources function that services international is a good place to start.

If no reasonable leads exist for determining where overseas R&D is located, the human resources function that services international is a good place to start. Its data will usually be only approximate but may be the best starting point available. Regardless of the source of early leads for R&D locations, however, it is critical to identify the prime point of contact for that location, along with voice and fax numbers. Initial contacts can help establish what languages are spoken by the R&D staff—another early constraint in developing a worldwide R&D plan—as well as a recent history of collaboration with other R&D locations.

Site Visits and Capabilities Profile

A program of visits to each R&D site is the next logical step in establishing worldwide R&D coordination. Care should be taken to establish early written communication with the point of contact describing the nature of the visit. Since that person may operate in a much more hierarchical environment, the formal notification to superiors that may be required will be easier with a letter than with strictly on-the-phone arrangements. In addition, an unprecedented visit from headquarters could easily be construed as having an agenda much more threatening than the actual one, and a letter can be drafted to put some of these concerns to rest. Finally, written communication may help avoid misunderstandings that can arise when either party is not working in its native tongue.

Sending a list of questions in advance of the visit will go far to make the trip not only a productive one but also a pleasant experience on which to build a lasting relationship. Prior knowledge of the kinds of information that will be needed gives the contact time to assemble portions that he or she may not already have while avoiding the potential loss of face that might accompany the person's inability to supply what was requested when he or she has had no prior knowledge. By seeing the general nature of the information being requested, the contact may become more comfortable with the actual nature of the trip.

Typically, the information that should be covered in a capabilities profile includes the number of scientists and technicians in the lab, the discipline and level of education of each scientist, an inventory of major equipment, estimates of throughput for appropriate tests (e.g., 100 heavy metal analyses per month or 50 GC runs per week to qualify lots of fragrances), and areas of special interest or capability. Types of computers and software are useful to inventory as well, along with language capabilities of the staff. It is also useful to know whether a plant is on the same site as the lab and, if so, what production lines or pilot plant facilities are available. Other information peculiar to the specific industry may of course need to be added.

The capabilities profile that is developed from a site visit is of critical importance when a worldwide R&D strategy is being created. The profile can help identify what R&D work can be conducted at each site, where redundancies and gaps exist, and likely candidates for consolidation.

A final comment relevant to initial site visits and early interaction with an offshore R&D site generally is that they are cross-cultural experiences in which there is great potential for misunderstanding and misinterpretation. Particularly for early contact before a personal relationship has been established, visitors should try to be as sensitive as possible to colleagues' expectations and protocols. Compared with that of most other cultures, US business etiquette is much more informal, fast paced, and outcome oriented. The extra effort required to sense the process-related details that colleagues are observing will go far in ensuring the proper outcome.

The Annual Worldwide Meeting

Once a clear picture of the capabilities of each R&D site has been developed and at least a rudimentary relationship established with each R&D manager, a logical next step is to gather all the managers together for several days or a week. The purpose is two-fold: to exchange technical information for mutual benefit and to develop rapport among the R&D managers (both headquarters-to-offshore managers as well as among offshore managers themselves). Although there is no doubt that the opportunity for in-person exchange of technical information would of itself justify the expense of such a meeting, the rapport developed during a week together is invaluable in easing the barriers to picking up the phone when a technical problem arises after everyone is back home. The author's experience is that headquarters-to-affiliate communication develops first, followed by affiliate-to-affiliate contact. The author believes both are critical to a successful worldwide R&D operation.

Developing a successful and relevant agenda for a worldwide R&D meeting is a simple task. Generating a preliminary list of topics is easy, particularly with input from colleagues in adjoining offices. Some fairly ubiquitous topics are recent new product development activities at each R&D site, competitive product news, local regulatory trends, developments in quality assurance methods and management, packaging issues, and project management techniques. The author's company has had good success in sending the preliminary list of agenda items to all participants well in advance of the meeting and asking them to do two things: to indicate their level of interest (high, medium, or low) in each topic proposed and to suggest additional topics not on the list. A quick review of the responses to this survey can generate the final list of topics, which will have the built-in input and buy-in from the meeting attendees. Care should be taken to have presentations and discussions balanced between headquarters R&D staff and affiliates; suggesting specific topics for which the affiliates prepare in advance may be helpful in ensuring that such a balance exists.

> **A program of visits to each R&D site is an important step in establishing worldwide R&D coordination.**

If the formal agenda is the obvious place for technical information to be exchanged and for interaction to occur in a familiar and comfortable manner, the evenings are the place for relationships to be explored and true rapport developed. The comments offered previously regarding cross-cultural exchanges apply equally well here. Suffice it to say that entertaining overseas guests in the evening is something that they expect and value, that it can generate the glue that holds a widely dispersed technical organization together, and that a small amount of planning and orientation of colleagues can go very far in making after-hours activities supportive to the larger goals of the meeting. R&D managers should consider purchasing and distributing to headquarters staff some of the excellent books on the topic of cross-cultural communication and behavior that are available in most good bookstores.

The Worldwide R&D Quarterly Report

If an environment in which technical information flows freely between labs is successfully developed, meeting face to face once a year becomes inadequate to keep everyone up to date. Even with the occasional visits that may occur outside the realm of the annual meeting, it is still difficult to keep people in touch. One way to supplement the yearly gathering is to generate a worldwide R&D quarterly report that describes the major projects at each site. Such a document can go far in promoting technology exchange and is valuable for headquarters staff concerned with limiting redundancy in the worldwide R&D portfolio and increasing productivity. This document can also drive worldwide R&D coordination in cases in which a large project has been separated into several components, each assigned to a different lab consistent with its technical specialties.

The author has found that the common practice of simply including each lab's input to the report in its own section and distributing the resulting document has limited value. The author's company has had better results when input is assembled into sections of common technology, so that a specific technical need results in the scanning of a specific section.

One last point of this topic is that no matter how comprehensive a lab's description of the work it is conducting, an interested party will invariably need to follow up for more details. The author encourages input to the quarterly report to be as brief as possible, with a clear point of contact for each project. This policy reduces the time required by each lab to prepare the project summaries and also reduces the time required to read them. Shorter is better. Hands-on points of contact are critical.

Developing a Worldwide R&D Strategy

There is a wealth of literature devoted to the topic of developing and implementing R&D strategy. Assuming that an R&D strategy has been developed and that it recognizes the necessary linkages to marketing, manufacturing, and the rest of the company, the logical next step is to increase its scope to worldwide rather than just domestic R&D.

To ensure that the entire R&D portfolio supports the R&D strategy, managers obviously need to know the composition of the current portfolio. The steps described previously should help.

These steps should also develop a good sense of what each lab's capabilities and limitations are—from a human resource, hardware instrumentation, and organizational perspective. (A cost perspective is also useful, if harder to elucidate.) This knowledge, combined with the larger R&D strategy, presents the opportunity to move the current R&D portfolio to one that is strategically better suited to company goals, delivers results faster or less expensively, develops and retains critical R&D talent, and allows repatriation of more affiliate profit through opportunities for creative accounting procedures.

Inherent in developing such a worldwide R&D perspective is developing a sense of the role that each lab should play in the overall R&D structure and moving each site closer to that ideal. Research may be limited to a single site and engineering of a new production equipment to another. The concept of centers of excellence may be appropriate to the organization. Alternatively, geographical variation in raw material supplies, regulatory constraints, or unique customer requirements may dictate the use of broad-based minilabs in many locations. Some labs may have only minimal value when a global perspective is developed and could be candidates for rationalization. Regardless of the specific arrangements that make sense for the situation, however, a clear sense of the role of each lab is fundamental to the success of a worldwide R&D strategy and the R&D coordination that is dictated by that strategy.

EVALUATING R&D PRODUCTIVITY

Harold G. Wakeley

This chapter looks at three aspects of R&D productivity: a short history of methods used to measure and improve R&D productivity, organizational issues that influence R&D productivity, and the development of measures of R&D productivity.

Present-day companies could not survive using Thomas Edison's organizational methods, as tempting as they may sometimes be. As an R&D manager, Edison was a total failure. His lack of scientific mathematical skills made him uneasy when working with people of recognized talents and his lack of management ability prevented him from effectively delegating work to people who had the skills he needed. Finding Edison's managerial style inhospitable, those good scientists who did come to the laboratory moved on quickly. In the Edison laboratory, R&D management consisted of Edison assigning projects to individuals and then making rounds to ensure that his directives were explicitly followed. A researcher's value was based largely on the salary paid, rather than on scientific output.

A direct consequence of this form of management was that rarely did any project other than the one that Edison was personally interested in proceed smoothly, and more than two projects were never sustainable. Even though his procedures earned him more patents than any other person (1,093), his methods caused him to miss important discoveries—for example, the development of synthetic rubber—that could have been accomplished in his laboratory. Organized research, such as the work done at Westinghouse and General Electric, very easily outdistanced that of Edison's laboratory.

World War II was the watershed period for R&D management as a means of controlling highly complex research. It became clear to government and to industry that some research approaches exceeded the capacity of any single person to understand completely and handle competently. In combat, the development of the task force changed warfare forever by focusing the energies of people with a wide range of competencies on a single goal. Similar approaches were applied to the development of weapons systems, the most conspicuous being the atomic bomb, for which widely disparate units contributed to a end.

This experience with organized R&D as it applied to warfare showed that some methods could yield better results than others. Since World War II, industrial companies have also taken many approaches to find better methods of organizing R&D and measuring and improving R&D productivity.

This chapter looks at three aspects of R&D management and productivity. First, a short history of methods used to measure and improve R&D pro-

---

HAROLD G. WAKELEY, PhD., is a principal scientist at the Human Factors Research Group in Winnebago IL.

ductivity will be presented. Second, some of the organizational issues that influence R&D productivity will be examined. Third, the development of measures of R&D productivity will be analyzed.

A SHORT HISTORY OF METHODS TO MEASURE AND IMPROVE PRODUCTIVITY

Frederick Taylor, with his theories about rigid structure in workplace design, marked one of the earliest points in the development of approaches for increasing productivity in manufacturing. At roughly the same time, Thomas Edison employed a similar approach to improving R&D productivity—using cadres of laboratory technicians. Edison's approach was particularly effective in applying new technology to the solution of problems. His application of vacuum pumping systems to the development of the electric light is an excellent example of this approach. Edison's ability to use new technology in practical applications allowed him to achieve very high levels of R&D productivity at a given level and within Edison's abilities.

The most profound changes in the ways in which R&D was conducted occurred during World War II. Until then, aside from Edison's laboratories, little had been done to apply the results of basic research to the development of usable products. During World War II, it became clear that a more effective method was needed for applying the new scientific and engineering knowledge being developed in universities. The previous approach to applying scientific and engineering knowledge—teaching students, who would then go into industry—was no longer working.

One of the most important mechanisms for applying this new knowledge was the development of the not-for-profit research center. Research carried out under contracts with research centers became a practical means for industry to obtain the knowledge needed to develop new products. The contracts described the problem concretely and limited the scope of research to variables likely to yield a useful product.

R&D Productivity Becomes Imperative

Contract research, however, was not sufficient to meet industry or government needs during the 1950s and 1960s, when the development of highly complex products required the coordinated efforts of many different technical disciplines. The push into space marked a further watershed in the way R&D was conducted, with hundreds of companies all peforming research that was used in systems and subsystems that all served a single purpose. The enormous expense of this effort along with the continual pressure to meet ever-tighter deadlines was sufficient to spur a number of companies to consciously seek ways to improve the quality and quantity of their R&D. Whereas a few years earlier R&D laboratories had been content to develop a single product at a time, usually under the guidance of a handful of scientists, aerospace companies were now committing tens of thousands of scientists and engineers to the conception and development of hundreds of separate systems that had to work in harmony, with absolute reliability, while performing at their physical limits.

The most profound changes in the ways in which R&D was conducted occurred during World War II.

To gain and maintain a competitive advantage, a number of aerospace companies deliberately set out to study how organizations could be designed so that R&D would proceed most effectively. Even though management consultants, industrial engineers, and social science researchers for years had been studying the problems of people working in groups, it now became apparent that such work was more than just an exercise for academics. Managers of these aerospace companies realized that they could gain real improvement only if the principles of human behavior could be understood and accounted for in organizational design.

During the 1950s and 1960s, a number of studies concerning innovation identified two important factors affecting successful innovation: the flow of information among group members and focused decision making. Members of highly innovative groups typcally used peers as sources of new information. They also frequently asked supervisors to provide critical evaluation of their work. In addition, successful supervisors used information from outside the group to critique the group's work.

These studies found that successful supervisors helped integrate disparate aspects of solutions developed by the group. This role of the successful supervisor complemented the activities of the group, particularly those related to generating and exchanging complex technical information. Thus, the studies pointed out that in the production of highly innovative research, the supervisors should serve as an integrator of information by facilitating communication among group members, relaying messages and integrating ideas so that a single unified solution could occur. If the supervisor did this, the group members could function as idea getters without being inhibited by the fear of having their ideas continually evaluated by the supervisor.

Once the problem of improving R&D productivity had been recognized, the challenge of measuring it became significant. Although the neat and tidy social science studies of innovation had pointed out critical factors in innovation, they did not offer much advice about measurement. Furthermore, these studies could not be easily applied to the industrial environment, where demands for steady, productive work leading to useful products were unrelenting. In industry, innovation had to be considered as only one of a number of activities required to keep a company profitable. It would be challenging for a company to maintain an orientation toward innovation while accomplishing the myriad tasks required to run a successful business.

Studies concerning the measurement of R&D productivity commenced and reviews of the literature on scientific performance showed that there was little agreement about what constituted scientific output or what measures should be used to reflect that output.

In the 1960s, both General Dynamics/Astronautics and Martin Marietta tried to apply traditional time study methods to the quantification of scientific and engineering productivity. These methods were spectacularly unsuccessful. In the studies, random observations of the activities of scientists and engineers were recorded. Some of the measures consisted of talking to a peer, reading, writing, and drawing. Thousands of scientists and engineers learned during these studies that communicating information within the group was penalized, while working in isolation was highly rewarded. While these studies may have led to an increase in productivity of certain technical groups that were performing repetitive tasks, the productivity of the technical groups that were responsible for creating new subsystems plummeted. Shortly thereafter, the time studies were cut back and eventually abandoned.

Meanwhile, other analysts were using other approaches to measure the productivity of technical groups and of scientists and engineers. These analysts found that even though rating systems based on subjective factors contained certain biases, objective measures of R&D productivity left large amounts of variance unexplained. For example, correlations between the number of publications of a scientist with the actual contributions of a scientist were well under .39. Similarly, the number of patents of a scientist, when weighted for age and experience, was still not helpful in identifying the scientists who were extraordinarily productive.

Measuring the quality of a scientist's work was found to be just as difficult. Significance, relevance, timeliness, and other measures of quality could not be weighted and put into a useful algorithm.

Factor-analytic studies were proposed that would determine how to evaluate a scientist's written output, how to determine the quality and quantity of this output, how to evaluate citation counts, and how to measure the degree of originality or creativity contained in technical work. But these studies did not yield useful results, either.

Even though problems with measurement proved difficult to overcome, these studies of organizational and group effectiveness did show that one's rank and whether the person was a member of a work group strongly affected communication patterns. Although supervisors could play an important role as gatekeepers to information, there appear to be group contributors who were not necessarily high technical performers but who played an absolutely vital role in the communication within a work group.

After many of these studies of R&D productivity had been completed, a factor analysis was conducted to determine the effect of behavioral science variables on laboratory management. This analysis showed that the results of these many studies were not contradictory but would be difficult to integrate. For example, some studies focused on the variables that affectd the climate of the laboratory, such as leadership style, organizational structure, and work load. Other studies concentrated on physical aspects of the laboratory, such as equipment, support staff, funding, libraries, and physical layout. Still others looked at variables involving the scientists or engineers themselves, such as inherited abilities (as measured by various tests) and technical achievement or performance. Additional studies focused on the variables that a manager could control, such as awareness of company needs, management desires, and the activities involved in research (e.g., conducting experiments, reading literature, making visits to customers, holding technical discussions, and consulting with others). Finally, other studies considered measures of R&D productivity such as patents, solutions to problems, and the ability to conduct R&D projects on time and under budget.

Although drawing definite conclusions from combining the results of these studies proved difficult, they did indicate that the study of R&D productivity was becoming focused and that it was receiving serious attention as a means of relating R&D goals to the goals of the company.

Successful supervisors helped integrate disparate aspects of solutions developed by the group.

New Developments Regarding R&D Productivity During the 1970s

During the 1970s, the most impressive study related to improving R&D productivity (as opposed to measuring it) was conducted by analysts at Hughes Aircraft Co. Hughes Aircraft supported a complete review of the state of the art in R&D productivity in an attempt to improve its own management practices. The analysts who conducted the review put their conclusions in a practical notebook that was easy to use; it was published and distributed to thousands of its R&D managers. What was most impressive about Hughes Aircraft's notebook was that it presented a clear description of R&D productivity goals, criteria, measures, methods, and counterproductive practices, all in a form that every manager could use. The Hughes Aircraft analysts also consulted corporations with large commitments to R&D to assure that the notebook represented a summary of the best thinking and practice at that time. It became known among R&D managers as "the book of management rules that I always knew, but may have forgotten."

The practical nature of the notebook was exemplified by such chapter titles as "Improving Operational Productivity", "Improving Managerial Productivity", and "Improving Employee Productivity." The notebook also provided hands-on support for the user with checklists of specific activities and practices that would yield high R&D productivity.

Because Hughes Aircraft was committed to using the results of its study, it not only distributed the notebook to all of its R&D managers but set up a program of group meetings and critiques about the notebook that continued for years. Hughes Aircraft program was based on two principles: that employees take their cues from management and that the greatest improvements in R&D productivity occur when management takes a systems approach. Hughes Aircraft's approach to improving R&D productivity was professional and endorsed strongly by top management. Improving R&D productivity was not made an end in itself but was treated as a normal line-management responsibility. Formalization of methods to improve R&D productivity through rules and regulations were discouraged. Instead, the emphasis was placed on improving communication among technical contributors who were directly involved in carrying out R&D projects.

By the 1970s, therefore, the first generation of studies on R&D productivity had matured. The efforts of General Dynamics/Astronautics had launched company studies of R&D productivity. And the efforts by Hughes Aircraft were a vast improvement over the General Dynamics effort, both in terms of level of effort and how the new knowledge was applied. The common elements in the General Dynamics and Hughes Aircraft efforts were support by top management, a focus on specific goals for improving R&D productivity, and a program for applying this knowledge in company operations.

During the 1970s, the study of R&D productivity began addressing the problems of information overload. The volume of technical literature and data bases was growing so quickly that in some technical disciplines, the knowledge that students gained during their first year of university was obsolete by the time they had graduated. On top of this, students who had taken advanced graduated degrees often had more technical knowledge than many R&D managers.

Consequently, the problem of bringing new people into a mature R&D department came to light. Issues related to how scientists and engineers collaborate and communicate, how teams work effectively, how a key technical resource person could be used effectively, and how the performance capabilities of the new generation of R&D researchers could be matched to R&D assignments all had to be dealt with. As a result, by the 1970s the study of R&D productivity had developed many new dimensions.

R&D managers also had to use new tools to handle increasingly complex situations within laboratories. For example, they began using statistical studies and modeling approaches to match capabilities and resources to project needs.

During the 1970s, new members of R&D staffs frequently became dissatisfied when the demands of their jobs did not prove challenging enough. In order to keep new recruits motivated, R&D managers had to deal with new issues, such as varying assignments, granting autonomy, encouraging identification with a task, and giving feedback. The baby boomers wanted to be heard and respected.

Studies of R&D productivity had to deal with these new complexities within the laboratory. Several studies came to the conclusion that in order to manage this new environment effectively, R&D managers should push decision making down to the level of the technical contributor, reduce the hierarchy within the laboratory by expanding horizontally into larger teams, use team supervisors as communication facilitators rather than as the sources of innovation, and promote in every way possible the transfer of information between technical contributors.

> **Once the problem of improving R&D productivity was recognized, the challenge of measuring it became significant.**

Although the study of R&D productivity was moving in these new directions, the older and more pragmatic methods of improving R&D productivity were not yet abandoned. Representative of these older methods was a study of how an R&D laboratory methodically improved productivity by introducing a number of industrial engineering practices or tweaking the system as Taylor had taught at the turn of the century. The goal of the effort was to simplify the work, as opposed to helping the laboratory increase its standards. Specifically, the goals of the effort were to:

- Minimize hidden, lost time.
- Eliminate unnecessary work.
- Combine or rearrange job elements.
- Simplify laboratory test procedures.
- Move to automated or labor-saving devices.

Not surprisingly to either R&D managers or industrial engineers, it was established in this study that hidden, lost time was the principal area of inefficiency in the laboratory. Work loads were cyclical. To accommodate peak loads, however, adequate labor had to be provided. Management was encouraged to schedule the work load more carefully, to eliminate some of the peaks and valleys. During the peak work load, less-skilled staff from related areas were to be used. Managers were also encouraged to identify tasks that were not directly related to the R&D department's goals and to eliminate them. Programs were devised for gaining employee participation and for establishing survey teams to obtain special information. Through the implementation of these programs, the results of the study were applied in the real setting of the laboratory. Often enough, the measures adopted did not require significant investment or organizational change.

Studying R&D Productivity in the 1980s

A new trend in the study of R&D productivity began in the early 1980s. By then, computer-controlled experiments were being conducted in many laboratories. Most laboratories also had access to a variety of data bases. Information became so available, in fact, that access was less of a problem than sorting useful information from irrelevant material. A management information systems (MIS) department was established in many companies so that senior management could have access to even the most trivial data. This led some managers to argue that there was no longer a need to push decision making down to the level of the technical specialist because senior management could access and manipulate data about company operations immediately, with the aid of spreadsheets. Spreadsheet management became a game in which everyone could play. Now senior management also had a tool for dealing with the perplexing problem of evaluating R&D; it could use data about company operations to quantify R&D performance.

Auditing departments in companies also were interested in using these new information sources as a method of internal control. In 1947, the AICPA had explicitly limited the scope of review in an audit to accounting and financial matters. By the late 1970s, however, the AICPA concluded that operations were also a necessary part of an internal audit and by that time most internal auditors were examining all of a company's activities.

There were a number of reasons why the internal auditing department expanded its coverage to include R&D activities. Since World War II, the budgets of R&D departments had grown enormously. During the 1960s and the 1970s, moreover, the development and commercialization of new technology seemed less risky in the US because the expanding US economy seemed able to accommodate virtually every new technological development, from latex paint to semiconductors. By the end of the 1970s, however, this picture had begun to change.

As an illustration, during the 1950s, not-for-profit research institutes had proliferated and during the early 1960s, most of them embarked on major expansion programs. But with the Vietnam war going strong and with Americans' priorities shifting away from technological R&D, most of these not-for-profit research institutes found support hard to obtain.

Furthermore, the cost of doing research increased significantly during the 1970s. Senior management of many industrial companies found that cutting in-house R&D was an easy way to cut costs. They also sought new tools to help them make better decisions about R&D. The internal audit proved a useful tool for cost cutting. The auditor's principal charge was to conserve assets with a view to improving the short-term bottom line. In the 1980s, with more information available through data bases, the auditor was able to shift from just inspecting and auditing for compliance to playing an advisory role through recommendations for improving control sytems.

Internal audits of company operations have been controversial. Managers of various audited functions have charged that these inquests were dysfunctional, led to inflexible policies, and were a barrier to innovation. However accurate these charges, intenal audits have become a tool for review of individual corporate operations, evaluating whether their work is consistent with company objectives.

An internal auditor carries out his or her responsibilities by applying various weighted formulas, or algorithms, that rank both management and operational areas. Parameters may include book value of capital, years of experience in the area, technical complexity of the discipline, internal control, the extent to which management can influence outcomes, special fudge factors, and exposure to risk.

In industry, innovation had to be considered as only one of a number of activities required to keep a company profitable.

An auditor's efforts to evaluate R&D management performance, in particular, is fraught with pitfalls. Because the complex technical activities of an R&D department are beyond the understanding of most auditors, outside help should be brought in to the audit. Unfortunately, a computer specialist is generally the first person recruited, and because of company restrictions, no other external consultant is usually brought in. This lack of technical competence on an auditing team forces auditors to focus on examining what they understand best: mainly administrative controls.

Articles outlining how an internal audit such as this could be implemented started to be published in the early 1970s, and in the late 1970s and early 1980s, articles describing the actual application of these internal audits began to appear. These internal audits appealed to senior executives by providing them with objective measures for decision making and strategic planning. During the late 1980s, many companies hired major accounting firms to perform their audits. Little has been published about the effectiveness of these audits, however, because the consulting accountants were obliged to protect both their clients' confidentiality and their own reputations.

A far more productive approach to measuring R&D productivity was taken by the laboratories of the US Department of Defense in the 1980s. These studies involved a statistical analysis of the effect of R&D organizational and individual variables on innovation and R&D productivity in the laboratories of a highly formalized and structured organization. The objectives of the studies were to determine how to define and measure the performance of staff in R&D laboratories and what weights to assign to different measures of R&D productivity.

At this point, it was accepted by analysts of R&D productivity that ratings, despite their subjective nature, were one of the more appropriate means of assessing innovative output. In addition, many analysts accepted that the number of patents and publications of a technical contributor still provided the most satisfactory measure of R&D productivity. The analysts who conducted this study of military laboratories made one more critical assumption—that the research products really did contribute to the solution of actual problems of operational military units.

After making this asumption, the greatest statistical problem that the analysts faced was obtaining a method for weighting the various dimensions of scientist and engineer output. Several previous studies had shown that there are two types of R&D output: technical and contractual/managerial. The analysts accepted this idea and therefore measured technical output by considering technical reports, journal articles, computer codes, and technical presentations. They measured the contractual/managerial output by considering proposals to management, proposal evaluation, procurement evaluation, and management presentations.

The studies of military laboratories yielded many interesting findings. Some were not new, but the rigor with which they were supported was important because they validated and helped publicize many earlier ideas about R&D productivity.

For example, the studies found that education was not related to contractual/managerial productivity but that it was most definitely related to technical productivity. The studies also found that technical success was related to greater external communication, to having the technical contributors participate in goal setting, and to the presence of strong support and empathy from the technical supervisor.

Importantly, the studies also showed that some laboratories within the US Department of Defense, which in general is a rather homogeneous organization, were significantly more productive than others. The difference in R&D productivity was somewhat ambiguously attributed to changing priorities within the less productive laboratories and was not adequately explained, which in itself may be sufficient explanation.

One of the most important conclusions that came out of the studies of R&D productivity in the 1980s was that the essence of R&D productivity measurement is finding meaningful comparisons. Comparisons may be made between organizations or within the same organizations over time, but there has to be some means of establishing a base measurement. Once a significant difference has been found, the cause of that difference can be sought. The ultimate goal is to find a way to manipulate that cause to improve R&D productivity.

Studies of R&D productivity were now being designed more rigorously and confidence in the results of these studies increased. The importance of a technical contributor's tenure on a job was found to be extremely important. Specifically, R&D performance increases for the first few years of tenure and then declines steadily. This lapse in R&D performance has been associated with a decline in communication with both internal and external groups.

Overall, the job tenure of a technical contributor can be marked by three stages: socialization, innovation, and stabilization. New employees, or employees assigned to new groups, require some time to learn the norms of a group while also learning the technical requirements of the job. At some point they gain sufficient familiarity with the job to be able to devote most of their energies toward accomplishing important tasks. It is at this point in a technical contributor's tenure in a job that participative management provides significant improvements in R&D productivity, and participative management will also extend the time that the employee remains in this innovative stage.

After a technical contributor has had enough tenure in a job to feel comfortable, however, the demands of a job become more predictable to him or her. From this point on there will be a slow shift away from a high level of innovation. During this third stage of job tenure, a technical contributor may require more directive management or may need to be reassigned to a new job with new challenges.

In the 1960s, both General Dynamics/Astronautics and Martin Marietta tried to apply traditional time study methods to the quantification of scientific and engineering productivity.

In general, most of the studies of R&D productivity during the late 1980s have confirmed the principal conclusion of Hughes Aircraft's earlier study of R&D productivity: that skilled and responsible R&D management is the most important factor in improving R&D productivity.

THE INFLUENCE OF ORGANIZATIONAL ISSUES AND R&D PRODUCTIVITY

The Cornerstones of High R&D Productivity: Clear Goals and Information

As mentioned previously, Hughes Aircraft Co. is responsible for the most extensive study of R&D productivity ever performed. The study found seven factors that correlate with a high level of R&D productivity:

- Skilled, responsible management.
- Outstanding leadership.
- Organizational and operational simplicity.
- Effective staffing.
- Challenging assignments.
- Objective planning and control.
- Specialized managerial training.

Simply put, a high level of R&D productivity occurs when technical personnel effectively work toward a clear goal. Getting an individual technical contributor, much less an entire R&D department, to do this is more difficult than it appears. Moreover, even if an individual technical contributor or an R&D department does reach a high level of R&D productivity, there is still a great challenge involved in maintaining that level of performance.

One of the reasons that both individual technical contributors and R&D departments become less productive is that success spoils them. Once a strategy for selecting and carrying out R&D projects has worked well, they often take the prudent approach of maintaining the status quo for too long. A current saying to this effect is, "If it ain't broke, don't fix it." The problem is that by the time a technical contributor or an R&D department realizes that the system is broke, the damage may be quite severe. Constantly assessing whether conditions are still the same can help guard against risky complacency.

The only way to assess whether significant changes have occurred is to continually monitor the environment of the R&D laboratory. For an R&D department to maintain a high level or productivity, therefore, it must make sure that it obtains a wide variety of information and that this information is used to monitor changes in the R&D department's environment. Such information should include the business goals of the company, new trends in customer needs, technological developments, changes in the technical personnel, and other organizational developments in the company.

In its broadest sense, all of this information provides the R&D department with a method to reduce uncertainty. Using this information, an R&D department can never reach a high level of productivity if it directs its projects toward the wrong goals. Information about a laboratory's environment, therefore, is critical not only to carry out R&D projects successfully but to reach and maintain a high level of productivity.

Defining the Ultimate Goal of an R&D Department

Jim Harbour, formerly of Chrysler Corp. and now an independent automotive management consultant, is fond of saying "The goal of every company and its staff is to satisfy the everlasting customer." If you have not satisfied your customer, you have not been productive. Moreover, if others can do a better job of satisfying your customer, you have a serious productivity problem.

The difficulty with Harbour's statement is that customers often have differing needs. Furthermore, customers may not have spoken of or even thought about some needs. Their perceived needs may have little or nothing to do with the technical solutions that you are developing.

Consequently, the goal of satisfying the everlasting customer goes far beyond taking and filling orders. The ultimate goal is to develop a new product that will satisfy customers' real needs. By doing this expeditiously, an R&D department can achieve true productivity. Anything less may satisfy the written requirements for the product but will not truly satisfy the customer.

Despite some qualifications to Harbour's comment, he is essentially correct. The goal is to satisfy the everlasting customer. The task is to know the customer's tangible and intangible criteria for success and to take advantage of your R&D department's unique capabilities to develop products meeting these criteria. If your R&D department's customers are internal, such as the marketing or production departments, your objective is still the same. An inherent advantage in this arrangement is that R&D is closer to its customer and can maintain better communication.

A critical factor in reaching a high level of R&D productivity, therefore, involves understanding customers and how the R&D department can work most effectively to serve them. If you can quantify your R&D department's productivity, so much the better.

Matching People and Functions

Technical people, like most others, have different talents. Some are better than others at working with other people, hardware, or data. Because of this, R&D managers must come to understand the individual abilities of their staff members and assign jobs that match their specific abilities. For example, technical people who are particularly good with other people should be used to gather information from other members of the company and from customers. They also should be used in the later phases of a development project as it approaches commercialization. Technical people who are particularly skilled at working with data should be used in planning a project and in conducting various tests and market research studies. Those who are particularly skilled at working with hardware naturally should be used primarily for carrying out the technical tasks.

R&D managers also should take into account the degree of cognitive complexity possessed by the members of their technical staff. One aspect of cognitive complexity to identify is whether an individual thinks primarily in simple, concrete terms or in complex, abstract terms. Whereas someone who thinks in simple, concrete terms will view problems or solutions as being either black or white, someone who thinks in complex, abstract terms will see greater differentiation.

Broadly speaking, technical contributors who think in simple, concrete terms should be used on development projects. With these projects, the objectives are usually clear and the emphasis is on meeting technical milestones on time. On the other hand, technical contributors who think in complex, abstract terms should be used primarily on exploratory R&D projects. With these projects the goals cannot be defined as well—and also may have to be continually redefined—and the emphasis is on investigating the opportunity rather than on delivering a product.

Hughes Aircraft supported a complete review of the state of the art in R&D productivity in order to improve its own management practices.

Organizational Structure Influences R&D Productivity

In structuring their R&D departments to improve productivity, R&D managers should consider a variety or organizational issues. Five key issues are:

- Division of labor.
- Hierarchy of authority.
- Degree of formalization.
- Degree of centralization.
- Degree of complexity in activities.

Division of Labor. Six ways in which the members of a technical staff may be divided are in terms of function, number of employees, product, customer, geography, or process. For example, technical personnel could be organized in terms of their scientific or technical discipline, relatively equal groups (in order to keep the groups manageable), the company's products, the customers that the company's various products are directed toward, geographically separate laboratories, or the technical or production processes within a company. As one can easily imagine, if the technical personnel of an R&D department are organized in a way that is incompatible with the needs of the company, the technical personnel will not be productive.

Hierarchy of Authority. Issues related to authority continually arise within R&D departments. For example, what level of R&D management needs to approve an expenditure for new laboratory equipment? Or, how much authority should a leader of an R&D project have? What should be the level of involvement of middle-level R&D managers in the selection of R&D projects? When the appropriate balance is not maintained between the amount of authority that should reside with senior R&D managers and the degree of freedom technical personnel should have to take the initiative, poor decisions can result.

If authority is not delegated properly, two things can occur. Perhaps the required laboratory equipment is not bought because senior R&D managers, who had the authority, did not understand the need; or the wrong laboratory equipment is bought because technical personnel, who made the decision, did not understand the overall requirements of the laboratory.

Degree of Formalization. Formalization is related to hierarchy of authority, but it also involves issues. For example, how formalized should an R&D department's process be for selecting R&D projects? Should it be quite formal, so that outside pressures from a marketing department do not force R&D managers to initiate a project without carefully evaluating its potential? On the other hand, how much freedom should there be for R&D managers to adapt to new circumstances during the year by shifting resources from ongoing R&D projects to newly identified technical problems?

The degree of formalization is particularly relevant in relation to the size of an R&D department. How much formalization should a small R&D department have so that it is not only innovative but also efficient? Or, how can a large R&D department keep from becoming so regimented that it is no longer responsive to new market needs?

Degree of Centralization. Should there be one R&D department that serves all the business units, or should each business unit have its own R&D group, which reports directly to the head of that business unit. That question continues to be debated. Perhaps the best way to view the history of changes in this area is to consider the successive reorganizations of R&D departments as swings in a pendulum—first, toward greater centralization, then toward greater decentralization, and then back again. Currently, the trend for most R&D departments is toward greater centralization, and given the strong drive among business managers to make R&D more responsive to business needs, perhaps R&D departments will remain decentralized for some time.

> By the 1970s, the first generation of studies on R&D productivity had matured.

Degree of Complexity of Activities. During the last two decades, R&D departments have had to adopt far more complex tasks. Where once an R&D department might have been organized simply, now many R&D departments may have a matrix system of authority, many multidisciplinary project teams, new venture teams, informal centers of excellence in which technical personnel from different R&D groups cooperate, and technical task forces.

The pure matrix system consists of various technical departments in which the technical personnel reside for the purposes of training and evaluation and various projects on which the technical personnel work. Technical contributors report to the leader of a particular project. At any one time, technical contributors can serve on one or more projects.

Many R&D departments have developed their own variation of this pure matrix system. Some R&D departments use a watered-down matrix system. Other R&D departments have established multidisciplinary project teams that do not fit the mold of a matrix system, or they have set up new venture teams that are quite independent of the R&D department's regular activities. Centers of excellence can facilitate cross-communication between R&D groups and task forces can be used for specific developmental projects.

These new organizational changes have taken place to allow R&D departments to handle more complex and specialized activities. It is critical to determine how a new organizational design can enhance productivity and when it might actually detract.

Each of these five organizational issues—division of labor, hierarchy of authority, degree of formalization, degree of centralization, and degree of complexity of activities—has a great effect on R&D productivity. Efforts to motivate technical personnel and to give them the appropriate tools to improve productivity will be undermined if these issues are ignored.

How Organizational Issues Affect R&D Productivity

Two illustrations of how organizational issues can affect the measurement of R&D productivity can be drawn from the automobile industry. Automobile manufacturers have different organizational cultures, but all are striving to differentiate their products from their competitors' in order to gain a marketing advantage.

For example, Lee Iacocca and others at Ford developed the Mustang during the 1960s to seek such a market advantage. Even though the Mustang was based on the stodgy Falcon economy car design, it was meant to appeal to an entirely different segment of the market. The Mustang was successfully developed only because the R&D people who championed it overcame enormous management resistance. Even the success of the Mustang did not allay these conflicts, and management turmoil and firings ensued.

Decision making was highly centralized at Ford, and managers were evaluated on the basis of criteria set by the CEO. Even though the product delivered by the R&D people was highly successful in the marketplace, it did not meet the criteria of management performance that had been established by the CEO of Ford.

The development of the Miata at Mazda provides a contrasting example. Until the development of the Miata, Mazda had had an overall record of producing unremarkable vehicles, with the exception of the RX7, which was powered by the Wankel engine. The successful development of this engine, which General Motors and others had failed to carry out, demonstrated that despite its reputation, Mazda excelled in engineering and design.

Nevertheless, without the radical changes that were made in Mazda's organizational culture before the Miata was developed, the Miata never could have been developed successfully. Before the development of the Miata, decision making was so formalized that it would have been unlikely that a vehicle as different as the Miata could have been developed. (The Miata is a two-seat roadster.) Fortunately, a new president, Norimas Furuta, took over before the Miata was developed. Furuta deliberately chose to decentralize the decision making involved in new car design. Specifically, he kept senior management out of the design process. Consequently, despite the reservations that the senior marketing executives of Mazda had about the Miata, the car was developed and has been a runaway best seller. But the Miata never would have been creaed had the previously centralized and formalized organizational structure of Mazda continued to exist.

MEASURING R&D PRODUCTIVITY

Overview of R&D Productivity Measures

Measures of R&D productivity may be either qualitative or quantitative. Both types of measures can be used to evaluate a whole R&D department or to evaluate an individual. The disadvantage of qualitative indicators is that they do not provide the apparent precision of quantitative indicators. On the other hand, qualitative indicators can create an illusion of accuracy.

Qualitative Measures of an R&D Department's Productivity

Qualitative measures of an R&D department's productivity include:

☐ The ability to repeatedly win competitive solicitations for R&D.

- The development of simple, reliable, maintainable, and operable designs that meet customer requirements.
- The ability to meet commitments to other company departments or to customers.
- The ability to perform to meet stated criteria or requirements.
- The capacity to meet specific technical performance, cost, and schedule milestones with ease and enthusiasm.
- The ease with which R&D designs can be put into production.
- Product performance during its life cycle.
- The image the R&D department projects to its internal and external customers.
- Organizational morale, as expressed by employee motivation, attitude, and non-work-related communications.
- The quality and usefulness of ideas generated.
- The ability to obtain and apply new technical information and skills to solve R&D problems.
- The ability to contribute to the overall advancement of the state of the art in an R&D discipline.
- The creation of valid, reliable, replicable research.
- An ability to respond to emergencies and to peak or sustained demands.
- An ability to simplify and master problems without interfering with ongoing R&D.

> **Authority issues continually arise within R&D departments.**

Qualitative Measures of a Technical Contributor's Productivity

Qualitative measures of a technical contributor's productivity include:

- Performance relative to job requirements.
- Performance relative to co-workers or others doing similar technical work.
- The contribution of the individual's work to the objectives of the R&D project.
- The congruence between a technical contributor's goals and the R&D department's goals.
- Reliable and effective work habits.
- Approaches and methods used to accomplish tasks.
- The ability to apply new or more complex information and skills to the accomplishment of a task.
- Completing technical tasks on schedule.
- Frequency of problems and errors.
- Time required for mastery of new skills and complex tasks.
- Ability to communicate effectively and reliably with superiors.

EVALUATING R&D PRODUCTIVITY

- Repeated requests for an individual's technical services.
- Ability to acquire and apply complex information with minimal direction and support.
- Feedback from customers.
- Relationships with co-workers.
- Ability to manage personal life so it does not interfere with work life.
- Ability to innovate and create on the basis of information gained elsewhere that was obtained through personal initiative.

Quantitative Measures of an R&D Department's Productivity

Quantitative measures of an R&D department's productivity include:

- Percentage of technical proposals won.
- Proposal backlog in relation to the percentage of proposals won.
- Dollar value of proposals won versus dollars spent on bidding (unbillable charges) to win those proposals.
- Profits generated per R&D dollar spent.
- Dollar volume generated per R&D dollar spent.
- Profit per payroll dollar.
- Profit per employee (exempt and nonexempt).
- Sales per payroll dollar.
- Sales per employee (exempt and nonexempt).
- Any measure of earned value that provides a comparison of actual work completed and cost incurred with work scheduled and cost budgeted.
- The ratio of units R&D effort to units of R&D product delivered.
- Ratio of workstation time to units of workstation output.
- Error rates.
- Time per unit or document for processing designs through engineering drawing release.
- Drawing change rates per unit of time.
- Labor-hours per standardized test.
- Setup and calibration time.
- Computer hardware and software investment versus savings.
- Ratio of exempt to nonexempt staff.
- Ratio of administrative to technical staff.
- Ratio of facilities, service, and support staff to technical staff.
- Cost of service and support functions.
- New hire offers accepted, hiring costs, and new hires per unit effort.

- Absenteeism.
- Voluntary turnover rate.

Quantitative Measures of a Technical Contributor's Productivity

Quantitative measures of a technical contributor's productivity include:

- Units of work per employee per unit of time.
- Computer instructions programmed per programmer per day.
- Engineering drawing releases per clerk per day.
- Error or rework requests per unit of production.
- Time off work.
- Publications, presentations, and demonstrations.
- Patents and inventions.
- Project suggestions.
- Accident and injury frequency.
- Number of technical proposals prepared per unit of time.
- Proposal backlog.
- Billable hours.
- R&D dollars brought in per unit of time.
- R&D labor-years backlog.
- Margin between R&D costs and gross revenues.

> R&D managers must determine whether the validity and reliability of R&D productivity measures justifies the effort of obtaining them.

How to Measure R&D Productivity

The shopping list of R&D productivity measures presented here shows that the human mind can devise a comparative measure for nearly every activity. The trick is to apply these measures effectively.

One problem that invariably arises in using these measures is deciding whether the data are worthwhile. For example, a highly structured R&D department will have data bases that describe the activities of every staff member. R&D managers must determine whether the validity and reliability of the measures justifies the effort of obtaining them.

Interpretation. Although it is easy to rank all technical contributors or R&D groups on the basis of overhead, billability, volume, or backlog, the range of normal variation in the measures must not exceed the differences observed between technical contributors or R&D groups. In several cases, decisions by industrial companies to reduce R&D staff on the basis of billability were followed almost immediately by a need to turn down R&D contract awards because sufficient R&D capabilities no longer existed in the company. The loss of corporate memory that resides in the collective experiences of researchers who are laid off can be irreplaceable.

Using R&D Productivity Measures to Evaluate R&D. Measures of the productivity of R&D groups or technical contributors are not always reliable guides for determining whether a technical area is no longer productive or whether a technology will be displaced by a more sophisticated technology. Metallurgists all recall the immense interest that existed a decade or so ago in both reactor metallurgy and the nonferrous alloys used for high-performance aircraft. Once the customers' problems were solved in these areas, though, support for related R&D all but disappeared. Therefore, even though many scientific questions still remained unanswered and conventional measures of R&D productivity related to reactor metallurgy and nonferrous alloys still remained fairly high, there was no market for R&D in these technical areas.

Measurement. As discussed, R&D productivity measures are classified as either quantitative or qualitative, depending on how readily the subject under study can be divided into discriminable units. There are four levels of measurment:

- *Nominal.* This consists of naming or classifying.
- *Ordinal.* This involves ranking items on a dimension with unequal intervals between ranks.
- *Interval.* This involves ranking items with equal intervals between ranks.
- *Ratio scaling.* This entails all the characteristics of interval scaling but involves a zero point as well.

These four scales, in the order presented, provide increasing precision of measurement.

Most R&D productivity measures should be classified as either ordinal or interval, which means that the precision of measurement is not great. And the problems of measuring R&D productivity can be compounded by the type of R&D being measured. Exploratory R&D, by its nature, generally should be measured qualitatively, whereas technical support and product improvement are likely to involve quantifiable measures that can be used repeatedly for similar projects. Applied research and development, which generally can be studied with only nominal or ordinal measures, likes somewhere in between.

To determine whether R&D money is being spent effectively, the measurement of R&D productivity described earlier can provide good direction. To predict when any R&D will lead to significant technical breakthrough and in what direction, however, the user is better advised to investigate the application of catastrophe theory or of fuzzy sets. New approaches or objectives make old approaches obsolete, and this occurs in a way that is partly an extrapolation of what has gone before.

Stimulating Interest in R&D Goals

Quantitative methods to measure R&D productivity may be most effective when used to stimulate interest in the goals of an R&D department. Quantitative methods can force R&D managers to examine lines of communication within a laboratory and to evaluate how information is being transmitted between R&D groups. By looking at patterns of communication within a laboratory, R&D managers may influence R&D personnel to change their patterns of communication, where necessary, which could result in improvements in R&D productivity.

Exploratory R&D generally should be measured qualitatively, whereas technical support and product improvement are likely to involve quantifiable measures.

A case involving the Astronautics Division of General Dynamics illustrates this point. In the early 1960s, General Dynamics/Astronautics was developing the Atlas missile for use as an ICBM and as a booster for Project Mercury and other space launches. The Atlas missile project included every form of research, from the most basic studies of zero gravity effects to the development of new hardware and improvements to existing products.

Interestingly, General Dynamics/Astronautics was not concerned at that time about R&D productivity, as defined by numbers of items produced. What General Dynamics/Astronautics was concerned about was ensuring that every Atlas missile would perform reliably. Studies of failures established that physical causes were not a principal problem. Instead, the principal problem was found to be individual differences in human behavior, which were introducing variations in components and assemblies that eventually produced faulty equipment. The critical measure of R&D productivity for General Dynamics/Astronautics, therefore, turned out to be a function of the failure rate of parts in specific subsystems of the missile. This measure of R&D productivity was an ordinal measure.

In the course of trying to determine where the human error problems originated, analysts at General Dynamics found that whenever meetings were held involving the R&D personnel who contributed directly to a product, error rates declined. These meetings were designed to provide free communication between small numbers of R&D personnel working on related tasks, and participants in the meetings were asked to describe problems with their technical tasks and to propose solutions. Afterwards, the solutions were implemented. Analysis of data on failures showed that the R&D groups whose members participated in meetings had significantly lower failure rates for several months following a meeting, whereas R&D groups whose members did not participate in such meetings did not have any change in error rate. This increased communication facilitated free flow of complex technical information that usually did not get passed on within the formal system of communication.

The case of General Dynamics/Astronautics illustrates that an analyst of R&D productivity often must make a leap from a set of numbers representing productivity to attempt to identify the variables R&D managers can influence.

There is not a single formula for measuring R&D productivity. Instead, an analyst of R&D productivity must understand the goals of the R&D department to develop a measurement that fits.

Composite Measures of R&D Productivity

A more arcane approach to measuring R&D productivity is to combine qualitative and quantitative measures to yield a weighted index of an R&D project's contributions to business opportunities. The overall efficiency of an R&D project can be estimated in terms of opportunity generated per dollar expended. This approach can provide a means for making decisions about which R&D projects to fund. Through use of estimates of market potential and cost of development, it is possisble to work back to a price for the newly developed technology. With this information in hand, a return-on-R&D index can be obtained as a function of the total cost of carrying out an R&D project.

Another composite measure of R&D productivity combines such factors as potential annual benefit, commercialization possibilities, competitive technical status, and the breadth of the R&D contribution. This measure can be used to make gross judgments about an R&D program's value.

Methods for measuring R&D productivity range from bean counting to macromodeling of markets and technology advances. Each method represents an attempt to boost productivity in some aspect of R&D. Success with any method depends far more on the R&D manager's understanding of his or her environment than on the particular method used. R&D managers themselves must tailor a method of measuring productivity to suit their department's particular goals.

CONCLUSION

Reviews of studies of R&D productivity up to the late 1980s point out several practical criteria for measuring R&D effectiveness:

- Skilled, responsible management.
- Leadership that is effective in cutting through complexity.
- Methods for developing practical solutions to difficult problems.
- Communication of these solutions to other functions.
- Simple organizational structures.
- A staff capable of handling complex problems, particularly when a new staff member is being added.
- Assignments that challenge the capabilities of the individual.
- Effective planning and control.
- Management training.

It is more difficult to measure how well technical contributors and R&D departments are meeting these criteria than it is to develop the criteria. Even though data bases can be developed for every activity in an R&D department, managers need to decide just how much effort should be expended in a particular area. New data may not provide any more valid measures than that which is already at hand.

Factor analysis and modeling approaches have confirmed the validity of many of the R&D productivity concepts that have developed during the last 40 years. These concepts concern the need for communication between R&D staff members, with direct users, and with others in kindred research areas. An R&D department also should be organized so that the optimum technical challenges are provided.

With all of the false starts and failures, the R&D manager of today has better tools to manage than his predecessors. The R&D manager of today has a far better appreciation of which organizational issues affect R&D productivity and a wide range of measures to use in evaluating the department's work. Our challenge is to build on this base to help the R&D manager in the future.

The overall efficiency of an R&D project can be estimated in terms of opportunity generated per dollar expended.

R&D COST ACCOUNTING

Stephen Hellman

> R&D cost accounting (i.e., chargebacks) in a multiclient consumer products R&D group helped to align the mix of R&D projects with company business goals.

Many valuable technologies—ranging from project-by-project reviews to longer-range strategic planning approaches—are available to help R&D groups determine the focus of their portfolio of projects. A less obvious but very powerful tool for maintaining a tight focus in R&D projects is customer-based cost accounting (commonly known as chargebacks in many industries). This chapter, through the use of a case study, describes how a system of R&D cost accounting in a multiclient consumer products R&D group helped to shift the mix of R&D projects in a direction more in line with the company's larger business goals.

R&D PROJECTS FROM OVER THE TRANSOM

A multinational over-the-counter nonprescription drug and personal products firm conducted virtually all of its R&D at a site near its corporate headquarters in the US. Worldwide sales for this product segment were just under $1 billion, with approximately 2% of sales devoted to R&D. A senior-level reorganization had just been completed—basically to groom an executive from an international affiliate for a broader-based position by giving that executive responsibility for all consumer products worldwide. The executive, in participating in the activity of revitalizing consumer businesses, took an aggressive interest in the R&D being conducted to support the consumer businesses, the rationale behind the projects, and considered how to achieve more bang for the buck. One of the first meetings between the new management team and R&D management centered on such questions as:

- What percentage of R&D time is spent on support to existing products—that is, on cost reductions, qualifying alternate suppliers of ingredients, and different packaging?
- What percentage of R&D time is devoted to new products with high potential for sales and profit?
- To what degree do each of the consumer businesses contribute to the R&D budget, and what is their share of the actual work done?

The response to such questions was, predictably, something like "We'll study the questions and get back to you soon." The request was bounced downward

---

STEPHEN HELLMAN is the director of Technology Planning for Clairol, Inc., a division of Bristol-Myers Squibb. In this position, as well as others at multinational consumer products and consulting firms, he has specialized in developing and implementing processes for allocating R&D resources consistent with business strategy.

to virtually all levels of R&D managers, and the responses varied considerably. Most managers were not comfortable enough to provide even rough estimates; those that did rarely talked in increments of less than half or thirds of the total pie. Even these estimates were so inconsistent that it soon became evident that R&D had no idea of how it was allocating its resources. R&D apparently worked on whatever projects were requested by the customer and whatever projects of a more technology-driven nature they themselves felt were worthwhile. Each project was assessed on its own merits, with no thought given to what benefits would be reaped if the projects were part of an integrated portfolio. Because the concept of portfolio management was so foreign to the managers, no discussions had ever taken place regarding what an appropriate mix of projects might be.

COST ACCOUNTING IS INTRODUCED TO R&D

At approximately the same time that the new head of consumer operations was asking the aforementioned questions, a new president of R&D—with a staff of two—had been installed. The president and staff implemented a fairly standard system in which R&D personnel reported weekly, detailing which projects they had worked on and for how long. The system also tracked project expenses. Such R&D reporting is fairly standard for R&D groups and would have amounted to a standard, but uninspired attempt at accounting for costs and probing the questions that had been asked of R&D. In this case, however, three enhancements to the standard R&D cost accounting system were added:

- A field was included in the computer code identifying each R&D project (the project number) that identified the division for which the work was being done.
- A field was included in the project number that identified the type of effort the project represented (e.g., existing product support, line extension).
- Five additional fields (currently unassigned) were included in the project number that could be assigned any value later on and that could be the basis of computer-generated reports that would flag all projects and their associated costs with a particular value for the field. For example, inserting the code TAX in a field for appropriate projects would allow printing a report of all project charges that qualify for an R&D tax credit, thus simplifying cost accounting.

The basic advantage of this system was that reports that would describe how R&D was spending its time, broken out in virtually any conceivable way, could be generated in less than a minute. Questions of the sort asked by the head of consumer operations could be answered instantly. More important, R&D could track its own resource allocation and adjust its project portfolio to reach its goals in this area.

R&D FOR SKIN CARE GROUP IS MOSTLY SHORT RANGE

Once the system was fully operational and six months of data was online, a series of reports was generated as a means of determining how resources were allocated. The pie charts in Exhibits 1, 2, and 3 are representative of the information that was available.

Several facts are immediately evident from the pie charts. Exhibit 1 shows that R&D as a whole is devoting 30% of its effort to existing product support and 30% to line extensions. This was a shock to all concerned, because there was general agreement that existing product support should consume no more than 15% of the R&D group's resources. In addition, line extensions were being deemphasized in favor of new product introductions, which were thought to hold more profit potential while reducing cannibalization of existing brands.

Exhibit 2 shows that the type-of-effort breakout for the skin care group was heavily skewed to shorter-term projects (i.e., existing products and line extensions) with virtually no resources applied to exploratory work and new products. The skin care group's stated strategy was aggressively oriented to new products; although the topic was discussed extensively, obviously not a lot was actually being done about it within the R&D division. The fact that so little exploratory R&D work was under way indicated that R&D was not initiating the basic science and proof-of-principle work it should; the lack of new-product programs similarly indicted the marketing group for not distilling their market research and opportunity analyses to identify specific new product opportunities.

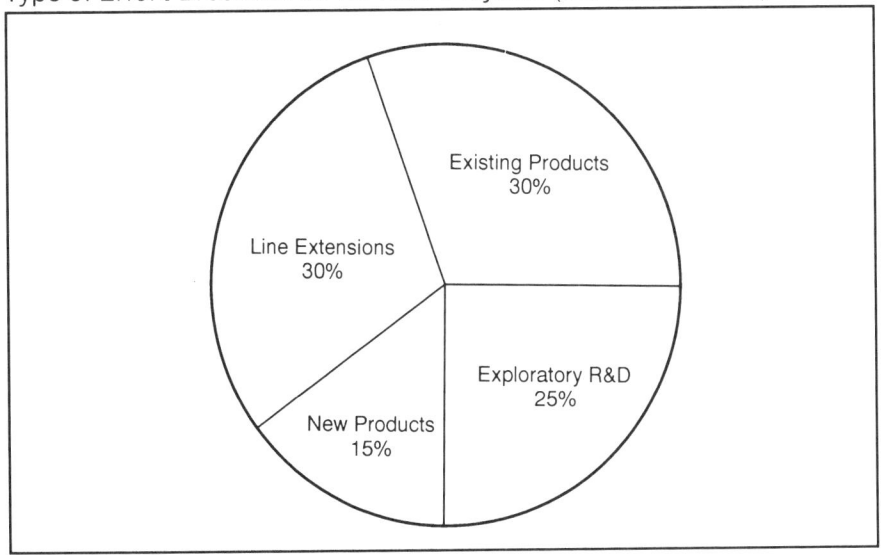

Exhibit 1
Type-of-Effort Breakout for All R&D Projects (six months data)

- Existing Products 30%
- Line Extensions 30%
- Exploratory R&D 25%
- New Products 15%

Exhibit 2
Type-of-Effort Breakout for Skin Care Business (six months data)

- Existing Products 60%
- Line Extensions 30%
- New Products 5%
- Exploratory R&D 5%

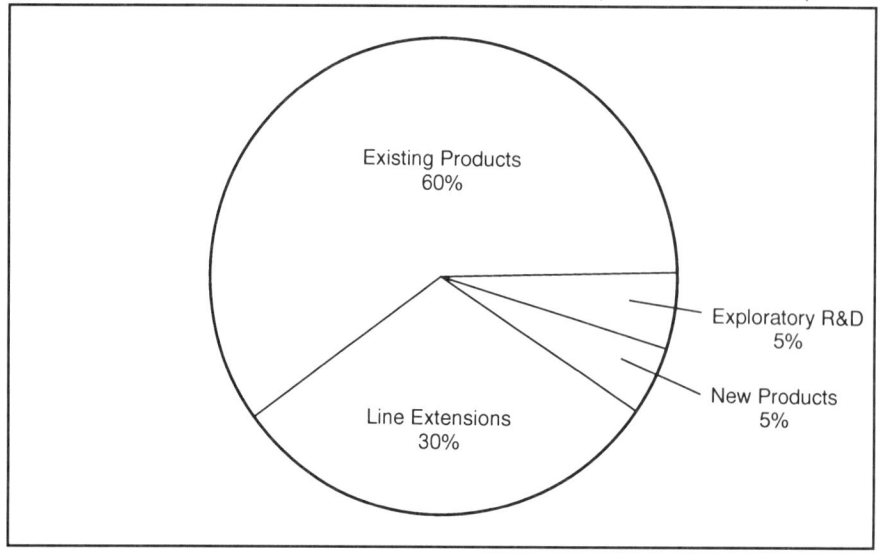

Exhibit 3
R&D Effort by Customer (six months data)

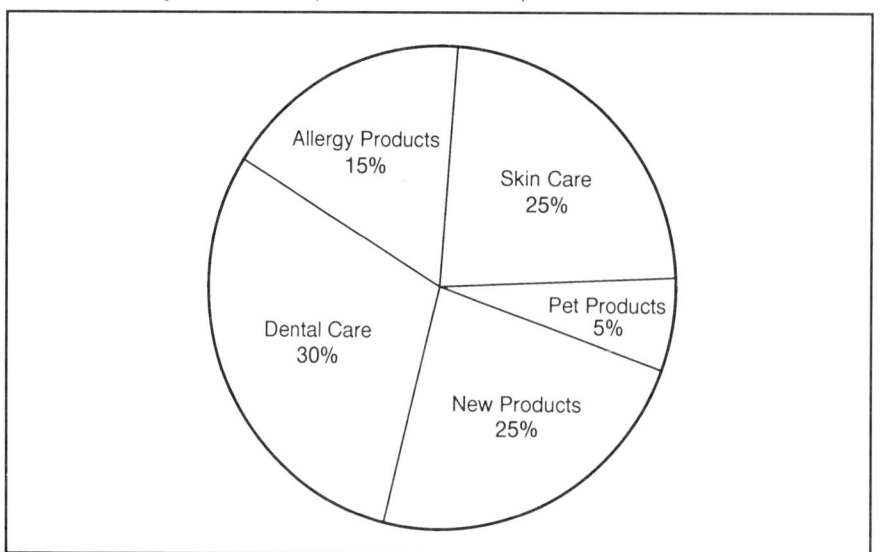

- Skin Care 25%
- Pet Products 5%
- New Products 25%
- Dental Care 30%
- Allergy Products 15%

MANAGEMENT OF R&D

Exhibit 3 shows that the dental care group is consuming the highest proportion of the R&D resource and that pet products are consuming the lowest proportion. These figures, of and by themselves, are meaningless. Therefore, a comparison with the revenues generated by these groups and the strategy (e.g., harvest versus invest) pursued by each was necessary. There was, however, some idea of how R&D resources were being allocated to the various businesses. Also, the effort-type breakouts provided a quick overview of how each business used its share of the resources.

The basic advantage of the cost accounting system is that reports that describe how R&D spends its time can be generated in less than a minute.

R&D FOR NEW-PRODUCTS GROUP IGNORES EXPLORATORY PROJECTS

Another change in resource allocation that was made after analyzing data produced by the cost reporting system pertained to the new-products group—a business unit dedicated to identifying new consumer product concepts, conducting consumer and technical proof-of-principle studies, and, ultimately, commercializing the most promising candidates. After about a year of operation under the new cost reporting system, an analysis of the type of R&D effort in support of the new-products group was undertaken. Initially, resource allocation seemed to be just where it should be: about half of R&D effort for the group was devoted to commercializing new products, and the other half was devoted to supporting various consumer tests of new and promising concepts. This latter half of the effort included making concept samples for focus groups, and making and packaging supplies for clinical tests and large market-research efforts. Absent, however, was R&D-initiated work for proof-of-principle of new technology that could be the basis of new products. New products were apparently receiving their appropriate share of the R&D resource, but it was the commercialization of concepts that were already in the pipeline that was getting attention; virtually no new technical concepts were being investigated at the early stages of development. If this trend continued, the new-products group would be without new concepts to commercialize in several years' time. The group was understandably concentrating on the shorter-term, revenue-producing commercialization projects and in the process ignoring the longer-term, technical exploratory projects that would be the basis of future products. This simple resource allocation analysis focused the situation instantly, and corrective steps were taken to reduce the dominance of shorter-term projects.

Another example of how the data generated from this cost reporting system focused management's attention on the need to redeploy certain R&D resources involved a business unit's use of R&D skewed very heavily toward high-risk exploratory projects. Contrary to the immediately preceding example, revenue-producing commercialization work had been totally ignored. Although all involved realized that there was an active exploratory R&D program in place, no one realized that none of these initial efforts were maturing into development projects. There had been a bias toward new and exciting projects that appealed to the business unit leader's sense of adventure and action; thereafter, special attention was given to moving at least a small number of promising projects through the normal development cycle. Once the status quo of R&D resource allocation was established in the various ways described, management eventually reached a consensus regarding the nature of the desired guideposts. For example, a goal of reducing existing product support to the 15% figure mentioned earlier was established—both for the R&D organization as a whole and for each individual business unit. The skin care group conducted a two-day, off-site meeting to cull expendable short-term projects and reach agreement as to the appearance of an ideal new-products program.

Once the goals for R&D resource allocation had been agreed to, periodic discussion occurred between concerned parties to track actual project charges. Rather than immediate compliance with goals being expected, trends in that direction were considered acceptable and unusual circumstances considered. The critical change was, however, that both R&D management and their customers had a new consciousness of R&D chargebacks and allocations. Bang for the buck was no longer a catchphrase of a senior executive; mid- and lower-level managers were now questioning the value of specific portions of the R&D project portfolio and trying to determine a reasonable balance of projects.

CANDLES SHOULD NOT COST MORE THAN THE CAKE

The two requirements for the activities described previously were:

- R&D personnel reporting of time spent on each project.
- Availability of versatile software to quickly generate resource allocations broken out in any way required.

The second requirement is critical, because the ability to generate data for resource and cost allocation on a fast, as-needed basis is a must. Numerous software packages are available to help accomplish such tasks. Customized application programs are also feasible, particularly in mainframe environments in which such high-level user languages as SQL place the management of data and flexible report-writing within the reach of most organizations. Only when the analysis of costs and resources described can be accomplished in a minimal time frame can these activities be fully justified from a cost/benefit perspective. Although the value of such analysis to the strategic direction of a company is substantial, it is difficult to justify the effort when weeks or months are needed to generate the information. In addition, current data is much more valuable than outdated data (learning in two months that the R&D mix for the last quarter was skewed too heavily in a particular direction is not nearly so helpful as finding out within 10 days).

OTHER BENEFITS OF COST REPORTING

Once a cost reporting system is in place, other benefits can be derived from it. Aside from the ad hoc analyses that management requests from time to time—which otherwise could take days of drudgery to respond to—occasional requests from corporate finance can be easily answered by using these cost reporting systems. Responses to requests for a listing of projects (with associated charges) necessary to comply with new EPA regulations, for example, are easily generated using the unused flag fields in the R&D project number. Similarly, applications for tax credits for R&D work done in support of new products or exported technology are easily supported by use of the cost reporting system.

R&D cost reporting is usually regarded solely as a mechanism both to track manpower and dollars spent against individual projects and to support chargebacks to internal business customers. A flexible system, however, can also be a substantial aid in shifting the mix of projects in the R&D portfolio to a balance more consistent with business goals. High-level software is critical to respond to the information needs generated in an unpredictable and changing business climate.

The availability of software able to generate data for resource and cost allocation on a fast, as-needed basis is a must.

RECOGNIZING AND REWARDING R&D PERSONNEL

Timothy J. Kriewall

> If the employees who are necessary for developing the company's future—R&D personnel—feel their worth is insignificant, then morale, enthusiasm, creativity, innovation, commitment, persistence, dedication, loyalty—in essence new product development and profitability—will become only empty words.

In running an effective company, management assumes responsibility for three fundamental attributes of the corporation: the business, the organization, and the individuals. Unfortunately, management can become too preoccupied with the business (watching the top and bottom lines) to give much thought to the organization and its individuals. By doing so, management fails to see how the individuals, effectively organized, will help them to attain the loftiest of business objectives. If the individuals (employees) who are necessary for developing the company's future feel their worth is insignificant in comparison to the business, then morale, enthusiasm, creativity, innovation, commitment, persistence, dedication, and loyalty will become only empty words.

The operative question here is how do the employees feel. How does management make its employees feel? Is that really the job of management? You bet it is.

Recognizing the worth of their employees is the second most important responsibility of management. (The first is attending to the needs of the customers.) Recognizing all employees' worth is important, but R&D personnel, by their very nature, seem to require this recognition more than do others in the organization. Perhaps this seems ironic because it is the R&D people who create the new technologies and products for the company; without them, no one else in a product-based company has a job.

Perhaps it is because their recognition is relatively short-lived compared with the recognition others may receive that management must work at formally and frequently recognizing R&D employees. For example, sales personnel can sell all existing technologies and products until the new ones come along. Sales forecasts can be set and exceeded daily, monthly, quarterly, and annually while R&D personnel may struggle for years to perfect their concepts and creations. Sales incentives can be set and met routinely.

When R&D personnel succeed in bringing their concepts into reality, several single-event things may happen. They may be admonished because they took more time than anticipated in getting their job done; they may be included in a one-time recognition ceremony after months of work and commitment;

TIMOTHY J. KRIEWALL, PhD, is a technical manager at Sarns 3M Health Care in Ann Arbor MI.

they may be asked to modify their creation to accommodate market requirements that have changed since the project's inception (i.e., revise what they did); or they may be asked to immediately move on to the next project that may already be behind schedule. Whatever the case, recognition and reward are short-lived compared with the commitment of work and time.

R&D employees, especially outstanding ones, usually have strong egos. They must have a confidence in themselves that will carry them through periods during which discovery is well beyond eyesight—all the while knowing they will be able to achieve success. So along comes management asking such questions as: How long will it take to get the job done? Do you realize how much money you're spending? Can you present a project review to senior management to keep this project alive? Do you think we should quit and cut our losses? Should we make a midcourse correction? Should we bring in a consultant? This type of questioning only fosters resentment, defensiveness, frustration, self-doubt, disrespect, paranoia, and anger. These are counterproductive, energy-sapping, defocusing emotions that stand in the way of accomplishment.

Instead, what R&D people may really like to hear is what athletes hear from their coaches: constant encouragement. They need to feel the emotions that empowerment, trust, confidence, commitment, support, and recognition bring. For best results, these emotions have to be fostered constantly, throughout the project. Herein lies the challenge for management, the time commitment for management, and the job of management.

ACTIONS OF RECOGNITION AND REWARD

Recognition and reward require overt actions or actions of commission. Disincentives are usually acts of omission. Exhibit 1 lists various ways to give recognition and reward; how each can be used is discussed in the following sections. Disincentives are discussed in subsequent sections.

Organizational Recognition and Rewards

As a general statement, with organizational awards who is recognized is not nearly as important as who is not recognized. That is, recognizing members of a team or organization who achieve certain goals or objectives is, indeed, of prime importance. However, the importance of the award soon fades.

In comparison, the team members or contributors who do not receive the recognition or reward will never forget the omission. It may be that these persons did not meet the criteria for inclusion, but if in their own mind they think they qualified, resentment will be carried for unreasonably long times. No one may ever know of these employees' feelings, because the individuals may not want to risk embarrassing themselves by admitting their slighted feelings only to find out the omission was intentional. It therefore behooves

Exhibit 1
Examples of Recognition and Reward

Organizational	Individual
Financial achievement award	Distinguished scientist society membership
Team recognition	Patent and publication recognition
Product plaque award	Performance appraisal
Off-site meetings	Technical excellence
Group training	Monthly lab awards
Enthusiastic leadership	Empowerment
	Personal notes
	Customer contact
	Remuneration and gifts

those who establish team awards to clearly establish inclusion criteria, and it may even be advisable to request self-nominations or peer nominations.

Certainly, there comes a time when the size of team awards must be restricted for logistical or financial reasons. Also, if all team members are given equal recognition no matter how small their role, their inclusion in the recognition may detract from the intended benefits for those who contributed yeoman efforts. This fact further reinforces the need to make the inclusion criteria clear and to hold to the criteria consistently.

Awards for Financial Achievement

At the completion of a project that results in a product, an effective way to recognize the achievement is for the company to host a meeting or banquet at which others are present to witness the recognition. The criterion for receiving this award may be that in the first or second year of the product, it exceeds stated sales and profit milestones. For large companies, there may be numerous examples of such products. A single annual event at which the same criteria are used for all recognized product teams seems to work quite well.

Those present beside the team members themselves should include senior management to the highest level possible. In addition, the recognition will be much more meaningful if a spouse or designated other is included. Being able to share with a nonbusiness partner the feeling of success that comes with such recognition imparts a long-lasting benefit to the event. If the award ceremony is successful in its purpose, both patience and encouragement will persist on the homefront as the award recipients work on their next projects.

Financial achievement awards can be used effectively for noncorporate R&D teams as well. For example, those who win research grants from funding organizations would be logical recipients.

At these ceremonies, management should make a sincere effort to demonstrate appreciation. Extravagance is not required, but cutting corners sends a strong negative message. Listing individuals' names in the program, providing quality food (if food is served), having entertaining speakers with succinct messages, and presenting a quality gift to commemorate the event and the specific accomplishment are all important ingredients of successful recognition.

Team Recognition

Team recognition may be for achieving goals intermediate to final business goals or for achieving goals not measured by product sales. For example, in this era of time compression and concurrent engineering, cross-functional development teams may be rewarded if they adhere to their time schedule. This could be done annually for long-term projects so that encouragement is given along the way. Clearly, for this to work, objective measurable milestones need to be set early in the project planning. This works best when the team itself sets its own milestones.

Where there is a multitude of projects ongoing within one R&D laboratory, group rewards may be appropriate if some percentage of all projects remain on schedule. All project schedules need to be visible to all the people within the organization and updated regularly (e.g., monthly). This offers a double benefit. First, it fosters a focus on overall laboratory priorities so that if one or more projects begin to slip, support resources can be given to accelerate the projects in jeopardy. Secondly, it spawns a sense of friendly competition that encourages all the teams to avoid being the one team that keeps the laboratory's goals from being achieved.

In the world of R&D, discovery and accomplishment cannot always be attained. Therefore, a reasonable goal for success might be that a certain percentage of projects or teams meet their milestones by year's end. Modest quarterly rewards may be appropriate to let everyone know management is interested and is encouraging success. Assuming the year-end goal is met, a two- or three-day off-site working and recreational retreat for all the teams, even the unsuccessful ones, is a strong statement of encouragement to the R&D personnel.

The retreat may involve a mixture of educational and training seminars appropriate for the group. Team-building exercises, design of experiments, technology updates, and giving effective presentations are examples. In addition, a good share of the day should be devoted to relaxing activities.

As a footnote, a real temptation is to downsize or cancel altogether a retreat or recognition activity if circumstances change in the subsequent year—for example, money becomes tight, business suffers, the economy falters, or the majority of projects subsequently fall behind schedule. Although no one would want management to ignore signs of imminent disaster and carry out a costly reward program if conditions were really precarious, it would also be better if no reward program had ever been established if management is not willing to stand behind it. Making excuses to cut back or curtail in moderately difficult times is not acceptable. The budget should be established and protected in the year of measurement so the commitment by management can be fulfilled.

Product Plaque Awards

Management can become too preoccupied with the business to give much thought to the organization and its individuals.

Peters and Wateman in their book *In Search of Excellence* credit William Manchester as saying "a man wouldn't sell his life for you, but he'll give it to you for a piece of colored ribbon." Although people do not take life-threatening risks for ribbon, once they have survived the risk, the ribbon is worn proudly or kept for a lifetime.

Peters and Wateman go on to say that in doing the research for their book, they "were struck by the wealth of nonmonetary incentives used by the excellent companies. Nothing, they say, is more powerful than positive reinforcement. Everybody used it. But top performers, almost alone, used it extensively."

Thus it is with plaques. R&D people do not work for a plaque—they work to make a better product or make a new discovery. If they receive a plaque for an accomplishment, some may only throw the plaque into a drawer to be forgotten. Still, many display plaques and certificates proudly. I am reminded of an individual who was quiet and unassuming, certainly not arrogant or an exhibitionist. He unexpectedly died while still an active member of the work force. While visiting this man's wife at their home to express his condolences, his manager was shown into his study where every certificate and plaque that he ever received was carefully framed and hung. The walls were covered. The certificates and plaques were a legacy to his family of his accomplishments.

A form of plaque that seems to work well is an achievement plaque to which can be added nameplates or attachments with specific product or accomplishment names. For really productive people, the plaque can become a litany of successful accomplishments.

The awarding of these plaques needs to be done carefully for the reasons already discussed. If someone is overlooked, feelings of being shunned do not seem to fade. If management does not take a personal interest in the awards, all benefit is lost. If weak contributors are included, or worse, if the awards are politically motivated, those who toiled will feel cheated.

Off-Site Meetings

Off-site meetings are good for clearing the head and taking the group out of its normally structured environment. These meetings can be used at the beginning of projects for brainstorming, norming, or forming exercises; they can be used midproject for team-building, motivational, or inspirational purposes; they can be used at project completion for rewarding, recognizing, and just saying thank you for a job well done.

In the latter case, if it can be done, inclusion of spouses or significant others will make an enormous impact on future attitudes. Maybe this is too personal a feeling, but the effect I have witnessed is simply amazing. What used to be jealousy toward the company when evenings and weekends were consumed with work turned to unbelievable patience and acceptance. And the only time my wife was included in such a three-day, off-site meeting was ten years ago. The effect can be lasting.

Group Training

Rewarding R&D personnel can take shape in the form of education and training. Discovery is clearly an important part of R&D professionals' lives, and involvement in formal training methods, processes, or tools is a motivating activity. When groups are taught together, the R&D members feel a kinship and sense of security in that others share in their new-found knowledge.

As an example, training an R&D group in a new computer tool such as a word processor, data acquisition package, or statistical package can create a unified feeling within the group. In contrast, consider what happens if group members are left to select different tools and learn on their own. Work products will be chaotic and probably incompatible because there will be different levels of ability and nontransportable code.

Enthusiastic Leadership

Rewarding and recognizing R&D personnel begins and ends with the manager's attitude. Have you ever noticed how the overall attitude of an organization will emulate the personality of its leader? If the manager is doleful, untrusting, angry, challenging, and reclusive, the organization will be sour, paranoid, abusive, in conflict, and uncommunicative. An enthusiastic leader, on the other hand, tends to exude confidence, energy, success, commitment, and perseverance, and the organization will follow.

Enthusiasm must be coupled with ability, though, and must be directed for the benefit of others to be effective. R&D personnel are not motivated by a court jester, or a lottery. Their leadership should express an enthusiasm for the team and its work.

INDIVIDUAL RECOGNITION AND REWARDS

Management has a wide variety of ways in which recognition and rewards can be given for personal achievement. Recognition can be based on objective criteria—for example, patents issued—or on more subjective criteria, such as attitude and helpfulness to others. Recognition and rewards should be tailored to the achievement. An element that adds real credibility to recognition is having selections come from peers.

Distinguished Scientist Award

For life-long scientific contributions to an organization, a prestigious award would be election to a select organization or group of distinguished scientists or engineers. An objective review body can be established to review nominations on an annual basis. The society may be named after a company founder or a scientist who made a significant contribution during the early formation of the company or institution. A balance needs to be made between having an objective set of measurement criteria and maintaining an openness that allows a variety of professionals to be considered. Nominations should be solicited annually. The longer the organization of scientists remains as an institution, the more prestigious and meaningful the nomination.

> Recognizing the worth of its employees is the second most important responsibility of management.

Recognition for Patents or Papers

Patents are a means to protect the competitive position of a company; they have demonstrable business benefit. Therefore, any company would be remiss if it did not recognize patent recipients. One obvious way to recognize the recipient is to publicly present the patent certificate in a meeting of peers. Another is to have a banquet for all laboratory or corporate winners.

For a nonprofit institution, papers accepted for publication in peer-review journals may be an alternative measure of demonstrating accomplishment. Promotions can be based on publications, but promotions are too infrequent to be the only way to recognize these important accomplishments. If an institution wants its R&D people to publish, then it would benefit the institution to encourage such efforts on a relatively frequent basis (e.g., annually).

Performance Appraisals

There is probably no more effective way to recognize the accomplishments of R&D personnel than through annual performance appraisals. Done correctly, they take a good deal of preparation and planning by management.

A good performance appraisal presents objective information not only of recent accomplishments but of ways to improve performance. That is not to say a manager needs to point out every mistake and failure of the past year. In fact, unless discussing the mistakes and failures can be instructive, it is counterproductive. (I cannot recall the reference, but I remember hearing that it takes about 42 compliments to overcome the effect of one criticism.) R&D people are told they are to make mistakes; it is OK to fail trying. Revisiting the mistakes will surely discourage risk taking in the future.

Instead, performance appraisals should offer ways in which performance can be improved through paradigm shifts, education and training, mentoring, attending professional meetings, or other positive actions. This presumes the

R&D manager knows, therefore, what it takes to improve performance. The end result of an appraisal should be feedback and direction for success.

The more R&D managers are respected and recognized for their own accomplishments, the more valuable the performance appraisal will be. Conversely, precocious, inexperienced R&D managers will have a difficult time presenting constructive recommendations to senior R&D personnel.

Technical Excellence

Within our company, an annual recognition program for technical excellence involves a peer nomination process. All R&D personnel are asked to nominate a peer for work performed in the previous year. The nomination should include detailed and specific examples of why the work of the individual should be recognized for its technical excellence. Not only are the professional (degreed) employees candidates for nominations, but all support people are eligible, including technicians and clerical personnel.

In general, this program has been well received. Nominations are screened by a committee of peers, including some previous winners, and the finalists are announced in a departmental meeting. Many feel that simply receiving recognition from their peers is reward enough, but out of the entire group one or two are chosen to represent the organization in the corporate recognition program. These people, their spouses or designated others, and the laboratory manager are invited to a large banquet recognizing all nominees. They have an opportunity to choose a gift from a gift catalog created for the occasion and at the banquet, a group of corporate winners are selected from those present and recognized before the entire audience. All nominees are given artistically crafted awards commemorating the recognition and those receiving the ultimate corporate awards are invited to a three-day retreat at a north woods lodge.

An interesting phenomenon connected with this process is the psychology that seems to surround it. Most R&D people feel they deserve at least to be nominated in any given year, but they are reticent to write a nomination for someone else. Part of this is because writing a nomination takes a fair amount of time (two to four hours). However, part of the reticence may also be due to selfishness or arrogance.

David Myers reported that in a psychological study, 60% of those in the study rated themselves in the top 10% of the population as leaders, 25% rated themselves in the top 1%, 70% rated themselves in the top quartile in leadership, and only 2% rated themselves below average as leaders. Maybe it is because these attitudes still persist that it requires management to cajole, coerce, and coax nominations from the R&D personnel. Recipients are indeed grateful for the recognition, but the nomination process is not spontaneous.

Monthly Awards

Our laboratory manager instituted a monthly awards program, similar to an employee-of-the-month award, that has been well received. Nominations are solicited at the end of each month for two awards: one for the lab member who has made the most significant contributions to the lab, and one for the nonlab person who has assisted the lab the most. Nominations are solicited before ballots are disseminated, after which all members are given the opportunity to vote for their choices. Recipients are announced at that month's lab meeting and are awarded a gift certificate from an area restaurant.

> **The team member or contributor who does not receive the recognition or reward will never forget the omission.**

Empowerment

It seems with each passing year, catch words and phrases come into vogue in the business literature. Currently, empowerment is one such word. Since its sudden popularity, however, I have seen no change in behavior to reflect its impact or true meaning.

However, what I think empowerment means is that management gives decision-making authority to those who are directly affected by a decision and who have the most knowledge of the decision's impact. For empowerment to work, there are two requirements: a clear vision of what management wishes to achieve, and trust.

Without trust, all efforts of recognition and rewards are fruitless. Professional trust is based on management's belief that those reporting to them are both able to do their jobs and able to make good decisions. Upon reflection, if employees cannot make the right decisions, then they should not be in their professional positions. Anyone who has worked for someone when trust was not reciprocated knows how oppressive working conditions can be.

There is, therefore, no higher form of professional recognition than trust. With it, morale and productivity will be at their highest. Without it, management should cut its losses and throw in the towel. The organization will quickly become totally dysfunctional.

Personal Notes

Have you ever received an unsolicited, handwritten note from your boss or a colleague that simply says you did an outstanding job on something and offers gratitude or accolades for the work? It feels great, doesn't it? You probably read it two or three times and then put it into a personal file for posterity. Maybe you never reread it, but you never forget that it is there.

This is an excellent way to give recognition. It does not take much time, it does not cost anything, but the impact is unbelievably positive. If an R&D manager cannot find the time to do this spontaneously, that manager should block out one hour each week to reflect on the accomplishments of his or her employees over the past week, select an exceptional effort, and write a note to the employee. The response will be gratifying.

Customer Contact

Customer contact has a double benefit; it allows R&D personnel to know what their customers' requirements are, and it denotes trust, because they are being allowed to represent the company to the customer.

Having R&D personnel speak with customers is usually anathema to sales staff. The reasons are many (e.g., unprofessional presentation, complex lexicon, promising more than can be delivered, excessive detail), but the feelings are, indeed, sincere. R&D personnel must be aware of the sales staff's concerns and make an effort to demonstrate that they can be trusted to conduct themselves accordingly. Sales should work with R&D to develop an appropriate company presentation, and R&D personnel should be encouraged and allowed to get into the field to meet customers. This is a form of recognition that will not only benefit R&D personnel but will benefit the business or research organization through increased customer contact.

Remuneration and Gifts

Monetary gifts or salary increases are a clear way management can show its recognition of the worth of its employees. Although this method of recognition and reward may not be available for frequent use because of limited resources, it must be used nonetheless. Even if all the previously described ways of recognition are used, salary increases that are limited to cost-of-living increases will give R&D personnel a very strong negative message about their worth. Conversely, if pay increases are generous, R&D employees will feel it is a reflection of their worth to management.

BEHAVIOR IN OPPOSITION TO RECOGNITION AND REWARD

Management may want to give abundant and appropriate recognition to its R&D staff. However, certain actions and behavior can be more detrimental to reaching the company's R&D goals than any overt action taken to show recognition. Exhibit 2 lists behavior counterproductive to recognition and reward that unfortunately gives negative messages as compared with the positive actions listed in Exhibit 1.

Business-Only Attitude

If managers are interested only in profit and other basic financial metrics, their business will surely be at risk. They must balance their time between attending to customer requirements, running their business, and looking after their people. If little to no effort is directed toward their R&D people, creativity, innovation, and timely introductions of new products will suffer.

Dual-Ladder System

Dual ladders are intended to give R&D personnel the opportunity to receive promotions, and thereby salary increases, commensurate with their achievements

Exhibit 2
Behavior Opposing Recognition and Reward

>   Business-only attitude
>   Dual-ladder system
>   Short-term vision
>   Threats
>   Finger-pointing memos
>   Politics

and responsibilities. However, promotions on the technical side of the ladder can be arbitrary. Consequently, review boards are established to review the promotions, and they can become accomplished reviewers—so accomplished, in fact, that no one who is not able to turn lead into gold seems to be promoted.

Meanwhile, nontechnical people, and especially managers at higher levels, are the apparent movers and shakers who receive substantial promotions. If this happens within an organization, R&D personnel will at least become disgruntled or, more likely, will join a competitor's organization.

Short-Term Vision

Nothing is more discouraging to an R&D person than to have his or her projects terminated because management has decided to curtail funding. I personally know quality investigators whose careers are at a standstill because they have been assigned to truly difficult tasks only to have the projects canceled before the discovery is made or the project has had the opportunity to end successfully.

One or two such cancellations in a career are probably acceptable, but for some, this is a never-ending scenario. Management might think that invention can be planned or that discovery can be completed within a half-year, one-year, or two-year time frame, and when it is not, management will accuse the investigator of being too slow. A technical manager may break up a task into achievable segments or may mix assignments so that development tasks, generally more easily attainable than research tasks, can be used to support promotions.

Threats

Threats are typical of the desperate manager who feels the need to manage by intimidation. Going back to the coaching analogy mentioned earlier, can you imagine the response of a team to a threat of failure? Decathlon hopeful Dan O'Brien's coach did not threaten punishment if he did not make the pole vault. Dan did miss the pole vault, and he did not make the cut to represent the US in the 1992 Summer Olympics. Reebok may not be able to recapture the advertising coup they had all but sewn up. Should his coach or Reebok have made a threat? Would it have made a difference? Dan will never forget his failed attempt, and he will always be the better competitor for it. We learn too much from our professional mistakes to be threatened for them. There is no place for threats in the R&D workplace.

Memos and Corporate Politics

Writing memos to set the record straight of who is to blame or who is responsible if milestones are not met is an excellent way to discourage good R&D personnel. Good researchers are too focused on task at hand to worry about the organization as well as the individuals.

I once read in a prominent weekly news magazine that the only people who get hurt by corporate politics are those who do not play the game. If that is the case, most good R&D personnel will be hurt, or their productivity will be so negatively affected that few milestones are met.

If R&D people are effectively managed, if the organization is supportive, positive, and enthusiastic, if there is a can-do attitude on the part of all the people within the organization, R&D will attain its goals. If there is finger pointing, blame shedding, and cynicism within the organization, R&D productivity will grossly suffer. It is important to weed out cynicism and naysayers before the cancer they spread is incurable.

THE RESPONSIBILITY OF R&D RERSONNEL

Obviously, the role of effectively recognizing and rewarding R&D personnel is not only that of management. The R&D staff has an important responsibility as well.

First, R&D staff members should be gracious recipients. Recognition efforts that are shunned or discredited will hinder any future attempts for awarding good work. I know one individual who refused to attend a ceremony given on his behalf because he said he did not think he did a good enough job to deserve the recognition. No amount of imploring persuaded him—he refused to come, and the event took place without him. His absence, unfortunately, really hurt his career.

In another case, a large group of people contributed to the success of a major product development effort. For a number of reasons, only a small set of contributors were able to receive recognition at the primary award ceremony, and a secondary event was planned for the balance of the contributors. One young engineer who felt he had made extraordinary contributions was so offended by not being selected for the primary event and by the nature of the gift of recognition that was given in the secondary ceremony that he keeps the award in his desk drawer and gets angry every time he sees it. I advised him to throw the award away inasmuch as he is wasting a lot of emotional energy reliving his feelings. He refuses.

Second, R&D staffers should control their jealousy of others. A sincere congratulatory statement to another award recipient, a pat on the back, an immediate handshake, will make the award process a long-lasting one. R&D nonrecipients need to have confidence that their moment in the spotlight will come.

> In the world of R&D, discovery and accomplishment cannot always be attained.

Third, humility is a favorable attribute. It seems that many R&D people, especially those who do a lot of work but never seem to get anything accomplished, need to blow their own horn frequently. It is hard to get close enough to someone to present an award or recognize achievement when a trumpet is being blasted in your face.

CONCLUSION

Recognizing and rewarding R&D personnel is not easy. People desire recognition in varied ways. Some like it done quietly and personally, some like much fanfare. Two people will look at the same award in absolutely opposite ways: one may consider an award an insult because of its nature, size, or timing, while another may be grateful for receiving the same award. There is no single award or method of recognition that should be used. A wide variety has been presented and many more exist; they should be tailored to the occasion and the person. Recognition and awards must be sincere, planned, deserved, and meaningful, and they should be awarded by people who have earned the respect of the recipients.

GUIDELINES FOR MOTIVATING SCIENTISTS AND ENGINEERS

Stephen J. Fraenkel

R&D personnel have different management requirements than other company employees, particularly in their general allegiance to their technical professions and in their need to integrate their highly specialized technical work into a general business environment. R&D managers who effectively address these issues foster productive, profitable technology development.

Scientists and engineers have many attributes that make managing them different from managing production, sales, or marketing people. For example, scientists and engineers are often more committed to their professions than they are to their employers. They also have more specialized skills than do most other company members.

Because scientists and engineers differ professionally from others in a company, R&D managers should take into account two basic issues in managing them. First, they should consider how R&D employees' professionalism influences their work and attitudes. Second, R&D managers should help their personnel integrate their technical work into their company's business. The 18 tips presented address both issues comprehensively.

DON'T ALLOW YOUR R&D STAFF TO BECOME OVERSPECIALIZED

R&D personnel need to maintain a balance between specialization in a field (e.g., electrical engineering or polymer chemistry) and intellectual flexibility. Unfortunately, technical personnel too often err on the side of overspecialization in one field. An R&D manager should prevent R&D staff from becoming too narrowly focused.

One way to achieve this balance is to recruit technical personnel who do not show a tendency to overspecialize. Moreover, a person who is a fast learner in a related technical area (e.g., an electrical engineer who can deal with questions related to physics) can often help a development laboratory more than a person who knows a great deal about only one technical area.

Furthermore, highly specialized technical personnel present two problems. First, it is difficult to keep such individuals motivated if their long-term

---

STEPHEN J. FRAENKEL, PhD, is founder of Technology Services Inc, a Northfield IL consulting firm specializing in R&D management and materials engineering. Before founding Technology Services, Fraenkel had a distinguished career in industrial R&D management.

professional aspirations are not fulfilled. Second, it can become difficult to use these people's technical skills if the needs of the company change.

As an R&D manager, I have seen many situations in which a technical person who was a fast learner did an excellent job. For example, I once had the task of overseeing the development of a process for rapidly coating a metallic web. The problems posed by the challenge of developing this process involved many technical areas, including heat transfer, mechanical metallurgy, process stability control, and chemistry. Unfortunately, in our laboratory at that time we could not assign—or even find—all of the experts in the various areas who were needed for this project. In addition, we were concerned that a committee of experts would not be able to solve the problems. Instead of forming a committee, we appointed a physicist to be in charge of the whole endeavor. This individual was a quick learner and knew just enough about all the aspects of the problems to ask the right questions and find the right technical assistance; therefore, he was able to organize the search for a solution, and the process we needed was successfully developed.

RECRUIT R&D PERSONNEL WHO KNOW WHEN TO STOP TRYING TO IMPROVE A NEW PRODUCT

In recruiting R&D personnel, the company's goal is to hire scientists and engineers who are interested in and capable of improving its products. On the other hand, it is important to ensure that they know when to stop improving a new product. That may be a surprisingly tall order. An understanding of this aspect of product development comes from experience. No college course teaches when to stop improving a product or processes. Many scientists and engineers are perpetually dissatisfied with the performance of their innovations and downplay issues related to delivering a new product to the marketplace as quickly as possible. Although scientists and engineers are not the only ones in a company who should recognize trade-offs between technical performance, cost, and time, they still should take them into account in their technical work.

USE COMPUTERS AS TOOLS

R&D staffs face few temptations more widespread, inviting, and available than the computer. Of course, few tools are more useful than the computer, and few are more fascinating in the variety of tasks to which they can be put. However, R&D personnel can become enraptured by the computer and view its use as an end in itself. Such fascination, accompanied by the inability or even unwillingness to recognize the computer as only a tool, no matter how powerful, can entice R&D staff members from their objectives and toward quantitative overkill, delaying project completion. Clearly, a measure of awareness of computer operations can be expected of all R&D staff. In certain instances, the presence of experts can even be mandatory. However, it is the generation and the judicious appraisal of novel ideas that make or break an R&D operation—not the size or sophistication of its computing resources.

As tools, computers are merely analytical instruments. In a successful R&D operation, foremost emphasis is on conceptual thinking, especially by the supervisory staff. Once the ideas are there, commensurate computer use will follow.

> R&D personnel should not be allowed to think of the process of discovery as an end in itself.

IMPRESS UPON THE STAFF THE NEED FOR R&D TO SERVE THE COMPANY'S BUSINESSES

Technical personnel often consider issues related to cost, schedule, safety, and producibility to be peripheral to their primary functions. From the perspective of a business manager, however, the results of an R&D project are purely academic if these issues are not considered from the beginning. If an R&D engineer indicates tolerances that are not producible, specifies materials that are not procurable, or does not propose methods to dispose of the by-products of a manufacturing process, a business manager has limited use for that person's work.

Consequently, the technical staff must be oriented toward serving the company's business needs. First, technical staff members should be evaluated to see whether they operate well in the business world, not just in the laboratory; that is, they should understand the company's business in its entirety and understand how such social concerns as safety and pollution control issues affect the company's businesses. Naturally, some are not interested in this broader perspective and consider the world of science and technology the only worthwhile arena. If these individuals cannot be shaken from this opinion, they may be better off working in either a university or a very large R&D organization that has a substantial organizational cushion between R&D and the management of a company's businesses.

Second, R&D managers should instill in technical staff a sense of urgency. Although not all R&D projects should be crash programs, even the most relaxed endeavor has to be undertaken as quickly as possible. There is never time in industrial R&D for a leisurely approach. Moreover, most business managers will accept approximate answers and near-optimal solutions if the situation is urgent.

Third, technical staff members should understand that they must product results. R&D personnel should not be allowed to think of the process of discovery as an end in itself. Business managers do not really care about the intellectual elegance of a technical approach; they want useful answers to their problems.

LOOK FOR TANGIBLE EVIDENCE OF CREATIVITY

Every R&D manager would like to have a very creative technical staff. To build such a staff, however, R&D managers must do more than wish for good luck. In recruiting candidates, an R&D manager should seek tangible evidence of creativity. For example, does a candidate understand and use known techniques to stimulate creativity? These techniques include synectics, which is largely based on analogies; brainstorming; morphological analysis, in which a search for the structure of a problem is conducted; and value engineering. Evidence of creativity can also be found in published papers and patents.

INSIST THAT R&D PERSONNEL REALIZE THE LIMITS TO THEIR TECHNICAL COMPETENCY

No technical person can be expected to know everything. In most cases, R&D staff members need to seek advice or guidance or to discuss their technical problems with their peers. Volunteering to seek advice reflects positively, not negatively, on the person who seeks information. To facilitate this process, R&D managers should insist that technical personnel recognize the limits of their techical competency and work cooperatively with other specialized members of the department.

DO NOT NEGLECT ISSUES RELATED TO EMOTIONAL STABILITY

R&D managers should pay close attention to the emotional stability of their staff members. The most important elements of an R&D project are the scientists and engineers involved. Most crises that arise in R&D projects stem from interpersonal conflicts rather than malfunctioning equipment, late deliveries, or unreasonable customers. For example, in Harvard Business School's collection of business case studies, those related to R&D show a striking number of problems that stem from technical personnel who are emotionally unstable, react defensively to criticism, or do not tolerate dissent. The emotional well-being and interpersonal relationships of R&D staff members can be crucial to the outcome of the project.

EXAMINE PERSONAL QUALITIES CLOSELY WHEN EVALUATING R&D STAFF MEMBERS FOR PROMOTION

When evaluating technical staff members for promotion, personal qualities as well as technical capabilities should be carefully noted. For example, how much supervision does the person require? A properly motivated technical person should not need—and indeed should resist—constant supervision. Conversely, how does that person react to supervision when the situation requires it?

The intellectual curiosity of staff members should also be taken into account. One evaluation technique is to ask questions that stray from the narrow confines of a person's technical specialty. The R&D manager should note whether that person has a record of continuous learning and whether the person can accomplish more in a greater variety of areas than, say, five years ago.

BE AN EFFECTIVE MANAGER IN ADDITION TO BEING TECHNICALLY ADEPT

Effective managers of other technical personnel must be able to manage their own activities effectively. For example, the ability to set sensible objectives is important. One of the most important of these is setting high standards for personal performance. In such intellectual endeavors as R&D, it is difficult to develop quantitative measures to evaluate how well such performance standards are met. Nevertheless, the R&D manager should insist that the staff be intellectually honest and, moreover, be perceived as such.

In addition, realistic work objectives must be established; those objectives should factor in the resources that are available. It is also important to set deadlines and hold the R&D staff accountable for meeting them. For example, if a project review is scheduled on the third Wednesday of every month, that review should be held as scheduled even if it takes only five minutes. If laxity is allowed there, laxity will also creep into other matters that may be more important than a project review meeting.

Finally, the R&D manager should delegate authority to the lowest possible management level. In this way, the whole staff participates in activities and has a stake in making those activities successful.

HAVE THE R&D STAFF FIND OUT WHAT THE COMPANY'S CUSTOMERS NEED AND WANT

> If efforts to put the R&D staff in contact with the company's customers cause conflicts, these conflicts should be resolved.

Depending on the nature of the company's businesses, it may be necessary to reach beyond the company's customers to formulate a market-driven R&D program. For example, the R&D staff I managed at a packaging-products company found that the input of our customers' customers was necessary to make technical work useful. What ultimately sold a package was not the opinion of the company that bought and filled the package but the opinion of the person who opened it and used its contents. In addition to having my R&D personnel visit such consumer-product companies as Procter & Gamble, Lever Brothers, and Stroh Breweries, I had the staff arrange its own consumer focus groups to evaluate new package designs before the packages ever reached our company's customers.

At our packaging-products company, we also developed many new products for industrial applications, including containers for hazardous substances. In these instances, my R&D staff dealt with the users of those containers (i.e., chemical companies) as well as with the producers of the hazardous substances. The staff found for the most part that the practices and preferences of users were the key considerations in developing a new container.

If efforts to put the R&D staff in contact with the customers' customers cause conflicts with the marketing or sales department (which happened to us occasionally), these conflicts should be resolved rather than avoided. Experience shows that the R&D staff—not the marketing or sales departments—will pay the price if the new products are unsuccessful.

ACCOUNT FOR ALL FACTORS WHEN ASSIGNING R&D PERSONNEL TO SPECIFIC JOBS

When considering an R&D staff member for a job, the R&D manager should take all factors into account. For example, not only the nature of the technical job should be considered but also such secondary aspects of the job as the amount of communication required, the types of resources to be managed, and the commercial issues to be addressed. Ensuring that a technical person can deal effectively with the nontechnical aspects of a job can be as important as ensuring that the technical requirements are met.

As R&D manager at a paper-producing company, I observed a case involving a glaring mismatch between a technical person's abilities and the needs of the situation. Even though the person assigned was an expert in metallurgy—one of the basic technical requirements—he lacked communication skills. Therefore, when he was assigned to monitor the paper mills for corrosion, he worked alone rather than in coordination with personnel from the mills. First, he wandered around the mills on his own, taking photographs, then withdrew to his office and developed an action plan for the mills without consulting anyone. As it turned out, the mills rejected his advice and his whole effort failed.

DO NOT SET GOALS WITHOUT THE STAFF'S INPUT

The technical staff should be involved in setting goals. Few activities increase staff commitment to effective R&D more than actually setting departmental goals. By allowing staff members to participate in this important activity, the R&D manager bolsters their professional status and their commitment to success.

When the technical staff is involved in setting goals, such financial goals as overhead charges and R&D budgets should be included along with R&D project goals. Staff members must be made to understand budget requirements and constraints—although they may not like these restrictions, they cannot then complain that the restrictions were imposed without explanation.

When I was the director of a laboratory, I always met with section and department heads to discuss their budgets for the following year. I asked them to make the first draft of their own budgets and encouraged them to explain their budgets to their staffs. This method of developing a total R&D budget worked extremely well. All members of the laboratory were kept informed of the budgetary constraints within which they would have to work.

DO NOT MAKE ALL OF THE PRESENTATIONS AT MEETINGS YOURSELF

The R&D manager should not be the exclusive spokesperson for the R&D group; if that is the case, the professional status of the R&D staff is demeaned, albeit unintentionally. Therefore, technical staff members should be given a chance to speak publicly, particularly about ideas they originate. Assuming this role provides many benefits, including:

> The underlying challenge is to keep the R&D staff intellectually flexible rather than to prevent technological obsolescence.

- Acquainting business managers with the staff.
- Enhancing the morale of the staff.
- Minimizing misunderstandings that arise when technical ideas are explained by someone who is not intimately familiar with them.
- Providing the staff with an incentive to learn how to communicate with business managers.

ENCOURAGE R&D AND NON-R&D STAFF TO ROTATE JOBS

The more the R&D staff and non-R&D personnel in the rest of the company understand one another's capabilities and needs, the better the R&D projects will serve the needs of the company's businesses. One way to hasten such mutual understanding is to implement job rotation among members of the R&D staff and non-R&D personnel.

Unfortunately, in most companies, R&D staff members are seldom assigned to the marketing or finance departments. It is even rarer for a marketing or finance person to be assigned to an R&D department. There is no reason, however, that R&D personnel cannot function effectively in planning, marketing, and finance departments. These individuals are skilled in analyzing phenomena in quantitative terms—a talent that can benefit those departments as well. Temporary transfers of non-R&D staff to an R&D department are more difficult to arrange but not impossible.

As R&D manager at the packaging-products company, which was also an operating division of a large oil company, I had the opportunity to compare the two scenarios because the parent company rotated jobs and mine did not. In the parent organization, engineers worked in the planning and finance departments and finance personnel worked in the R&D department, particularly in R&D planning. This experience led me to conclude that no disadvantages stemmed from rotating R&D and non-R&D personnel; on the contrary, there were many advantages, especially in terms of integrating R&D into the company's businesses.

BREAK DOWN BARRIERS BETWEEN R&D AND TECHNICAL SERVICE PERSONNEL

In laboratories, barriers can develop between R&D and technical service personnel. Sometimes, these rifts occur because R&D personnel regard their technical services colleagues as operating at a lower technical level and having fewer capabilities. The R&D manager should try to prevent such an attitude. This view of technical services personnel's capabilities is not necessarily accurate and can destroy any chance of fostering the much-needed cooperation between the R&D and the technical services staff.

USE A DUAL LADDER FOR ADVANCEMENT OF R&D PERSONNEL

In many companies, R&D personnel who want to advance are forced to become managers, even if they are not suited to the position. Using a dual ladder to advance R&D personnel solves that problem by allowing such individuals to climb a technical ladder if they are not suited for or are not inclined to pursue a management career path.

For a system of dual career ladders to operate effectively, qualifications for positions in both paths need to be spelled out as specifically as possible. This is usually easier to do for positions on the management ladder than it is for positions on the technical ladder.

Problems with a dual-ladder system generally arise for two reasons. First, such a system is sometimes misused by R&D managers who use the technical ladder as a place to put R&D personnel who have failed as managers. Second, a dual ladder system does not work well if no genuine equity exists between corresponding positions in the technical and management ladders.

HELP YOUR R&D STAFF COMBAT TECHNICAL OBSOLESCENCE

Although technical obsolescence can be a problem in any company, it presents particular difficulties for companies whose technologies change rapidly. In many organizations, however, the main technical problems involve applying existing knowledge rather than making cutting-edge technical breakthroughs. Nevertheless, the problems of applying existing technology can still be complex and difficult. The underlying challenge is more often to keep the R&D staff intellectually flexible rather than to prevent the staff from becoming technically obsolescent.

An R&D manager can ask several questions to evaluate the intellectual flexibility of the R&D staff, including:

- Can the person reason by analogy from a biological system to a mechanical one?
- Does the individual have and apply the knowledge that the same differential equation describes both current flow and mechanical vibrations?
- Can a wear problem in a wood-based product be dealt with by using experimental results from mechanical metallurgy?

The R&D manager can decrease technical obsolescence and increase intellectual flexibility by helping the R&D staff gain skills with broad applicability. One such skill is the use of statistics to analyze and interpret research findings. It is amazing how many highly trained R&D professionals grind out reams of data but have little knowledge and make no use of statistical techniques, probability theory, and inference methods.

> R&D managers should encourage their staffs to write clearly.

Statistical techniques are applicable to all branches of science and engineering, and they are increasingly useful in business as well. In addition, statistical techniques provide insights into the meaning of data that cannot otherwise be gained. As manager of an R&D department, I required everyone on my technical staff, including technicians, to take a training course in statistics, and R&D project leaders were required to take a course in experimental design.

TEACH THE R&D STAFF HOW TO COMMUNICATE

R&D managers should encourage their staffs to write clearly. The litmus test of clear writing is whether it can be understood by the most senior managers to whom it is addressed. Because senior managers are often business managers with minimal technical education, technical jargon should be avoided.

Furthermore, R&D managers should encourage R&D staff to be more effective in their oral communication. One way to do this is to have a capable instructor conduct a short course on making oral presentations. As an R&D manager, I regularly had instructors teach my staff how to give speeches and make presentations at meetings. These short courses on public speaking helped my staff members develop their stage presence, self-confidence, and ability to think on their feet.

ACHIEVING DEDICATION AND EXCELLENCE

The points included in this chapter are based on my personal experience as an R&D manager. They worked in my situation, and I believe that they are effective rules of thumb that can be applied in almost any technical development environment. Because all R&D staff members are individuals, as are their managers, the generalizations made here will not apply in all cases; however, for the sake of offering some useful advice, a few generalizations can be made. The basic premise is not to characterize R&D personnel as lacking in particular areas but to show that encouraging commitment, pride in work, team spirit, and respect for oneself and one's peers encourages dedication and technical excellence.

SECTION 3

PROJECT MANAGEMENT AND TEAMWORK

Project management and teamwork are an important part of R&D management that, depending on the company, can also affect an R&D department's relations with the rest of the company. Because a R&D project usually involves more than one person, someone must set goals for the two or more technical specialists who work on it and track their progress. In some cases, formal teams are created to carry out these projects. Personnel from non-R&D functions may serve on them. Managing such teams requires special finesse.

In the first two chapters in this section, Robert Szakonyi tackles the broad subject of project management—how to select, plan and carry out R&D projects. James A. Ward explains how managers can reduce project risk by understanding and controlling project variables—work, resources, and time.

In a subsequent chapter, Stewart L. Stokes, Jr. suggests guidelines for implementing self-managed teams. Robert Szakonyi explains how to establish a cross-disciplinary team for product development. Ronald A. Martin offers tips on motivating teams, and John Van Atta discusses the dynamics of cross-functional teams. Finally, George Manners poses the questions organizations face in implementing a successful total quality program.

ESTABLISHING DISCIPLINE IN THE SELECTION, PLANNING, AND CARRYING OUT OF R&D PROJECTS

R&D managers need lots of discipline to ensure that R&D projects are targeted to meet customer needs and that are completed on time and within budget. This chapter describes an R&D evaluation system that helps impose this discipline.

Robert Szakonyi

In the selection, planning, and execution of industrial R&D projects, an R&D manager is caught in the middle. On one side are the company or customer needs that R&D projects should be targeted to meet. The manager must ensure that an R&D group is working to provide what the user requires. On the other side are the scientists and engineers, each of whom has specific technical capabilities and individual goals. The R&D manager must ensure that these specialists meet the technical objectives of their projects on time and within budget.

Because an R&D organization is caught in the middle, it is useful to speak of the two faces of discipline needed for R&D projects. Discipline is needed to ensure that R&D projects are targeted to meet company or customer needs. Discipline is also necessary to ensure that R&D projects actually do meet their objectives on time and within budget. Both kinds of discipline are required for R&D projects to pay off.

What R&D managers overlook at times in their efforts to develop a new business or a new manufacturing technology is that discipline is an essential element of innovation. Discipline serves as the gyroscope when an R&D manager, an R&D staff, and their company push into new directions.

This is not to say that there is no such thing as over-discipline in R&D projects. There is, and quite a few companies suffer from it. A company's marketing or manufacturing department, without fully appreciating technological opportunities, may overspecify what an R&D group should accomplish. Although discipline may be established in this situation, a company's technological potential can be wasted. R&D managers themselves can overmanage their own projects, treating an R&D project as if it were the construction

---

ROBERT SZAKONYI, PhD, is the director of the Center on Technology Management at IIT Research Institute, Chicago. He has consulted for many companies in various industries and has written two books and more than 40 articles on technology management.

of a building, not the execution of genuine science, or engineering research or development.

Most companies, however, do not suffer from the problems just mentioned. Instead, the questions that gnaw at most R&D managers are: what R&D should be selected and how can the department keep R&D projects on track? Both questions relate to discipline.

PROBLEMS INVOLVED IN TARGETING R&D PROJECTS TO MEET COMPANY OR CUSTOMER NEEDS

The crux of the problems involved in targeting R&D projects to meet company or customer needs is the great difficulty of getting marketing information to accompany available technical information (or manufacturing-related information if a new technology needs to be directed to a manufacturing plant). Some technical information is very inadequate, particularly if an R&D group is pushing the state of the art. In most cases, however, the disparity between technical information (which often ranges from good to very good) and the marketing information (which often ranges from bad to fairly good) is great indeed. To use a metaphor, an R&D group, which is pitching, often lacks a catcher's mitt toward which to aim.

Valid marketing information truly reflects a customer's needs. In general, the more that a company has had experience with a customer, the more valid its information is about that customer.

Consequently, the problem of targeting R&D projects to meet customer needs is most accurate when a company moves into new areas of business. For example, one oil company made three major diversification efforts over 15 years, and they all failed because the company had poorly defined objectives for developing new businesses (and the senior business managers lacked the patience needed for developing new markets). Similarly, a business manager at an electronics company once observed that his R&D staff was always much more successful in developing new technologies to meet their existing customers' needs than in using existing technologies to develop similar products for potential new customers.

An R&D group can also lack valid marketing information if the needs of customers change, and the company is slow to recognize their new needs. Without valid marketing information, R&D groups either continue to solve an older generation of the customer's needs or try to solve the customers' new needs without knowing what those needs are.

Getting marketing information channeled appropriately is also important. Management issues are also involved in ensuring that valid marketing information is available and that this information is used in discussions about R&D selection and planning. Two important management issues are the establishment of an effective marketing (as opposed to sales) group (or a technical marketing group if the customer is the US Department of Defense or industrial companies) and the creation of effective working arrangements between R&D groups and marketing groups.

PROBLEMS INVOLVED IN PLANNING AND MANAGING R&D PROJECTS

> The problem of targeting R&D projects to meet customer needs is most acute when a company moves into new areas of business.

The crux of the problems involved in planning and managing R&D projects is the difficulty that R&D managers face in holding their staff and themselves accountable. Although R&D project planning and management serve many purposes within a company, they are, first of all, tools for R&D managers to run their group, department, or organization effectively.

Effective R&D project planning and management must be different from, for example, the planning and management of a project to construct a building, because carrying out R&D is different than constructing a building. Nonetheless, in any R&D project, certain technical objectives should be met, schedules kept, and budgets observed.

Resistance to R&D project planning and management comes mainly from scientists and engineers but at times from the R&D managers themselves. For example, the senior R&D managers of an aerospace company continually have to engage in a tug-of-war with researchers and R&D managers to get them to specify the technical objectives of their projects.

An R&D manager at a computer company described the same problem from another perspective. According to him, the researchers in his company either do not want or do not know how to think about their technical work in the context of project objectives. They are primarily interested in substantive technical questions. Issues related to the practical technical results of their work—not to mention to the time and resources that it takes to get those results—are, at best, of secondary importance.

CREATING A PARTNERSHIP WITH OPERATING UNITS TO SELECT R&D

Given the difficulty of getting valid marketing information, many R&D groups resort to selecting R&D mainly by themselves—and developing a technology and pushing it to commercialization mainly by themselves. The problem that R&D groups and their companies encounter through this approach are often monumental. Because great effort is necessary to develop and commercialize a technology, company operating managers can withdraw their support, causing the technology to die before it is even marketed. Even if a technology is commercialized, the new product may be targeted poorly.

For example, R&D managers at an electronics company had directed the development of a new technology, the design of a consumer product based in this technology, and the introduction of this product into marketplace. Unfortunately, these R&D managers misread the needs (and wants) of the consumer, and their company eventually had to pull this new product from the marketplace. At an agricultural machine company, R&D managers developed and commercialized a new energy-related technology. However, they misdirected their efforts because they did not know what made commercial sense. Because of their mistakes, their company lost its entire investment in a major facility that was created to exploit this new technology.

There are other R&D managers, of course, who have excellent marketing sense. Some of them are entrepreneurs who leave their companies and start new ones. Most R&D managers, however, do not have great strengths in marketing or manufacturing—just as their counterparts in marketing and manufacturing do not have great technical capabilities. What is needed, therefore, is a partnership between R&D groups and company operating units.

Most R&D managers accept the idea—at least in theory—that an R&D group and operating units should create a partnership for selecting R&D. The major problem for them concerns not the idea of partnership but the issue of how to create it.

Creating a partnership between an R&D group and operating units is a set of multilayered problems. They consist of differences in time horizons, differences in opinions about what is needed to evaluate the worth of a new technology, potential shortages in the skills required for creating bridges between R&D and operations, potential conflicts over control or over whose options should be kept open, and a lack of genuine interest in creating a partnership.

Differences in Time Horizons

R&D managers and operating managers live, at least in part, in different worlds. Depending on their industry, R&D managers are concerned about research that will pay off in 3 to 20 years. Operating managers, on the whole, are concerned about quarterly and annual results.

This gap is obvious to most R&D and operating managers. However, in most cases, an R&D manager needs to close this gap more than an operating manager does. If the two managers do not discuss potential commercial application of a new technology, the R&D manager risks misdirecting important research, but the operating manager will probably not suffer in quarterly or annual results if such a dialogue does not occur. The long-run potential of the operating manager's business may suffer, but this particular operating manager may not be in the same position when the future weaknesses are exposed. In short, it is up to the R&D manager to take the initiative in creating a partnership.

An R&D manager can take the lead by creating an analytical framework for such a dialogue. An analysis of the profiles of the company's technologies—such as how they were derived, what their successive applications have been, and when their potential will diminish—can serve as such a framework. A strategic plan involving both technical and business considerations also can serve this purpose.

Developing such a plan is difficult. Even in lieu of a real strategic plan, however, an R&D manager can take actions to create a partnership. Developing his or her own version of a working draft of a company strategic plan (which may be kept undisclosed if necessary or may be circulated as a strawman if possible) provides an analytical framework concerning the company's future needs. Then by discussing each of the components of this fictitious strategic plan separately with the appropriate operating manager, the R&D manager can start some constructive dialogues about the company's strategic business needs.

> Creating a partnership between an R&D group and operating units is a set of multilayered problems.

Differences in Mindset Related to Technology Evaluations

An R&D manager in the operating division of an electronics company once complained about the researchers in his company. Because this R&D manager had worked for many years in his company's research laboratory, he sympathized with the researchers' problems. Nonetheless, he faulted them for developing technical concepts to only a limited extent. On a scale of 0 to 100, where 0 is the point at which a technical idea is generated and at 100 it is commercialized, the corporate researchers often carried out technical work to point 10. The operating division of this R&D manager, however, could not start working on the development of a technology until it had reached point 20 or even point 30.

Naturally, part of the problem in closing this gap between research and an operating division is the matter of resources. R&D managers who work in a corporate laboratory or in an operating division do not need to be reminded of this. They wrestle constantly with the problem of resources.

The other part of this problem, however, has to do with the mindsets of researchers and operating managers. (R&D managers often do not take this part of the problem into account.) In general, people who develop a technology and people who use it evaluate the worth of that technology from very different perspectives. R&D and operating managers must be aware of this fact to avoid talking past each other.

Because R&D managers need a partnership more than operating managers do, they must take the initiative in creating one. Ways to do this include analyzing in depth the applications of their technical concept or sometimes carrying out technical work that is applications-oriented. Often, though, what is needed is just analysis based on interviews with potential users. In other words, R&D managers need to recognize a potential gap between their point 10 and an operating division's point 20 and then demonstrate the applicability of their technical concept.

Shortage of Bridge-Building Skills

Two building blocks that can facilitate a constructive dialogue between R&D managers and operating managers are a technology plan and market research.

When R&D managers present operating managers with a systematic analysis of the technical potential and the possible applications of their R&D, they do their part in creating a constructive dialogue. When operating managers present R&D managers with a systematic analysis of their company's future market needs, they, in turn, do their part.

The trouble is that in many companies the R&D group and operating units lack the skills necessary to make good analyses. Some R&D managers do not even understand what a technology plan is. Sometimes, market researchers lack important knowledge. A manager of commercial development at a chemicals company, for example, complained that although the market researchers in his department all had commercial (and sometimes technical) experience in their niche of the chemicals industry, most lacked the analytical breadth and the range of experience needed to evaluate diversifications into new markets.

The solution is improved training for existing managers and better recruitment for such positions as R&D planner and market researcher. The first step in this solution, however, has to be taken by R&D managers and operating managers themselves. Managers at the chemicals company admitted that the primary reason their company recruited such poor market researchers was that the company's managers did not understand what market research was in the first place.

Potential Conflicts over Control

Underlying many of the difficulties of creating a partnership between R&D and operating units is the problem of control. For a partnership to work, an R&D group has to keep an open mind about, for example, an operating unit's recommendation to cut particular R&D projects. For a partnership to work, an operating unit has to take risks with a new R&D technology. Both an R&D group and operating units have to give up some control for the greater good of a company.

Realistically, however, R&D managers and operating managers are not evaluated in terms of their contributions to the greater good of a company. Managers, on the whole, are judged according to how well their own group performed on specific functional objectives. In this situation, managers understandably constantly strive to keep their own options open. Allowing an operating unit to influence which R&D projects are cut appears to diminish an R&D manager's portfolio of possible winners. Supporting R&D before a technology is proved appears to hinder an operating manager's flexibility in responding to new business needs.

The answer is for both sides to give and take. Give and take involves negotiation. As in all negotiation, both sides must give a little until trust develops.

Lack of Genuine Interest in Creating a Partnership

Sometimes, R&D or operating managers are not interested in creating a partnership. Perhaps the fraction of R&D managers who have this attitude is smaller than the fraction of operating managers who do. Because R&D managers depend more on operating managers than operating managers do on them, they need a partnership much more to reach their goals. Many an R&D manager has had to tread water while waiting for a key manager in another function to retire.

Both R&D and operating units must give up some control for the good of the company.

FOCUSING ON THE HIGHEST TECHNICAL AND BUSINESS PRIORITIES

Assuming that an R&D group and operating units recognize the need for creating a partnership, the question that they face is: how do they focus on the highest technical and business priorities? This question is very hard to answer. At a consumer-products company, R&D and marketing managers spent two years in the early development of a new product, unclear about what the new product was actually supposed to be. The R&D managers were pursuing various technical avenues, all of which were supposed to aid in new product development. The marketing managers were pursuing a variety of product options. Finally, after successive stages of stagnation, some R&D and marketing managers met to hammer out an agreement on a few objectives. The meeting was a turning point. The R&D managers now knew which few R&D projects should have the highest priority, and the marketing managers now focused on a limited number of product options. Ironically, the agreement turned out to be quite simple. Why did it take two years to come about?

Companies fail to focus on their highest technical and business priorities because R&D and operating functions do not agree on some unstructured rules about how R&D should be selected. The first rule is: Always have a formally agreed-on starting point for an R&D project. For example, at a meeting at a building-materials company an operating manager charged quite correctly (for the R&D managers could not refute him) that many R&D projects just started on their own. No one seemed to know when these R&D projects had begun and what their justification had been. Without some clear and agreed-on objectives, these projects were almost bound to fail.

The second rule is: R&D managers should not slight commercial consideration in evaluating R&D proposals or ongoing R&D projects. An R&D manager at a nuclear reactor company revealed that his R&D group would always concentrate on various technical criteria when evaluating R&D and then throw in a few commercial criteria as an afterthought—until operating units pressured the R&D group to be increasingly responsive. The R&D managers soon discovered that the success rate of their technically challenging R&D projects went up.

The third rule is: In discussing R&D with operating managers, R&D managers should use a matrix consisting of rows of technical capabilities and columns of company and customer needs. (The particular scheme chosen is not as important as that R&D managers and operating managers be comfortable with it.) At an electronics company, such a framework was used to decide whether the company's technologies and products were differentiated from those of the competition. At an aerospace company, an analytical framework highlighted the company's strongest technologies, the most probable areas in which the US Department of Defense would request proposals, and the probability that the company would succeed in each of these areas. Whatever the scheme, R&D managers and operating managers need to examine systematically how their respective interests relate to each other.

The fourth rule is: Discussions of technical and business priorities should be tied to the issue of resources. For example, an R&D manager at a household-products company revealed that the actual budgets for a long list of R&D projects contrasted sharply with the priorities enunciated at an earlier meeting in which the other R&D managers and operating managers had participated. The managers at this meeting had talked about defining priorities, but their actions spoke otherwise. A multitude of R&D projects were being funded, though at a miniscule level.

Following rules, however, is not enough. R&D and operating managers must grapple with forces beyond their control. They must take into account the human side of management.

For example, a scheme for analyzing technical capabilities and company needs does not address the issue of what technical efforts or business activities should be undertaken in the first place. A case described by an R&D manager of a building-materials company illustrates this point. Before this manager took his position, the various R&D department heads had never assessed how well their laboratory's efforts as a whole were meeting the operating divisions' needs. Surely, these R&D heads had realized they were all serving the same company. Why had they not examined their combined efforts before? They had never seen themselves as part of a greater whole. Thus, although they may have been capable of following rules to examine their own R&D department's work, they needed more than rules to think about the laboratory as a unit.

When R&D managers and operating managers do systematically evaluate technical and business priorities, they must do so according to unspoken goals. For example, at a producer of health care products, the R&D managers were aiming to make their company like a pharmaceutical company, whereas the marketing managers were aiming to make it like a toiletry-products company (through their respective recruiting). Moreover, neither the R&D managers nor the marketing managers realized which direction their counterparts were going in. These managers' conflicting but unspoken goals would have sabotaged any discussion based on a scheme relating various technical capabilities to various product areas. Their goals would not have been incorporated in such a scheme. The R&D managers and operating managers, therefore, might have followed the rules but would have talked past each other.

PROJECT MANAGEMENT AND TEAMWORK

Both R&D and operating managers should understand that following rules serves the best interests of both R&D and operating managers in bad as well as good times. When a business runs into trouble, panic sets in. Agreed-on rules, analytical frameworks, and strategic plans are usually put on the shelf. The best way out of trouble, however, is not to work harder while analyzing less.

> **Discussions of technical and business priorities should be tied to the issue of resources.**

The case of a chemicals company provides an important lesson. When the company ran into trouble, managers put aside their commitment to planning in order to address the current difficulties. Little did the company realize that market conditions were going to get much worse. When this did occur, the company had no strategy to follow. It has now been divided and sold.

ESTABLISHING R&D PROJECT PLANNING AND CONTROL

Because many scientists, engineers, and R&D managers are much more interested in science or engineering than in R&D management, many R&D managers meet significant resistance in establishing R&D project planning and control. Their opponents maintain that R&D personnel cannot be held accountable as other company personnel are and that R&D project planning and control squash the creativity required in R&D. Opponents of R&D project planning and control usually charge proponents with ignoring the role of serendipity in R&D (e.g., the development of nylon and xerography).

A good R&D project plan is not a procrustean blueprint into which complex technical issues must be forced. It is true to the spirit and content of a company's R&D.

The critical questions in developing a good R&D project plan are: What is the technical objective of a project (or two or three of its most important technical objectives)? What are the critical technical problems that will be encountered in carrying out the R&D? How can proof of positive technical results be demonstrated or measured? What are some logical technical milestones along the way? When should they occur? How much will it cost to reach each of these milestones? Which scientists, engineers, and nontechnical personnel will contribute to the project? What will they contribute? During which phases of the project will they make their contributions?

Finally, at the heart of good control of an R&D project are one or more project milestones as well as informal inquiries about technical progress, schedules, and costs that a project manager should make between these milestones. At each milestone, a project manager should evaluate a project in terms of such criteria as the quality of the experimental data gained and the validity of the original technical objective. Rather than waiting for formal meetings to reveal problems, a project manager should closely follow a project's technical progress, time lines, and resource use to recognize problems quickly.

Given the risks involved in R&D, however, many scientists, engineers, and R&D managers are threatened by such methods for holding themselves accountable. An R&D planner at a metals company said that R&D managers at his company would not invite him to their meetings because he might ask whether their projects were meeting their technical objectives. A technical manager of an oil company revealed that a proposal to establish a position of R&D planner (for which he was to be nominated) had to be withdrawn because R&D managers did not want anyone raising questions about the R&D they had selected. An R&D planner at an automotive company related that the resistance to his planning group's efforts to establish improved R&D project planning and control was so powerful that not only was he forced to take an early retirement, but his whole group was disbanded.

R&D project planning and control can lead to something being cut—an R&D project. Similarly, R&D project planning and control can lead to unnecessary difficulties if it is carried out by the wrong people (e.g., accountants). The company personnel who can best understand the vagaries and uncertainties of R&D—the scientists, engineers, and R&D managers themselves—should consequently establish technical objectives and milestones.

There are many good reasons for R&D managers to use R&D project planning and control:

- To go quickly from the development of a new technical concept to the introduction of a new product.
- To avoid being forced to deliver a new technology before it has been adequately tested.
- To speed the process of making decisions about the introduction of new technology into operations.
- For solid business plans, which are developed and taken seriously by operating managers and which can give an R&D group clear ideas about where operating units want to go in the future.
- To preserve an R&D group as a vital and effective company contributor.

Getting the Product Quickly to the Marketplace. After R&D project teams at an electrical-products company defined clearly their project objectives and means for reaching them, the teams dramatically cut schedule slippages and cost overruns. An R&D manager of an instrument company claimed that training his engineers in engineering design paid off significantly: they improved their project planning and completed in several months projects that they had previously taken one year to finish.

Not Delivering R&D Before It Is Ready. R&D project plans cannot keep operating managers from forcing an R&D group to meet unrealistic schedules. Nonetheless, an R&D project plan—along with evidence that a project is being controlled properly—provides ammunition that an R&D manager can use to persuade operating managers to loosen unrealistic deadlines.

Speeding Decision Making. R&D project planning and control cannot prevent operating managers from vacillating when it is time to introduce a new technology into operations. Still, by demonstrating that a new technology has been moving on track, R&D managers improve their chances of persuading operating managers to support its introduction. Although operating managers are very sensitive about risks, they also want successes. R&D project planning and control helps demonstrate that a new technology is a winner.

> R&D project planning and control can lead to something being cut—an R&D project.

Having Good Business Plans. All R&D managers want solid business plans to help them evaluate their R&D. There are no magic methods for persuading operating managers to plan better, but effective R&D project planning and control can set an example.

Preserving the R&D Group as a Vital Player. An early warning sign that an R&D group is in trouble is an effort by a non-R&D organization to provide it an analytical framework for improving project management. It is unfortunate that R&D groups in some companies have to be given such warning signs. It is even more unfortunate that some R&D groups ignore them. R&D groups throughout the US and Canada are being closed down or cut in size. Although much of this restructuring is occurring because of market forces, some of it must be happening because R&D groups are not holding themselves accountable. In other words, one of the best ways for R&D managers to preserve their R&D group as a vital and effective player in a company is to hold scientists, engineers, and themselves accountable through good R&D project planning and control.

DEVELOPING AND USING AN R&D EVALUATION SYSTEM

Another tool is an R&D evaluation system, which R&D managers can use to keep track of how well R&D projects are progressing. R&D managers should develop their own systems, remembering to keep them simple and to focus on only the most critical issues related to project progress.

The R&D managers of a pharmaceutical company learned these lessons the hard way. Several years ago they installed a complicated computer-based R&D evaluation system that they eventually had to eliminate because of resistance from almost all of the company's scientists, engineers, and R&D managers. Learning from this experience, the R&D managers who originally installed the system created a simpler one to facilitate discussions among R&D managers about ongoing projects. This new system also did not work well, however, because it highlighted schedules. Rather than focusing on the critical technical problems that surfaced as R&D projects proceeded, R&D managers argued about weather a milestone had been met or about who was to blame for slippages in schedule.

The heart of an R&D evaluation system has to be the critical technical issues of R&D projects. These are the same critical issues that should be taken into account in planning and controlling an R&D project. Questions that should be addressed in an R&D evaluation system include:

- Is technical work proceeding according to the technical plan?
- Have unexpected technical problems cropped up? If so, how good are any new technical approaches being followed for dealing with these unexpected technical problems?
- Are there new estimates of the probability that the technical objective of the project will be met under current schedules and cost projections?

Obviously, a few other questions concerning schedules and costs should be addressed in an R&D evaluation system. These questions would be answered in the context of the technical work. Just because questions concerning schedules and costs can be addressed more easily that technical questions can does not mean that they should dominate the agenda.

Finally, because an R&D evaluation system gathers information about R&D projects so they can be considered in relation to each other, system, senior R&D managers, and R&D planners can use this system to see how a collection of R&D projects is proceeding. This tool can have at least eight different uses:

- It can aid in R&D project planning and control. At a consumer-products company, an R&D planner used a simple reporting form for every R&D project to get project managers to plan their projects. This system of reporting forms, which evolved into an R&D evaluation system, forced R&D project managers to think about important project issues.
- It helps improve the selection of subsequent projects. For example, the R&D planner at the consumer-products company discovered that a system of tracking R&D projects provides a kind of corporate memory that can be drawn on when new projects are proposed. Now, when he discusses the advantages or disadvantages of a potential project with new marketing managers (who usually hold their position for less than two years), he can refer to progress or lack of progress of R&D projects that were carried out several years earlier.
- It can be used to optimize R&D resources. After the implementation of an R&D evaluation system, an R&D manager of an electrical-products company could finally point to inefficiencies in how his company managed R&D projects. Many tasks were being run in sequence when they could easily have run simultaneously. Project managers of several projects were all planning to use the same limited resource at the same time.
- An R&D evaluation system can document the need to reallocate R&D resources as existing technologies mature and new technologies develop. R&D

managers can use this information when asking for resource reallocation.

> The heart of an **R&D evaluation system has to be the critical technical issues of R&D projects.**

- Some large companies in the US and Canada are coordinating the efforts of previously independent R&D laboratories. Frequently, one or more of these R&D laboratories is part of a company that has been recently acquired. R&D managers responsible for improving this coordination usually find an R&D evaluation system that applies to all laboratories is a very useful tool for building teamwork within an R&D organization.

- An R&D evaluation system documents the overall progress of R&D projects. Although this system does not measure R&D output or R&D productivity, it can provide evidence that an R&D organization is moving in the right direction. R&D managers can use this information in discussions with operating managers about what the R&D organization is doing.

- R&D managers can use an R&D evaluation system to discover patterns of weaknesses in the management of R&D projects. For example, a food-processing company had a shortage of process engineers. Although it had many engineers it called process engineers, most of them specialized in constructing new manufacturing plants. This lack of skills, which came to light only after many R&D projects had suffered, could have been exposed much earlier if the company's R&D managers had used a system for evaluating projects. The system would have shown why many were running into trouble when transferred to manufacturing.

- Senior R&D managers can use these systems to promote an R&D's identity, which can be weak when R&D groups within an R&D organization try too hard to serve the needs of operating units. For example, the head of R&D at a natural-resources company could not control his R&D managers' work. The managers realized that, in practice, they were beholden to the managers of the operating units, who wanted them to serve only the operating units' immediate needs. The senior R&D manager needed to create an overall identity for his R&D organization that could counter the R&D managers' loyalty to the particular operating units. An R&D evaluation system can be used to promote an R&D identity.

HOLDING CRITICAL, BUT FAIR, PROJECT REVIEWS

In a project review, R&D managers, scientists, or engineers who are not participating in a particular project meet to evaluate its progress. They closely examine the procedures through which R&D managers hold the project manager and team accountable. In addition, because R&D managers are ultimately responsible for any R&D project, they must also hold themselves accountable through these procedures.

Holding project reviews is the most important way to establish discipline within R&D projects. Nothing better focuses the efforts of R&D personnel than knowing that at midpoint in an R&D project they will face a critical, but fair, review of their accomplishments. Unfortunately, few R&D groups conduct such reviews. For example, R&D managers at a chemicals company met to discuss projects, but no one addressed such critical issues as whether a project was meeting its technical objective or whether this technical objective was still appropriate. Instead, the managers wrangled over who was responsible for certain milestones not being met or what technical approach to follow for the remainder of a project. They found they could avoid examining critical issues related to an R&D project by concentrating on the mechanics of implementing that project.

Listed in Exhibit 1 is a set of issues that R&D managers must address to hold effective project reviews. Alongside each issue is one or more questions that R&D managers face in trying to resolve these issues. There are no easy ways for managers to resolve these issues, but the better they understand them, the more skilled they become in holding reviews. These issues are briefly considered in the following sections.

Exhibit 1
Project Reviews: Issues and Questions

Issues	Questions
Review focus	Is the review a technical evaluation? Is the project plan, schedule, or costs its focus? Implications of the technical results for, say, the eventual development of a new product?
Review agenda	Are good technical results being produced? What are the implementation methods? What are the project's objectives?
Level of detail examined	How much detail should be examined?
Stages at which a project should be reviewed	When is the right time to review a project?
Staff members representing a project at its review	Who besides the project manager should review a project? If project team members should be involved, who should they be? If R&D managers who are not on the project team should be involved, who should they be?
Make-up of the reviewers	Which technical disciplines should be represented among the reviewers?
Management level of the reviewers	Should the reviewers come from more or less the same management level or from different levels?
Authority of the reviewers	Should they comment, advise, or make strong recommendations that senior R&D management or business management will consider seriously?
Relative permanency of the reviewers	Should reviewers stay with a project at different stages?
Supervision or coordination of the reviews	How is the review process supposed to function? Does it need to be managed?
Following-up after a review	What should happen after a project's review? Should procedures be established to ensure that something happens?
Evaluating the review process	How can the evaluation process be improved?

Review Focus. Project reviewers must agree on the focus of their evaluation, for there are many aspects to an R&D project (e.g., technical accomplishments, schedules, and costs), and they are all intertwined, For example, modifying the technical objective of a project usually affects both the project's schedule and cost. Unless reviewers agree on what they are focusing on, they talk past each other. Reviewers also need to agree on whether they are concentrating on an R&D project or on the implications of the technical results for, say, the eventual development of a new product.

> Holding project reviews is the most important way to establish discipline within R&D projects.

Review Agenda. Are reviewers just concerned about the technical results or are they looking deeper into the project, evaluating the methods of implementation? Perhaps they are questioning the original project objective and wondering whether the project should be continued. Reviewers should agree on the agenda of each review.

Level of Detail Examined. How deeply should reviewers probe into a project? Probing in great detail is time-consuming, but it may be warranted if reviewers think that serious problems exist. On the other hand, if reviewers consider only major questions and allow a project manager and team great leeway in finding answers to those questions, the reviewers may overlook less obvious problems in a project.

Stages at Which a Project Should Be Reviewed. Reviewers should know when to assess a project's technical progress. If they do not, they may wind up asking the right questions at the wrong time.

Staff Members Representing a Project at Its Review. Besides a project manager, who should represent a project at a review? Probably one, two, or a few members of a project team, depending on its size. Possibly one or more R&D managers who are responsible for technical work that the project team needs to use. Choosing the staff members who represent a project at a review can be a ticklish matter that can also affect the validity and fairness of a project review. If made well, the choice of these staff members can help clarify difficult questions of authority and responsibility that may not have been resolved among the project manager, team members, and other R&D managers who are not on the team. If made poorly, the choice can bias the review itself.

Make-up of the Reviewers. If the choice of reviewers is not made fairly, the review will be biased from the start. It is not easy, however, to decide which technical disciplines and which members from these disciplines should be selected. Not only is it difficult to measure an individual's capabilities against those of another individual, it is difficult to determine whether someone from a somewhat related technical discipline brings a fresh point of view to a review panel or simply insufficient expertise in the technical area being evaluated.

Management Level of the Reviewers. There are two key issues here. The first is deciding how to achieve the right blend of technical competence and management authority. The second is preventing great disparities so junior members are not inhibited by senior ones.

Authority of Reviewers. What should the reviewers be authorized to do? Comment? Advise? Or should they make strong recommendations to senior R&D or business management? The answer depends, of course, on the management level of the reviewers. It also depends, however, on what senior R&D management or business management expects from the reviewers.

Relative Permanency of the Reviewers. Is it better to have reviewers who are familiar with a project because they had evaluated it at an earlier stage? Or might these reviewers need to preserve a project that they were once involved with? Each case must be looked at individually.

Supervision or Coordination of Reviews. Can project reviews run on their own once reviewers have been designated? Or will reviewers, all of whom have regular line responsibilities, come to project reviews unprepared if no one supervises or coordinates review activities?

Following Up After a Review. Although a project review is necessary, it is only part of a solution. The other part is follow-through—making sure, for example, that reviewer recommendations are followed. Issues related to follow-up procedures must be addressed during project reviews.

Evaluating the Review Process. Regardless of how effectively a company is managed, its process for conducting project reviews probably can be improved on.

CONCLUSION

Establishing discipline is difficult in any situation, but it is especially difficult to impose on R&D projects. However, with determination and time, it can be done.

CONTROLLING PROJECT VARIABLES

James A. Ward

Effectively planning and controlling three variables—work, resources, and time—is what ensures project success. This chapter points out the problems and opportunities associated with project control. By understanding the dynamics and interrelationships of these variables, R&D managers can better manage resources, balance time and people, and keep projects on course.

Every R&D manager has been involved in projects that were less than totally successful. This chapter offers a prescriptive approach to project control.

KEYS TO GETTING STARTED

The key activities that occupy the project manager at the start of the project are planning and scheduling. These activities must also be repeated when attempting to salvage a project that is heading for disaster.

Project Requirements

Projects can and do fail even when requirements are known, but there has never been a complete disaster when all requirements were fully defined, documented, understood, and agreed on by all involved parties. Many projects fail when resources are devoted to doing the wrong things.

Project scope, objectives, and requirements must be completely defined, documented, reviewed, and approved. This definition must take place before the initiation of a project or as the first activity undertaken during project execution. When a project is in trouble, requirements must be revisited, restated, and reapproved.

In addition to defining the scope and objectives of the product to be developed, the project manager must also define the project in detail. This definition addresses the work to be done, the resources devoted to that work, and the time that the effort will take. Management must review and approve this project definition to ensure that resources are being applied productively. Productivity results from the ability of the project manager to produce the greatest amount of work with the least resources in the shortest possible time.

JAMES A. WARD is an independent consultant specializing in systems development project management and implementation of TQM.

Work. The work to be performed should be defined in such a way that when it is accomplished, the project will be successfully completed. Although it may sound elemental, this is not always the case. It is unlikely that any two project teams, attempting to develop the same product within the same organization, would perform an identical set of tasks or activities in the same manner or sequence. It is also unlikely that they would produce the same results. Each project defines its own work.

Resources. Resources include the efforts of the personnel assigned to the project, computer time, software, supplies, management and support time, tools, and methodologies used.

Time. Time is defined for this chapter as the elapsed calendar time from the inception of the project to its successful completion.

PROJECT PLAN AND SCHEDULE: RELATING THE KEY VARIABLES AT PROJECT INCEPTION

After defining the work to be done, the project manager can apply resources over time to do that work. Intuition suggests that if one variable is held constant—in this case, the work to be done—the other two variables can be adjusted in opposite directions. More resources to do the same work ought to take less time.

Intuition is correct, but only within certain limits. Unless project managers understand the dynamic interaction of the three variables of work, time, and resources, the negative effects of these dynamics during project execution will drive the project to disaster.

Resources Versus Time

Resources assigned past a certain level will not shorten project time. Beyond a point, more resources will actually lengthen the time it takes to develop a product. Adding resources to a late project makes that project later. Even at project inception, this holds true.

Usually, no more people can be productively used than the square root of the total amount of estimated worker-months needed. If a project is estimated to take 100 worker-months (8½ years) of effort, more than 10 people assigned to the project will not shorten the calendar time it takes to complete the project. Excess resources become wasted or counterproductive.

This dynamic holds true because of the nature of project work. Certain tasks require the output of other tasks. Tasks must be performed in a specified sequence. There is a limit to the number of concurrent tasks that can be performed effectively. After a certain point, tasks cannot be effectively broken down and assigned to more than one person. In addition, the more people assigned to a project, the more overhead is incurred in communications, management, and coordination of activities. If one person had all the requisite skills to complete a 100-worker-month project, that person could probably do so more efficiently and use fewer resources than a larger project team.

> **The second person assigned to a project contributes less than the first, and the third less than the second.**

The Law of Marginal Utility

The law of marginal utility operates this way: the second person assigned to a project contributes less than the first, and the third less than the second. Although each new person makes a positive contribution and thereby reduces the overall calendar time it takes to complete the project, at some point the marginal utility curve turns, and the next person's contribution becomes negative.

The actual contribution of the eleventh person in the previous example will probably not be negative, but the marginal contribution of each person after the tenth will drop dramatically. Therefore, a 100-worker-month project should take at least 10 calendar months to complete no matter how many resources are assigned to it.

Because the productivity of each additional person added to the project will be less than the preceding person's, the average productivity of the entire project team is reduced. In allocating resources, management must achieve the optimum-sized project team that balances resources against time to achieve the most productive mix.

Overall Elapsed Time

Projects that will take more than one calendar year to complete probably should be redefined. They can be broken into phases or multiple projects. A 144-worker-month (12-year) project is about the largest discrete systems development project that should be tackled.

There are some compelling reasons not to undertake very large projects. First, an organization loses its attention span at some point and other priorities intervene. Resources tend to disappear, through either attrition or reassignment.

Second, business conditions change. If a project cannot be completed within one calendar year of its inception, this project, when delivered may be obsolete. Competition is moving more rapidly all the time. The government regulatory climate may change (this applies to virtually any industry).

Third, the larger the project, the greater the risk of failure. Even the best project managers and the best project management methodologies and tools become strained when attempting to control projects that are too large. Degrees of error in planning or estimating that would be easily correctable on small projects can be overwhelming on large ones. A 50% overrun on a project of one worker-year can be handled. The same 50% overrun could prove fatal to a 20-worker-year project. Management must have the visibility and the ability to control overruns and cut losses, and this is much more likely on smaller projects.

The chance of success for a large project is significantly enhanced when a project is carved up so that a major implementation occurs at least once a year. This method also provides ongoing visibility and organizational commitment to the project.

The Role of Estimates

Every organization has a way of estimating projects. Whether this involves drawing lines on a Gantt chart, applying sophisticated estimating algorithms, or just stating what the dates must be, the project manager either develops or is given an estimate of how long a project should take.

Estimates that are based on fact and are accurate can be a tremendous advantage in project control. Estimates that are no more than a guess or someone's wish can be extremely detrimental. Unless estimates have their basis in fact and are developed by the individuals actually assigned to do the work, they are best ignored, except when reporting progress to management.

The value of estimates is in productively allocating resources and in coordinating task interdependencies. Estimates should not be used as evaluation tools. The project manager must be free to adjust plans and estimates on the basis of actual project feedback without having to explain why initial estimates were not totally accurate.

Plans are guides. Estimates are just that—an educated guess about what should happen. The reality is the actual project work that is being performed. If the reality does not always conform to the plan, it is likely that the plan may need some modifications.

Task Splitting

By assigning personnel to more than one project, management may think that more things are being accomplished or that resources are being used more productively. In fact, the opposite is true. Task splitting has a negative effect on productivity and efficiency, and this reality should not be ignored.

A person can devote 100% of his or her time to one project. That same individual will be productive only 40% of the time on each of two projects or 20% on each of three projects. Deciding among tasks adds coordination and decision time. Time is lost in switching from one task to another. Task splitting is a notorious resource stealer. Dedicated resources are always the more productive.

PROJECT STATUS REPORTING: MONITORING PROGRESS AGAINST THE PLAN

For the project manager, project execution involves monitoring progress against the plan and schedule on a regular basis, recognizing deviations, and taking appropriate corrective action. The project manager is chiefly responsible for ensuring that the work meets all quality standards and that it conforms to requirements and specifications. Providing high project visibility to users and management is also a primary project management task.

The project manager must receive regular and formal status reporting. All project team members should report progress against the plan and schedule weekly. This reporting should be done at the lowest task level.

Each week, the project team members should answer the following questions about their assigned tasks:

- Is it done?
- If it is not done, when will it be done?
- If it is behind schedule, what are the reasons?

Estimates based on fact can be a tremendous advantage in project control; estimates that are no more than a guess can be extremely detrimental.

The intent of status reporting is to chart real progress and at the same time to verify the efficacy of the project plan, schedule, and estimates. It is best to avoid reporting percentage of completion on any task, however, because this reporting is invariably overly optimistic and conveys no real information that the project manager can use. In any event, the tendency to report percentage of completion usually indicates that tasks are too large to be accomplished in a time period in which they can be effectively controlled.

Providing High Visibility

When project managers report status to management and users, they should emphasize deliverable products. Managers should report progress against major milestones and major deliverable products as defined in the project plan.

A project cannot be too visible. High visibility ensures management support. Management meddles in projects when it does not know what is going on. When and if the project hits rough sledding, the project manager will need the support of management to take action. This support must be nurtured through the confidence that comes from keeping management informed.

Frequency of Status Reporting

Each member of the project team should have at least one task due for completion each week and on which to report. Under no circumstances should a team member ever have more than two weeks between task completion dates.

Management is not interested in weekly task-level reporting, however. Status reporting that is too detailed (or too frequent) usually obfuscates rather than clarifies project status.

A major milestone or deliverable product should be scheduled each month. A project should never go more than two months without some significant event. The completion of a project phase with a formal report, including submission of the plan for the next phase, dramatizes progress most forcefully. The submission of a major deliverable product that requires management and user review and approval is also a critical event. If at least one of these events does not happen in more than two months, the project should be restructured and replanned.

Recognizing Deviations Early

Equipped with accurate project status information, the project manager can assess progress and detect any deviations or problems at the earliest possible point. An axiom of effective project management is act early, act small. Corrective action can be instituted in small doses and in ways that will not be disruptive.

On receiving weekly project team status reports, the project manager must post actual progress against the plan and note any deviations. If the project or some members of the project team are consistently ahead or behind the estimates, the manager should think about adjusting estimates accordingly. Minor adjustments in estimates, scheduled task completion dates, and task assignments should be done weekly as the project progresses. These minor adjustments need not be communicated beyond the project team, as long as they do not affect the scheduled dates for major milestones or deliverable products.

Identifying the Causes of Deviations

In even the best planned projects, deviations will occur. The first place to look for corrective action is in the plan itself.

There will be times in the course of any project when monitoring progress and making minor adjustments to the plan will not be sufficient to keep the project on schedule. This may happen for any number of reasons, and the reasons will undoubtedly influence the actions the project manager may take.

The work may change because of changes in project scope. Resources may be lost to the project. Technology may be poorly understood. The project team may have problems interacting effectively. When projects deviate significantly from plan and schedule, the dynamics of the key variables of work, resources, and time will determine the likelihood of success of any action that is taken to get the project back on schedule.

Leaving aside those occurrences that significantly alter the work to be done (such as changes in scope or incorrect requirements definition), the most frequently encountered problem is severe or persistent schedule slippage. The easiest course of action is to admit that estimates were overly optimistic and that the project will simply take more time to complete. However, persistent schedule slippage is more likely to be a symptom of underlying problems than the cause of the problem. Extending project time will not cure these problems and may only allow the project team to dig a deeper hole for itself. It is lack of time that forces most project teams to face the reality of failure.

> **Task splitting has a negative effect on productivity and efficiency.**

Instituting Corrective Measures

If the project manager is thoroughly convinced that the problem is simply caused by overly ambitious estimates and schedules, then the schedule should be altered. Otherwise, under no circumstances will increasing resources against the same overall schedule be successful in and of itself.

The project manager must subject a troubled project to detailed analysis, usually with the help of an independent party. Unless the underlying causes of problems are eliminated, the project will experience greater problems as the project manager (and others) attempt to apply corrective action. Management must be supportive of this process.

Once appropriate corrective action has been initiated, the project manager must go back to the beginning and prepare a formal project status report. Bearing in mind that the dynamics of work, resources, and time will be much different from what they were when the project was initially planned, the manager can then develop a new project plan and schedule.

CONCLUSION

A predictable pattern results when project managers fail to understand the dynamics of work, resources, and time. First, schedules are extended, usually more than once. When this does not work, resources are added, making problems that much worse. Finally, in an attempt to bring the project to a conclusion, the work effort is cut back, often to the point at which the resulting product no longer meets the requirements it was originally meant to address.

Heroic efforts to meet the original schedule by working large amounts of overtime for extended periods will not work. Error rates will soar, project communication will become increasingly difficult, and teamwork will be severely strained. Often this drives the project into a never-ending sequence of testing and error correction. These consequences often result because of management's inability to distinguish between effort and productivity.

By understanding and managing the interaction of the key variables of work, resources, and time, the project manager can productively plan and control systems development projects. Potential failures can be turned into successes.

AN ACTION PLAN FOR SELF-MANAGED TEAMS

Stewart L. Stokes, Jr.

Thawing old attitudes and behaviors may be one of the most difficult tasks R&D managers face when they try to move to self-directed work teams. A plan for phased-in implementation can ease the changeover to this new work style.

Self-directed (also known as self-managed) work teams are groups of employees with all the technical skills and authority needed to manage themselves. They perform such management functions as planning, organizing, staffing, leading, integrating, controlling, and measuring. They may also be authorized to hire and fire employees on the team, establish team budgets, purchase supplies, track expenses, conduct performance reviews and appraisals for team members, reward performance, and discipline employees on the team for unsatisfactory performance.

Not all self-directed teams enjoy these levels of authority and responsibility. There is no one best model; in fact, the latitude and charter of a self-directed team is determined largely by cultural and political considerations. For example, some organizations invest their self-directed teams with the freedom to plan, organize, lead, and control but stop short of allowing them to budget, make purchases, and hire and fire employees. Some organizations cross-train everyone on a team so that everyone can perform everyone else's job; they then rotate people through the positions. Other organizations maintain their former, more traditional job descriptions and training practices.

Almost any business discussion today that has anything to do with human downsizing (or rightsizing) is sure to include mention of empowerment and its application through self-directed work teams as a strategy for increasing motivation and productivity. In some organizations, empowerment is merely a new spin on an old concept that may have been equally vague: employee participation. The opportunity to participate in problem-solving and decision-making activities has become a core value in most enterprises, although the extent of that participation varies from organization to organization and manager to manager. The conventional thinking goes that if participation is good, empowerment must surely be better.

For those enterprises and departments that wish to take empowerment seriously and attempt to put it into practice, self-directed work teams offer the ideal vehicle. They provide the context, the encouragement, the accountability, and the resources for empowerment to flourish.

STEWART L. STOKES, JR., is senior vice-president, QED Information Sciences, in Wellesley MA.

QUALITY ISSUES AND TEAM STRUCTURE

Empowered, self-directed teams make a major contribution to the success of total quality management (TQM) programs. According to one survey, more than 75% of all companies claimed quality management to be a priority goal. That is the good news. The bad news is, there seems to be a widening gap between goals stated and goals accomplished. Other studies show that the patience of executives is beginning to wear thin with the amount of time it takes to achieve quality breakthroughs. The problem lies not with the inspiration but with the execution.

To be successful, TQM programs require a high level of sustained team energy. Teams are the engines that drive quality programs. Teams can also bog down the process. When teams become top-heavy with people, the result is that participation declines, commitment sags, and energy drains away. As these conditions occur, the constant drive for continuous improvement that is the hallmark of every successful TQM program begins to suffer.

There can be an inverse relationship between the number of people on a quality improvement team and the results achieved. To be effective, a TQM team needs representation from those business units or sectors in which the quality issues are important. Participation and commitment are vital. The more people on the team, however, the more time, attention, and energy (personal energy as well as team energy) must be given to managing team dynamics. More time may be consumed, for example, in resolving intrateam conflicts than in researching, implementing, and measuring quality improvements. One guideline to follow is to encourage the representation needed to deal with the quality issues at hand, but to understand—and be prepared to deal with—the trade-offs.

TQM is as much a philosophy of organizational behavior as it is a process, and the philosophy rests on a constant, unyielding focus on customers. The cornerstone belief of successful TQM programs is that everyone is someone else's customer. If this fundamental insight is not understood and practiced by everyone in the organization, TQM programs cannot achieve their true potential. Team members must realize that they, too, need to perceive one another as each other's customers.

Empowered, self-directed teams are frequently the vehicle through which organizations implement and sustain customer-focused TQM programs. Teams can focus the knowledge and insights of their individual members, but only if the individuals know who their customers are, understand what is important to them, and know what internal processes affect these important issues. Self-directed teams that harness and concentrate the energies of their individual members in turn encourage an environment within which *kaizen* (the Japanese concept of constant improvement in every aspect of a person's job) can flourish. These benefits result only if the teams are championed, trained, coached, and rewarded for what will be, for many, a totally new organizational experience.

A NEW ORGANIZATIONAL MODEL

Self-management is a new model of organization and of work. It presents a host of new problems and challenges, not the least of which is a long and involved training agenda. The costs of self-management are considerable (as are the benefits), and organizations are well advised to examine costs closely before embracing this new organizational model.

> Self-directed teams are ideal vehicles for R&D departments that want to put empowerment into practice.

Major entries in the cost column include those for training, education, and coaching. All are critical success factors for the move to self-direction. These costs are tangible and can be budgeted. Less tangible are the costs of time and energy required to get self-directed work teams operational and fully functioning. Underestimating these costs can be a mistake. Self-direction almost always takes more time to implement than most people think it should. When this happens, frustration and impatience occur, and it may be a struggle to sustain the initial enthusiasm and support.

Characteristics of the New Model

R&D managers are accustomed to fulfilling their responsibilities in enterprises organized into vertical functions with work being handed off from those in one function to another (i.e., from marketing to sales to production to quality control to shipping). These vertical functions have come to be called silos or smokestacks, and from the early part of this century they have provided organizations with efficient and predictable structures.

The managerial model in smokestack organizations was a command and control model, with a vertical organization chart providing the hierarchy within each functional area. Senior or upper-level managers were at the top of the chart, middle-level managers were in the middle, and lower-level managers were near the bottom. Sometimes supervisors were shown on the organization chart as below the lower-level managers, and the nonmanagerial employees were shown at the bottom. One of the interesting things about these organization charts was that the enterprise's customers did not appear at all.

Customer Focus. The emerging enterprise model injects several fundamental and profound changes into the mind-set of organizations and the people in them. First of all, the new organization chart may show customers at the top, rather than the president or chief executive officer. The message is that the customer, not the most senior executive, is king. The corollary is that everyone in the organization is there to serve the customers, not to serve that individual who happens to be above them on the organization chart.

Organization Around Processes. The new enterprise model is also organized around processes rather than functions. A process is a complete unit of work, such as the creation of a product or service (an output) for either an internal or external customer. This output can be produced by a team of employees representing several functions, work areas, or task groups, all working together to create the product or service. The form of organization in these cross-functional teams is horizontal, not vertical, and the goal is a seamless flow of activity in the production and delivery of products or services.

Absence of Hierarchy. The new model is characterized by an absence of hierarchy, and for control purposes relies on the people on the teams to control one another from the inside out, so to speak. This "inside out" attribute means that R&D managers and professionals must have the ability to influence. The more self-directed these cross-functional teams become, the more important the ability to influence is as a survival skill.

New Reward Programs. A final difference between the old vertical model and the new horizontal, or flat, organizational model is the issue of reward. This is also the issue that leads to the malfunctioning of many self-directed teams. In many organizations, reward and compensation programs lag behind changes in organizational design and work flow. Until they become more aligned, it will be difficult for self-directed teams to live up to their true potential.

Many organizations still base their compensation programs in part on the number of people in the individual's span of control—on how many people report to an individual. This practice is supposed to signify a person's degree of importance. The more people that report to a given individual, the more important that person must be to the enterprise. This line of reasoning ignores the issue of contribution, or what it is that a given individual actually contributes to the work unit, group, department, and organization.

The R&D field and its highly skilled technical professionals still largely buy in to this span-of-control fallacy. Dual career paths (one career path for those who desire managerial responsibility and a separate but equal career path for those who prefer technical assignments) remain the exception rather than the rule. Many R&D professionals would prefer to climb a technical ladder but find it necessary to move into management to advance in their organizations. The result often is that the R&D organization loses skilled technical people and gains poor managers.

CRITICAL SUCCESS FACTORS

The move toward self-direction may sound like the answer to many of the problems organizations face: out-of-control costs, too many managers, administrative and production bottle-necks, bureaucracy, and convoluted chains of command. However, this move may cause more problems than it cures, if four critical success factors are not in place:

- Sustained vision and commitment to self-directed teams on the part of senior and line executives.
- Specific and unambiguous goals, objectives, and limits of authority for the teams.
- A compensation structure that rewards team-oriented behaviors and supports team goals and objectives.
- Extensive training, education, and coaching before, during, and after implementation of self-directed teams.

Organizing self-directed work teams takes time ... and time is today's scarcest resource.

The lure of self-directed work teams and the promise of fewer managers and fewer bottle-necks seem to be more than many organizations can resist. The expectation that cutting the fat out of the middle will result in improved financial performance and smoother work flow is difficult to argue with. The problem is time: it takes more time than most organizations bargain for, and time is today's scarcest resource.

Unless the work processes of the rightsized department or organization have been restructured, the remaining employees do the same amount of work that was done before. When doing more with less becomes the goal, the anticipated benefits of rightsizing are overtaken by problems engendered by overload and burnout. Gaps in the organization are created that manifest themselves in poorly conceived and executed projects. A byproduct of this collapse of managerial effectiveness is frustration, faultfinding, finger pointing, and conflict.

It is usually at this point that a major organizational redesign is announced and managed work groups are replaced with self-managed work teams. For self-management to work, however, the organization cannot neglect to specify the outcomes expected and the training and education that will be provided.

The number-one critical success factor for self-managed work teams is training and education. Self-management is frequently resisted because of the perception that the organization will not provide the training and education necessary to enable the self-managed work teams to succeed. This perception, when accompanied by the expectation of unrealistic results, is certain to get the move to self-management off on the wrong foot.

Another prerequisite for successful self-directed work teams is a highly placed executive or manager (or, preferably, several such people) who will not only visibly and vocally champion the move toward self-direction but who will also back up the move by changing the reward structure of the department or organization to reflect the new roles, responsibilities, and skills required of self-directed work-team members. Reward structures are among the first places within an organization to which employees turn for signals about what really counts in the organization. Organizations have typically rewarded individual performance, and competition among people and departments has been a preferred organizational style. With the move toward self-direction, however, internal competitive behaviors need to be tempered by more collaborative behaviors, and new reward mechanisms must be created.

Fully functioning self-directed work teams benefit from reward structures that enable them to measure, appraise, and reward team-centered contributions of team members. This requires that team members have the authority to reward (and discipline) themselves on the basis of their own appraisal of individual, team-centered performance. The issue ties directly to the other critical success factors for self-directed team success: clear and unambiguous goals and objectives, and the availability of significant amounts of training, education, coaching, and counseling. Individuals and organizations must invest in training and education involving both hard and soft skills—before, during, and after the formation of self-directed teams. Merely taking a group of people, calling them a team, and providing a brief team-building course is not enough; this can actually do more harm than good. A quick training experience raises questions and expectations that may, if not met through in-depth training, coaching, and mentoring activities, result in a collection of confused and even angry individuals.

A COMPETENCE-BASED APPROACH TO TRAINING

For best results, training for self-directed work teams should be competence-based—that is, it must be customized to the specific needs and objectives of the team, including the climate and culture within which the team operates. The culture of the organization is a key variable for self-directed teams, for the teams themselves become cultures within cultures. They can support existing organizational structures and values or, more likely, they will run counter to them.

One of the most important training topics for self-directed team members is creating and managing change. Self-directed team members are not only agents of change; they are the change. They will benefit from a thorough understanding of the dynamics of change, including how to reduce resistance to change.

A competence-based approach to training self-directed work teams focuses education on the specific knowledge, skills, attitudes, and values that are required for successful self-directed work teams. These competences can be categorized into three domains:

- Technical (i.e., specialized knowledge and skills).
- Interpersonal, intrateam, and interteam (i.e., human relations and group dynamics knowledge and skill).
- Conceptual (i.e., how teams fit together within the larger enterprise).

Technical training includes training in those job skills for which the team is responsible. Each team member must be cross-trained in every other person's job. The purpose of cross-training is to ensure that all team members are comfortable and competent filling in for each other during times of absence, such as for illness, vacations, or personal time off. It may require team members to work extra hours. One of the least-mentioned but most significant aspects of the move toward self-direction is the realization that being a team member is a responsibility. The investment of personal and team time and energy will be greater than that required of participants on traditional teams, in which supervisors and managers customarily work the longer hours and make the decisions.

> **The culture of the organization is a key variable, for self-directed teams become cultures within cultures.**

Interpersonal, intrateam, and interteam skills training includes training on topics that enable team members to understand themselves and others better, and to function more effectively in both collaborative and contentious team environments. This training in process skills is essential for success, because self-directed team members have to solve their interpersonal difficulties on their own; there is no manager there to help them sort out their grievances.

In some organizations, internal consultants are available to help teams deal with their interpersonal issues and problems. Other organizations have eliminated these positions as being nonessential to the core business practices. If an organization is moving toward self-direction, both internal and external consultants become valuable resources. They bring expertise to the teams and help team members learn and practice the human relations and group dynamics skills essential for success. Group dynamics skills include:

- Effective listening; giving and receiving feedback.
- Problem finding and problem solving.
- Consensus building and decision making.
- Conflict management and resolution.
- Influencing and negotiating.
- Planning and coordinating.
- Running effective meetings.
- Making informative and persuasive presentations.
- Screening and hiring.
- Training and coaching.
- Performance review and assessment.
- Disciplining.
- Termination.

Conceptual competences include in-depth learning about the businesses of the enterprise, including information about products, markets, customers, channels of distribution, and competitors. Teams can perform periodic analyses to identify specific departmental or organizational strengths, weaknesses, opportunities, and threats. Whereas managers may have done such analyses previously, the self-directed team members must now assume the responsibility.

Two other areas of conceptual competence vital to team success are those of quality requirements and organizational design—specifically, how teams fit together and are linked within the enterprise as a whole. Because team members must establish and maintain dialogue with other team members throughout the enterprise, they require a working knowledge of the design of the organization. Teams must be active participants in the enterprise's TQM program, and the more training they have in the specifics of TQM, the more effective they become.

Finally, team members may also need to prepare budgets and do financial forecasting and cash-flow analyses. Although interteam coordinators may perform these functions, these coordinators may also be team members and may be rotating through this linchpin role, so the need for financial knowledge and skill exists for all team members.

PUTTING THE CONCEPT TO WORK

To make a smooth transition to self-direction, R&D managers should:

- Be certain that the move toward self-direction is motivated by sound business reasons and is not in response to a short-lived whim or fad.
- Communicate these reasons early and often to those who are being asked to become members of self-directed teams.
- Select a pilot area within the R&D organization to field-test the self-directed concept and practices. Avoid the pressure for an all-or-nothing approach; learn to walk before running.
- Understand the depth of personal and organizational commitment required for self-directed teams to become successful.
- Expect that the move to self-direction may be resisted and be prepared to counteract this resistance through constant communication, training, and support.
- Review existing compensation structures and adjust them to reward new individual knowledge and skills and team-oriented behaviors.
- Determine the specific new knowledge, skills, and objectives that are necessary and will be rewarded and implement the required training and coaching.
- Reinforce the new skills and behaviors by helping team members learn from their mistakes instead of being punished for them.
- Reward successes and focus on what the team members do that helps make them and their teams successful.
- Prepare for the long term; self-directed work teams do not happen overnight.

It may be only a matter of time before the R&D department is asked to respond to a request from senior management to investigate or implement self-directed teams. To help when that moment arrives, here is a list of do's and don'ts to guide the move to self-direction.

What Not to Do

In implementing a team approach, the first piece of advice is not to leap into action unprepared. Likewise, it is a mistake to try to deny that change is needed and wait for the situation to blow over. In addition, R&D managers should not:

> It is important not to try to turn every group into a team overnight.

- Assume that the decision to move toward self-direction was carefully researched and thought through. It may have been, but assume nothing.
- Get caught in the activity trap of trying to turn every group into a team overnight.
- Make a premature announcement that beginning on this date, all managers will become known as team associates or team consultants, or team facilitators, or even team members. Egos are fragile.
- Immediately engage a vendor to come in and give everyone a team-building course. The same training material will not work for everyone.
- Create a planning task force that consists entirely of training specialists. Although they will be key players, they cannot speak for the R&D personnel who will be personally affected by the change.
- Expect attitudes and behaviors to change overnight, even if the move toward self-direction has been mandated. There may be public statements of support for the move but covert actions to sabotage it.
- Demand new behaviors without establishing the reward structures and providing the education required to support them.
- Forget that the move to self-direction is a major change. Time, energy, and support are needed to allow old attitudes and behaviors to thaw before new ones are put in place.
- Be discouraged by false starts and mistakes; more important, refrain from punishing those who may have difficulty acclimating themselves to self-direction.
- Take the bumper-sticker approach to self-direction. "Just do it" may be a great advertising slogan, but it is a poor way to bring about major organizational change.

What to Do

The best advice is for R&D managers to:

- Proceed as with any new project, and plan before executing the change.
- Learn as much as quickly as possible about the reasons behind the dictate to move to self-direction. The more people can learn up front, the easier the transition will be.
- Organize a pilot team or teams. There may already be people in the organization who work closely together and realize they are interdependent. They are already moving toward self-direction. Their experiences can suggest what

works, what does not work, and what is needed to help things work better. Self-directed teams are also learning teams, because they must adapt and change continuously.
- Communicate, with deeds as well as with words, an intention to build self-direction and personal empowerment into the R&D organization. The R&D manager should be prepared to relinquish some authority and demonstrate to others a commitment to empowerment.
- Organize a small task force of people who can work together to plan the implementation of self-directed teams. People responsible for R&D management and training should be included.
- Expand the task force to include the managers and staff who will be affected. Obtaining buy-in requires a great deal of personal selling. This can usually be done more effectively in a small-group setting than with a large group of people.
- Build competence models for the self-directed teams—that is, what team members need to know, understand, and be able to do and what attitudes and values are necessary for their roles and responsibilities.
- Customize training and education around these competences and to fit corporate and departmental cultures.
- Be supportive and communicate this support continuously. Developing self-directed teams takes time, patience, and commitment; communicating this support helps hold teams together.
- Be certain that all team members know who their customers are, what processes and outcomes are important to their customers, and what levels of performance constitute customer satisfaction. Teams can then concentrate their energy on improving these processes, outcomes, and levels of performance.
- Encourage team members to learn from their mistakes. Always examine what happened and determine how to improve it next time. Insist on continuous improvement. This is TQM in action.
- Show trust. True empowerment through self-direction can exist only within an atmosphere of trust. Trust requires respect, restraint, and responsibility—three qualities that exist within successful teams.

CONCLUSION

These guidelines should help R&D managers and staff formulate action plans when faced with a dictate or mandate to move toward self-direction. This change in organizational design and expectations can be daunting to those who have grown up in traditional command and control structures. With planning and phased-in implementation, empowered self-directed teams can release the personal energy, drive, and motivation to help organizations reach new levels of accomplishment.

ESTABLISHING A CROSS-DISCIPLINARY TEAM FOR DEVELOPING A NEW PRODUCT

Cross-functional teams hold tremendous potential for improving the effectiveness of new product development efforts. However, many obstacles can block the path to successful implementation of a cross-functional team.

Robert Szakonyi

Managers often attempt to solve problems in the development of a new product by establishing a cross-disciplinary team. This makes sense, because the results produced by a cross-disciplinary team typically exceed those produced by the independent efforts of individuals from various disciplines. To improve the likelihood of success, however, managers must understand the difficulties involved in establishing and managing a cross-disciplinary team.

This chapter examines the difficulties that two cross-disciplinary teams faced. One of these teams was established in a technology-driven company, called Company X. The other team was established in a marketing-driven company, called Company Y. Although the orientations of Company X and Company Y differed from one another, the two teams faced many of the same difficulties.

One of those difficulties involved the technical and business aspects of the new product developments for which the teams were responsible. In both cases, the team was responsible for a new product development that lacked clear direction. The two cross-disciplinary teams also faced management-related issues, such as determining who should lead the team, deciding which disciplines should be represented on the team, and ensuring open communication and trust among the members of the team.

DEVELOPMENT-RELATED ISSUES

The cross-disciplinary team at Company X was given responsibility for a development effort that held great promise but was not proceeding well. Because the company's senior managers could not agree on any of the proposed new technical approaches, the only technical activities that were being carried out in this development effort were those that had already been approved.

This development effort was not really a project; instead, it was a multifaceted technical effort for which extensive research was still needed. Although this development effort could have yielded various new products, it had spawned

---

ROBERT SZAKONYI, PhD, is the director of the Center on Technology Management at IIT Research Institute, Chicago. He has consulted for many companies in various industries and has written two books and more than 40 articles on technology management.

only one product. In essence, this development effort involved an embryonic business that was based on new technology.

Company X lacked a clear strategy regarding the results of this development effort. Research managers within Company X thought they knew which research questions should be answered, but the problem with this development effort did not involve the research strategy. Instead, the problem involved the lack of a technical and business strategy, of which a research strategy is only one part. In other words, Company X was proceeding with a development effort without knowing what new products were to be developed and commercialized.

Because of this lack of a technical and business strategy, the company's senior business managers could not decide how much to invest in this development. They were uncertain about the probability of success and the commercial potential of this development, so they were not fully committed to it. Senior managements's uncertainty about this development prevented the R&D and marketing managers from determining how to proceed. In other words, the lack of a technical and business strategy placed Company X in *research strategy limbo*, trying to do more than research, but less than full-fledged development.

The cross-disciplinary team at Company Y also was given responsibility for a development that was not proceeding well. Twice during the previous three years, management had almost terminated this development. Although Company Y had previously produced parts of the new product that was being developed, difficulties arose because the company had never produced the complete system that was now required.

The main obstacle to this new product development, however, was the lack of management consensus regarding the requirements for the new product. In fact, various managers disagreed about whether Company Y should even develop this product.

The company president supported development of this new product, because the product would fill a gap in Company Y's product line. The business manager and the marketing manager from the business unit that was to be responsible for the new product did not support the development effort, because they did not believe the product would be successful in the marketplace. During the previous three years, they had tried unsuccessfully to stop this new product development.

The company's R&D managers supported the new product. However, the marketing organization was not committed to the new product and frequently changed its definition of requirements. Consequently, the R&D organization wasted resources by adjusting to the frequent changes in marketing direction.

Finally, the managers of the manufacturing plant that was to produce the new product doubted that the company's business managers, marketing managers, and R&D managers were committed to the product. Although this manufacturing plant needed the new product, the manufacturing managers gave it tepid support, because they believed the other managers in Company Y were not interested in the product.

Thus, Company Y's new product development was faltering. After three years of development, the product lacked solid support throughout Company Y and it still was not clearly defined.

An effective team leader needs in-depth technical knowledge as well as managerial ability.

MANAGEMENT-RELATED ISSUES

The cross-disciplinary teams in Company X and Company Y had to deal with four management-related issues:

- Who should be the leader of the team?
- Which disciplines should be represented on the team?
- Do the various members of the team communicate effectively in terms of both technical and business matters?
- Do the members of the team trust each other?

Who Should Be the Leader of the Team?

Company X had no logical candidate for leading its cross-disciplinary team. Although one technical person had in-depth knowledge of the technical issues involved, he lacked both an interest in managing and the ability to manage people. Although a team was in place, the team did not function well, because it had no leader. The absence of a leader also caused many of Company X's managers to question the team's value.

A member of Company Y's Team was interested in being the team leader, and this person had many of the required management capabilities. However, other groups in the company did not acknowledge his responsibility for this new product development. For example, he was not always invited to meetings concerning the new product—in particular, meetings that involved middle-level managers.

This person's inexperience as a team leader also caused other problems for Company Y's team. Because he was more interested in the technical aspects of the new product, he performed many of the technical tasks himself and neglected some of the management tasks, particularly during critical periods in the new product development.

Which Disciplines Should Be Represented on the Team?

Company Y's cross-disciplinary team originally consisted of representatives from six disciplines:

- Process R&D.
- Corporate engineering.

- Quality assurance.
- Purchasing.
- Plant engineering.
- Business development.

The teams original mission was to transfer technology from the R&D organization to a manufacturing plant.

After the team was established, the team members quickly realized that the requirements for the new product were not clearly defined. In particular, the team did not know the company's marketing plans for the new product. Consequently, the team expanded its membership by adding a representative from marketing.

Despite the addition of a marketing person, however, Company Y's team still did not include all the necessary members. From the outset, the team should have included a representative from manufacturing. Throughout the development, manufacturing personnel believed that the engineer from the manufacturing plant did not adequately represent their interests. Because the team was developing a new product, a representative from the product R&D organization also should have been added to the team.

In Company X, two teams were responsible for activities related to the development effort. One was an informal team, with members from the research, development, clinical, and business development organizations. This informal team had overall responsibility for the development. The other team was a formal team that was responsible for completing the one project that evolved from this development. Because this project concerned clinical trials, the clinical member of the informal team also served as leader of the formal team. Other members of the formal project team were various technical specialists from the development and clinical departments.

Although Company X's formal project team functioned reasonably well, the informal team encountered several major problems. First, the informal team did not include a representative from the regulatory organization, because no one in that organization had the necessary knowledge of the technical issues involved.

Another problem confronting the informal team involved the team's representative from the research organization. This individual was not respected by his colleagues in research or by the other members of the informal team. To the key members of the research organization, this person's role on the team was that of a note taker. He was expected to attend meetings, take notes, and then report back to members of the research organization.

To further complicate matters, two groups within the research organization disagreed about how the technical work in the development should proceed. Because the team member from the research organization simply passed information back and forth between the research organization and the informal team, the team found it difficult to reconcile the research organization's often conflicting views of how the development should proceed.

> Communication problems often arise because of the differing orientations of team members from various disciplines.

Do the Team Members Communicate Effectively?

Company Y's team encountered communication problems between the various engineers and the marketing person on the team. The engineers on the team were most comfortable discussing how to develop the new product, because this allowed them to concentrate on the technical aspects of new product development. However, the marketing person could not add much to these discussions. He was most concerned with defining what should be developed. Consequently, he preferred to discuss the meaning of market research studies and how the team might deal with various market segments.

Although the engineers and the marketing person sometimes tried to bridge this gap, they usually found it easier to stay with what they knew best. Because the engineers outnumbered the marketing person, the team meetings typically focused on the technical issues of implementation.

In contrast, the communication problems among the members of Company X's informal team arose from a fundamental difference of opinion about the nature of the technical problems involved in the development. Various members of the informal team believed that some of their colleagues on the team simply did not understand these technical issues. This was not surprising, because Company X was breaking new technical ground. Nevertheless, the members of Company X's team found it difficult to function effectively in this environment.

These differences of opinion among the members of Company X's informal team resulted in disagreements about which team member should have the most control over various aspects of the development. Each representative on the team had legitimate reasons for seeking control over how the development proceeded. For example, the representative from the research organization sought control because the development involved research questions. The representative from the development group also wanted control, because the effort was primarily related to development. Finally, the representative from the business development organization wanted to have his say because the development had to be commercialized. Because various team members had legitimate reasons for seeking control, delineating responsibility for the various aspects of the development proved to be difficult.

Do the Team Members Trust Each Other?

A lack of trust limited the effectiveness of Company Y's team. The members of Company Y's team should have paid closer attention to the activities of other groups within the company, particularly with regard to how those activities affected the team's work. The team also needed to determine whether other groups within the company really were committed to the new product that the team was developing. With so many external problems affecting the team, the team members needed to discuss those problems and then determine how the team should handle them. Unfortunately, the team members lacked the mutual trust necessary for such open, frank discussions.

The members of Company Y's team also could have done more to resolve problems without management intervention. Although the solutions to some problems required the approval of senior managers within Company Y, the team members were capable of resolving many other issues.

The members of Company X's team also lacked the necessary trust in one another. With the exception of the team's representative from the research organization, the team members all had full responsibility for their respective disciplines' input on matters related to this development. As a result, the personalities of the various team members had a significant effect on the project outcome. Each team member's strengths, weaknesses, insights, and biases had a greater effect than would have been the case if several professionals from each discipline had handled issues related to this development. To complicate matters, the team members found it difficult to discuss their respective strengths and weaknesses. By avoiding this troublesome issue, they also increased their mutual mistrust.

SUMMARY OF THE DIFFICULTIES INVOLVED IN ESTABLISHING A CROSS-DISCIPLINARY TEAM

As these two cases show, a cross-disciplinary team for developing a new product may encounter several types of difficulties:

- The lack of a clear direction for the development effort.
- Problems in finding the right leader for the team or in gaining acceptance for that person.
- Difficulties in getting the right people on the team.
- Communication problems among the members of the team.
- A lack of trust among the team members.

The lack of senior management consensus and support hindered the product development efforts of both cross-disciplinary teams. In Company X, managers defined a research strategy but failed to define a technical and business strategy. Consequently, the company's senior business managers were uncertain about how much to invest in the development. Without this support from senior business managers, the R&D managers and the marketing managers did not know how to proceed.

Company Y's team encountered similar problems. The company president and the R&D managers supported the new product, but the business and marketing managers did not. Furthermore, the manufacturing managers questioned the other managers' commitment to the new product. Because of these problems, both cross-disciplinary teams found it difficult to define a clear direction for their development efforts.

> Overcoming the obstacles to establishing a cross-disciplinary team requires the efforts of a key senior manager and an outstanding team leader.

Both teams also found it difficult to find the right leader and then ensure that the entire organization accepted that person's leadership of the team. Company X did not have a logical candidate for the position of team leader. The person who best understood the technical problems was not an effective manager. On the other hand, Company Y's team leader understood the technical problems and could manage other people. However, this team leader was not fully accepted by other groups in the company. He exacerbated this problem by neglecting management issues.

Both cross-disciplinary teams failed to ensure that they included all of the right members. The cross-disciplinary team in Company Y lacked representatives from the manufacturing organization and the R&D organization. The cross-disciplinary team in Company X lacked a member from the regulatory organization. In addition, this team's representative from the research organization was not respected by his research colleagues and served only as a conduit of information between the team and the research organization.

Both cross-disciplinary teams had communication problems. The members of Company X's team disagreed about the nature of the technical problems their development involved. The team members also disagreed about who should have the most influence on the development. The engineers on Company Y's team did not communicate well with the marketing person on the team. Rather than address this problem, the engineers usually focused on technical issues.

Finally, both cross-disciplinary teams had problems related to the lack of trust among team members. Company Y's team members failed to discuss how they should deal with the other groups in the company. Rather than working together to resolve various issues, the team often asked management to intervene. Company X's team did not acknowledge the personal strengths and weaknesses of the individual members of the informal team.

OVERCOMING THE DIFFICULTIES INVOLVED IN ESTABLISHING A CROSS-DISCIPLINARY TEAM

No technique can overcome the numerous difficulties involved in establishing a cross-disciplinary team for developing a new product. Instead, successful implementation of a cross-disciplinary team requires two people who understand these difficulties and share a commitment to the success of the cross-disciplinary team and the new product. One of these people must be a senior manager, though not necessarily a general manager; the other must be the leader of the cross-disciplinary team.

Only a senior manager can overcome the first three difficulties mentioned in the preceding section: the lack of a clear direction for the new product development; problems in finding the right leader for the team or in gaining acceptance for that person; and difficulties in getting the right people on the team. A key senior manager can foster support for a new product within a company and ensure that the development effort has a clear direction. This senior manager can also ensure that the right person is appointed team leader and that this person is accepted by others in the company. Finally, this senior manager can ensure that the team includes representatives from all of the appropriate disciplines.

People in other disciplines tend to judge the value of a team in terms of the team's leader. The leader of the cross-disciplinary team must also understand all of the difficulties associated with establishing a cross-functional team. An effective leader can overcome difficulties involving the composition of the team. An excellent team leader can resolve the communication problems that often arise between team members from different disciplines. An excellent team leader also fosters trust among team members.

Of course, the major obstacle to overcoming these difficulties is finding both a senior manager and a team leader who not only understand all of these difficulties, but also will work together to solve them. Cross-disciplinary teams for new product development can succeed, with the guidance and support of a key senior manager and an outstanding team leader.

MOTIVATING R&D TEAMS

Ronald A. Martin

The enthusiasm of development team members is critical in keeping projects timely and on track. Project leaders must therefore understand the pressures that affect team motivation and performance. This chapter provides an analysis of these pressures to help managers better motivate R&D personnel.

The enthusiasm people have for a task or project is directly proportional to their anticipated gain. The most boring, unfulfilling tasks are those workers perform out of duty while doubting the value of the job. One example of an unfulfilling tasks is clearing up old work before starting something that is new or more exciting.

Several years ago, a research team had to wind down a drug development project and summarize its data before offering it to a third party. (This drug had failed to interest incoming senior management and was dropped in favor of more attractive ventures.) The wrapping-up process, though necessary, received a low priority because it did not seem to benefit transition team members, who therefore wanted to unload it onto anyone more junior than themselves.

By contrast, fulfilling tasks excite personnel and generate energy? Why? What determines the way personnel view a project? People are eager to participate in projects that benefit them. As in elementary physics, force (i.e., perceived benefit) is necessary to produce motion (i.e., project participation). Without the continued application of force, however, friction stops a body in motion. In project management, friction is reduced and motion is maintained by motivation.

Anyone can motivate through the use of authority or power. However, managers who motivate in this way use their energy, and not that of the team, to push projects along. What separates leaders from managers is the ability to get things done without coercion. Successful team leaders harness the enthusiasm of others. They make team members internally motivated.

Leaders motivate by offering such inducements as a cash bonus, the chance for travel, or the promise of a promotion. They also offer such intangible motivations as appreciation, the chance to improve self-esteem or visibility, or the opportunity to meet a challenge. Tangible motivations appeal to everyone; they are predictable and usually impersonal and can be controlled by others. Being the easiest for management to offer, they are its preferred tokens of gratitude. Intangible motivators tend to be unique to each individual. They

RONALD A. MARTIN is vice-president of drug development at the Parke-Davis Pharmaceutical Research Division of the Warner-Lambert Co.

involve personal growth; are less easy to predict; and cannot be budgeted for, or allocated on, any sort of ranking system. Providing intangible rewards is very hard work. Project leaders, however, must learn to deal in intangibles. Managers' success, and that of their projects, directly relates to their ability to recognize and fulfill intangible needs.

In discussing project management, this chapter draws from experiences in drug development, an especially complex area of project management because of the long time horizons, diversity of scientific skills, and regulatory considerations involved. Chemicals rarely move from discovery to marketing approval in less than eight years; it can take three to four years just to establish whether a drug is worth developing. The cost of developing a new therapeutic agent averages $250 to $300 million. Attrition rates are high, and the chance of any early-stage compound succeeding is around 10%.

Given the lengthy time frames and considerable risks involved, developing enthusiasm in team members can be as challenging as the projects themselves. However, developers must feel some level of passion about their projects. Without enthusiasm, a development becomes a collection of individuals merely going through the motions. A project loses creative momentum, and the quality of effort suffers accordingly. The following section examines some factors that impede team motivaton.

NEGATIVE FACTORS AND THEIR INFLUENCE

Financial rewards, or incentive bonuses, are a time-honored method of getting people to work harder and faster. This motivation is largely successful; if it were not, its use would have ceased long ago. However, bonuses condition people to work only for the next reward. They succeed in the short term in matters for which a longer-term objective is not relevant. For example, bonuses can be an important inducement for boosting yearly sales or reducing factory downtime. In the long term, however, economically based incentives lose their impact and can demotivate personnel working in a team, for when one member feels his or her compensation compares unfairly to that of other team members, animosity can develop that will destroy team cohesion.

Some have tried to democratize the reward system by rotating rewards, but this practice eventually leads to a meaningless award: something like team player of the month. Once people realize that the award is rotated, it becomes a right rather than a token of recognition. The cost of recognizing outstanding performance then inflates to reflect expanded expectations.

Monetary rewards can be appropriate to recognize outstanding individual performances when team loyalties or interdependencies are not involved. To be eligible, a person must have acted independently (or with distinction) in a manner that served the team, and this fact should be well-recognized by everyone involved. The danger, though, is that singling out one team member for a bonus, if substantial enough, does not encourage his or her teammates to pull with that team member again. Instead, it encourages independent action or, in extreme circumstances, vindictiveness. The alternative is to reward the whole team. However, if a team is rewarded for passing a significant but generic milestone, other teams will expect an equivalent reward when they pass similar milestones, and if these teams are not, their members will feel poorly treated.

> **Anyone can motivate through the use of authority or power.**

The issue of financial incentives is even more complex when a team includes technically skilled individuals. Scientists are driven by curiosity and the stimulation derived from solving problems. Money is not their principal motivator. Scientists receive bonuses enthusiastically and gratefully, but bonuses do not guarantee greater levels of creativity or sparks of intuition. In this sense, a bonus is a reward for past performance, not a generator of future breakthroughs.

Many motivators, if applied inappropriately, can backfire. For example, praise can be a good motivator if used in the right place and with conviction, but continuous praise of any single individual creates problems within any group. The author encountered a particularly subtle form of this problem several years ago when his organization was experimenting with a core team concept. To streamline the development process, four or five key members were chosen from a multidisciplinary drug development team of 10 or 15 individuals. The core team met frequently to deal with emerging crises and problems. The full team met less frequently to share information, review project status, and formulate strategy. However, the identification of core tam members made the remaining team members feel like second-class players. Their enthusiasm declined, and the projects suffered accordingly. This experiment was abandoned, and teams in late-stage projects were simply reduced to their core members. Today, internal consultants are called in to help teams deal with issues as they are needed. This current approach admits that certain individuals are more important to the successs of a project and dispenses with the political sham of a team within a larger team.

Any motivator, if applied without common sense, can lead to negative consequences. In the discussion that follows, the reader may wish to remember this simple fact.

POSITIVE FACTORS

Three broad classes of motivational factors exist: project related, individually based, and organizationally dependent factors.

Project-Related Factors

Challenge. The drug development system at the author's company has three distinct team phases (see Exhibit 1). Project teams guide the search for and identification of lead compounds, those agents with pharmacologic profiles that are appropriate for human trials. When adequate information is gathered to justify an investigational drug submission, the project team hands respon-

Exhibit 1
Drug Development Team Cycles

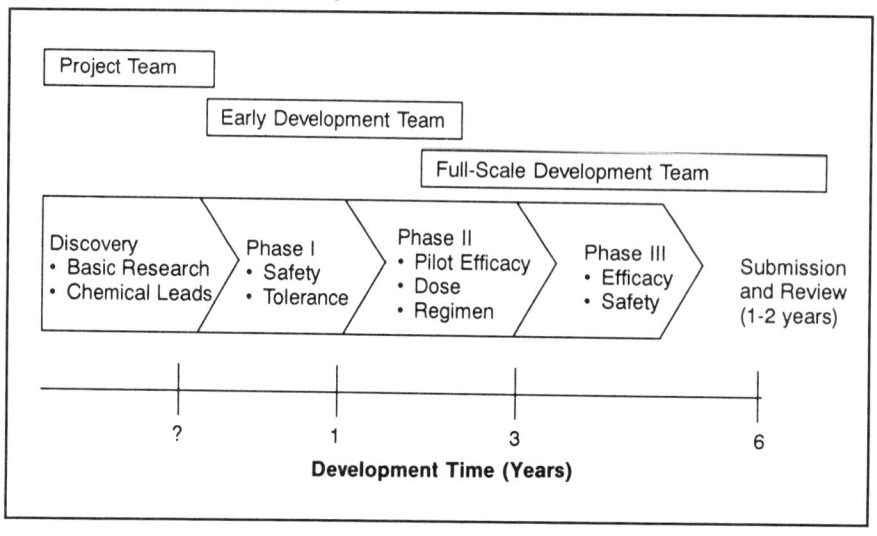

```
Project Team
    Early Development Team
        Full-Scale Development Team

Discovery        Phase I      Phase II         Phase III     Submission
• Basic Research • Safety     • Pilot Efficacy • Efficacy    and Review
• Chemical Leads • Tolerance  • Dose           • Safety      (1-2 years)
                              • Regimen

    ?            1            3                6
           Development Time (Years)
```

sibility to an early development team. This group files the drug, communicates with the Food and Drug Administration on the drug's unique development requirements, and establishes whether the drug works on humans. Both teams are headed by part-time leaders—scientists who maintain laboratory responsibilities, are experts in the necessary therapeutic areas, and are at the forefront of evolving technology. They can adapt new science to early development objectives and vice versa.

Once efficacy is established, the final step in the development process begins. The full-scale development team headed by a full-time leader who is a senior manager establishes the drug's efficacy (according to the criteria established by regulatory agencies around the world) and files for marketing approval and prelaunch scientific activities. Continuity is maintained from early development through full-scale development by the assignment of a full-time project manager. These information gatekeepers know the history of their projects and are the foundation of the team system. They guide less experienced team leaders at the early stages of the project, coordinate the development process, and work with line managers and team members to ensure that targets are met.

Each team has a stake in the process and a clearly defined role in the development cycle. Any project like drug development that takes six years or longer must be broken down into milestones. Otherwise, the project becomes too abstract, unfocused, and overwhelming. Dividing the development goal into bite-sized pieces enables team members to share in part of the success without worrying that the last person to touch the ball will get all the glory.

Manageable shorter-term objectives enable all, including management, to follow a project's progress, and a defined and manageable life span offers personnel an achievable challenge.

> **Bonuses condition people to work only for the next reward.**

However, a development team does not derive energy merely because it exists. To motivate team members, an identifiable enemy is useful. It provides a channel for team energy, encourages competition, and offers a concern to rally the troops. The enemy can assume a variety of forms, real or perceived. It can involve beating another company on the basis of quality or concept, ensuring that a critical event occurs at a particular time, or proving a point to a senior management. Whatever the challenge, it must be important enough to arouse key group members, because their enthusiasm is infectious.

Annually or quarterly, these challenges should be broken down into goals that are:

- Specific.
- Challenging.
- Realistic.
- Action oriented.
- Measurable.

The team should set its own goals and yardsticks; management should approve or adjust them. This division of responsibility ensures that all parties share equally in a project's success or failure; and the discussion it involves ensures that everyone understands how performance is evaluated. Because these goals are measurable, they can be managed objectively.

A Sense of Urgency. Because developing a drug takes a long time, development team members rarely remain at a drug company long enough to see a project from concept to product. Dividing development goals into a series of tasks performed by different groups ameliorates this situation because each group has a goal that can be reached quickly enough to permit management to use time as a key element in assessing the team's success or failure. Managers can consequently motivate a team by talking about the urgency of a task. However, using time as a motivator can also create conflicts if many projects must compete for the same scarce resources.

One of the worst things that can happen to a project and its participants is that it be considered a low priority. Team members want to feel that their careers are going somewhere; low-priority projects do not advance careers. They are annoying obstructions to spending time on more exciting activities that management is enthusiastic about. Low-priority projects are slowly starved for resources because of their place on the food line, and they just crawl along until management determines that it is time to put them out of their misery.

Much has been written concerning skunk works, whereby a group receiving few resources is challenged to show that an organization is wrong about something or to prove the feasibility a concept. With trimmed resources, a commitment from higher management to look the other way, and members with a vested interest in succeeding, the impossible can often occur. However, few projects can take on the aura of a mission without a leader or team members who are intent on swimming against the current. Without a high enough priority and universal commitment, a product has no urgency; without a time frame against which to measure productivity, it is doomed to destruction.

Strategy. Few people feel comfortable embarking on a major trip to an unknown destination without a map or itinerary. Why, then, should anyone feel comfortable participating on a project with only a slight understanding of its objective and a gut feeling for how long it should take? It is critically important to periodically determine a project's path and progress and to assess where a group stands in terms of its plan.

A strategic plan codifies goals, expectations, and risks. It defines the environment for a project, not for a team. It clearly lays out, for example, whether the project plan is low-risk, fast-track, or high-quality. It details how success will be determined, including time frames and standards for the accomplishment of key milestones. Using the strategic plan, each department can understand and negotiate its commitments so surprises do not occur.

The strategic plan should detail organizational and team responsibilities; tactical, or operational, plans should outline departmental or individual responsibilities. Team members should periodically review tactical plans to ensure that they meet current strategic needs. Management, in turn, should track the team's progress against goals in the organization's strategic plan. Plans, of course, can and should be altered as the development environment changes.

Individual or Personal Factors

Proficiency Level. Until members of a team get to know each other, uncomfortable periods of silence will occasionally arise in team meetings, usually when a decision must be made or an action taken on an issue. Sometimes, silence on an issue means general agreement; other times, it signifies members' shyness, lack of knowledge or experience, or unwillingness to commit to a position. The astute leader draws inexperienced team members out of their silence. Even if these members state only the obvious, that can be a valuable contribution. This leader also assesses team members' technical proficiency and self-esteem before construing their silence as agreement or disagreement with a position.

Seasoned team members tend to be aware of group dynamics. Having honed their negotiating and observational skills, they know the limits to which others can be pressured. They understand their company's political landscape. If they are recognized experts, they may be willing to risk their reputation by taking a stand on an issue. They are confident and usually willing to admit their errors in perception or judgment, (often with a show of humor). Having the confidence to admit errors, they gain added respect from their peers and encourage junior team members with more fragile self-esteems to become comfortable in admitting their own errors or technical problems to the group.

> **An identifiable enemy is useful to motivate team members.**

All leaders are willing to risk their images. Indeed, the greater someone's sense of leadership, the more likely he or she is to take on risks. Higher levels of risk tolerance come from greater experience and the expectation that something positive may happen, given enough time. Almost anyone will participate in a high-risk task when there is much to be gained. The challenge for team leaders is to convince team members that for a given project, the gain is worth the risk.

Respected Leadership. Not everyone should be put in charge of a project. A leader should have excellent interpersonal skills, including the ability to deal comfortably with people at various levels within an organization and skill in helping people with different interests or agendas to reach a common ground. The abilities and reputation of a team leader are always under scrutiny. A leader's failure to support the team will cause the group to lose respect for him or her. Failure to support management by cynically deriding its decisions will cause the group to question the value of its projects or to doubt their leader's credibility. A leader must display confidence and enthusiasm at all times—a tough job in times of stress or in an uncertain environment.

A leader must be clearly knowledgeable in technical areas, for without basic technical skills, the leader can not anticipate or, even worse, recognize problems and act with sound judgment. Certainly, team leaders can not be experts in all facets of their teams' projects, but they must know when and where to seek advice.

The leader must accept responsibility for team failures, enthusiastically acknowledge the contributions of team members, and be committed to the project at all times. The leader must also be prepared to fight with senior management over issues that the team feels strongly about. A team benefits an organization only if its opinion is appreciated; its leader must ensure that management hears and understands this opinion, recognizes its value, and acts on the opinion (but not necessarily in the manner the team proposes).

Finally, an effective project leader must confront and deal with performance problems. Avoiding this responsibility sends a signal that poor performance is acceptable. It also indicates that the leader is weak and has a little commitment to the project.

Opportunity. Most people feel rewarded when they are assigned a task, but that feeling does not last very long if they are not given the responsibility to succeed at it. Team members want to learn and expand their capabilities, not spend time on meaningless exercises. Managers must try to ensure that control systems, designed to provide information on products and processes, do not unnecessarily impede team members' efforts. Members must be allowed to make mistakes and gain experience from them. Management must delegate to team members those decisions that are reasonable and of substance and then stand back and carefully nurture the problem-solving process.

Organizational or Environmental Factors

Nurturing by Management. Teams, like other complex systems, must be carefully set up and continuously maintained. Teams operate in uncertainly areas, and many pressures affect their performance (see Exhibit 2). Periodic reviews are very important to leverage a team's value. By demonstrating interest and responsiveness, senior management conveys its support for a project to a team and to heads of functional management.

Reviews are an opportunity to acknowledge the successes of team members. Reviews enable a team to share new information, frustrations, setbacks, and proposed approaches. By offering its support and advice, management can keep a project on track or apply midcourse corrections before disaster strikes or positions become so entrenched that personalities and politics become overriding issues. There is more than one correct path to achieving a goal; some just take a little longer than others do or are less graceful; therefore, management must be patient and resist the temptation to closely supervise every aspect of a project. If management limits its involvement to issues of strategy and

---

Exhibit 2
Pressures on Team Performance

- Management Expectations/Support
- Team Aspirations
- Commercial Realities
- Regulatory Constraints
- Availability of Resources/Priority
- Individual Proficiency Levels
- Hidden Agendas
- Functional Workload
- Enormity of Task

resource allocation, its periodic reviews will provide a sense of stability and may even offer renewed life to projects that are faltering.

> **All leaders are willing to risk their images.**

Consciously or unconsciously, management decides who is responsible for a failure. A senior management review should emphasize that everyone shares responsibility for problems and that the team should work to find solutions. The review should focus on whether a proposed solution addresses a problem, not who is to blame for it. Because everyone owns the problem, problem solving must occur in a nonthreatening manner. When it does not, teams conclude that management can not handle bad news, and they pitch their presentations accordingly.

Communication Systems. Two factors that can limit the effectiveness of any team are tight resources and poor communication networks. However, even in the face of resource constraints, creative and enthusiastic people can meet goals and succeed at projects. Communication is a critical success factor. Communication must flow, for example, among team members, between management lines and levels, between the team and management, and from outside contacts to management or the team. Ideally, these lines all operate concurrently and with sufficient redundancy to ensure that all parties know what they must. A disconnection in lines sets up alternative communciation pathways—for example the grapevine.

Most crippling, not only to the team but to the organization as a whole, is any concerted effort to block the flow of information. Sometimes this effort is well-intentioned and perhaps necessary, such as when one group must first assimilate its data before sharing it with people outside the group. Other times, groups withhold information because they want to make one group look good at the expense of others; or, even more simply, information ceases to flow because one group simply neglects to inform another or because it innocently considers something to be an internal matter.

Organizations must carefully monitor their communication networks. They should step up systems to provide redundancy, for example, periodic reviews. Information must be current; networks must function smoothly and aggressively. The more data is shared in an organization, the easier it is to build commitment among parties to communicate easily and the less likely it is that political agendas will be a significant force.

Clearly Defined Expectations. Preschool children make up games for which the rules evolve as the games progress. Adults realize that they do not have a ghost of a chance of winning against a child, but play is play, and it is all good fun. In the real world, changing rules are not handled so amiably. Those who realize they are in a no-win situation either bail out, if they are career driven, or maintain a low profile and avoid boat-rocking. Either course is a loss to the organization.

Management must say how the success of a team will be judged, while remembering that a too-clearly defined mandate can restrict creativity. There are occasions when rules change or need redefinition, but all parties should agree on the need for change. Management should honor, as far as is practical, previously set rules.

As in all relationships between supervisors and personnel, both parties must demonstrate trust and earn credibility. Without mutual respect, neither party gives that bit of extra effort that distinguishes outstanding from mediocre performance. A team earns the respect of management by developing realistic goals and communicating its progress honestly. Management establishes respect by supportive, nonthreatening, and consistent behavior. Management must ensure that the company's overall strategy is clearly communicated. When teams understand their companies' plans, they know that what they are trying to accomplish will further management's goals.

FINAL COMMENTS

Before examining whether a team is motivated to solve a particular problem, management should determine whether that team is necessary in the first place. Organizations sometimes create teams to tackle problems when all they need is the effort of one or two individuals, or they assign so many teams and task forces to a problem that no one knows who has final responsibility for anything and activity is the only product. In other instances, the purpose of a team has long since expired, and the team exists for historical reasons only, adding little, if any, value to the organization.

Most people like to work productively and want responsibility. If management feels that a team is the answer to a need and delegates full responsibility to it, after the ground rules are clear, this demonstration of enthusiasm and support will go a long way to motivating the group. The rest is periodic fine-tuning.

AIR DROPS AND TOTAL QUALITY

George Manners

The implementation of a successful total quality management program involves answering four important questions. This chapter discusses these questions, how to answer them, and how the answers lead to a successful total quality program.

Contemporary management vocabulary has a fairly descriptive phrase—the air drop—to refer to the manner in which new programs are introduced into organizations to improve effectiveness. The image is pretty clear—the change program comes out of nowhere; at least, people cannot see it coming. It generally falls in too fast. It is not supported or attached to anything in particular. People often get hurt when it hits. It breaks into pieces. It becomes useless.

Simultaneously, the idea of total quality management has reached the stage of frenzy in the US, if not globally. People only need observe the agendas of recent professional meetings or the titles of articles in the R&D literature to realize the extent to which the R&D community is embracing the idea. But most corporate managers are struggling with just what total quality really means. Like most other conceptual constructs, the acceptance of total quality would be greatly aided by an operational definition that people could grasp. That has always been one of the primary problems of change efforts that require skill, attitude, and behavior change; people could not grab hold of any operational anchors.

Isn't there a reasonable operational definition of total quality? How does it fit with what has been learned during the last few decades? Are there not some timeless principles to keep in focus? Could well-intentioned total quality practices end up as another air drop?

The author has an answer to the last question first: absolutely. That answer is frightening, because the author happens to believe that total quality, as he understands it, is an intellectual windfall to the profession. He would like to say that it is a windfall to companies' competitiveness, but most people now understand that it is more a price of continued participation in the industry.

Thus, the author would like to share his opinions on the other questions mentioned previously. First, the author will offer what he considers to be an operational definition of total quality that is independent of jargon but is still, in his opinion, operational. Second, the author presents some conclusions on what conditions must exist in an organization for it to have a sustaining total quality culture. Finally, the author reinforces what most people understand as to the constancy of purpose.

GEORGE MANNERS is the director of services and facilities at James River Corp., Neenah WI.

WHAT IS TOTAL QUALITY?

The reason that the total quality movement is such an intellectual windfall is that the essence of it is the ability to answer four simple questions:

- Can the company demonstrate that it understands what its customers want?
- Can the company demonstrate that it understands its processes?
- Can the company demonstrate that its processes are under control?
- Does the company have a plan for continuous improvement?

The extent to which an organization focuses on answering these questions is the extent to which it can focus on what total quality is really about. Companies too often focus on total quality tools and air-drop training programs around those tools rather than focusing on total quality. There is a veritable plethora of total quality tools in the literature and offerings of consultants. If companies focus on the questions, employee buy-in is much more certain and the effort is in the right direction. In addition, the company will acquire the tools that it needs to get the answers.

If the entire corporation devotes its energy to the four questions, it does not need the many conceptual frameworks that confuse the issue. The management community expends an incredible amount of energy on the my-framework-is-better-than-your-framework exercise. The author is engaged in it right now, but he manages around these questions rather than wasting energy lecturing about them to the people in his organization. The following sections discuss each question briefly.

Can the Company Demonstrate That It Understands What Its Customers Want?

Experienced R&D professionals know that everything should start with this question. Thus, the author will share some insights that have come from trying to apply it. In the struggle to understand total quality, managers have begun to stress that total quality emphasizes speaking with data. So when the question is asked, managers require a data-oriented demonstration that the company is coming to grips with a real answer. At the business level, this could range anywhere from a highly sophisticated multidimensionally scaled market research study to a well-documented visit to see the products run on the customers' equipment. At the individual level, this has led to a much more pervasive emphasis on sound project contracting with a sponsor. At the departmental level, in addition to the contracting dimension, attempting to fundamentally answer this question has led to an ever-increasing use of internal-customer surveys and face-to-face feedback sessions.

Second, companies have begun to overtly avoid falling into the trap of wants versus needs. The operative word is *want*. It is first necessary to document that the company understands what the customer wants; then, the company might engage the customer in a possible upgrade to what the customer really needs. R&D organizations have too often projected a condescending image by ignoring what the customer

> Most corporate managers are struggling with just what total quality really means.

(generally represented by an operating division as opposed to the consumer) wants and working on what they think the customer needs. This activity has injured the careers of many R&D people. The folklore in the labs has a lot of success stories to justify this approach, but the number of R&D people who have been hurt by using it vastly exceeds the number of heros.

Can the Company Demonstrate That It Understands Its Processes?

In the company's drive to demonstrate that it understands its processes, the following outcomes seem to be occurring. First, R&D is realizing a much greater tie-in to manufacturing so that substantive technical-process characterization can take place. Second, this technical-process characterization is backing up into the pilot plant, which must have the same degree of attention in order to meet specifications.

Third, being driven by this question is forcing the company to come to grips with what its core processes really are. At the functional level, most companies have not given this question sufficient thought. It is now obvious that a clear articulation of what the organizational, departmental, and individual core processes really are is a minimum condition for getting total quality started. And again, once the company can specify its core processes, demonstrating that it understands them necessitates that it substitute data and logic for feel. Can a company really demonstrate that it understands how it selects projects? Can it really demonstrate that it understands how it plans and executes trials, appraises performance, or recruits talent?

Fourth, companies are more completely confronting the true multifunctional nature of business core processes. As the author writes this paragraph, he is also preparing to facilitate a third day-long meeting between R&D, engineering, and manufacturing on refining the processes for planning, executing, and analyzing multimill trials.

Fifth, being driven by this question has made companies much more aware of the role of infrastructure and well-documented systems for conducting their affairs. And this implies that companies believe that systems enhance, rather than inhibit, the creative process. Does this mean that companies should actually follow a system for selecting projects? Of course it does. Won't that inhibit creativity? On the contrary.

Finally, companies are increasingly aware that it is the question that is important, not finding a specific tool for answering the question. The human resources department within R&D has been judged as a total quality leader by the corporation's total quality vice-president. The primary reason for this is that the company has systematically documented every core process that it controls, using what it calls continuity folders. The folders are not elegant flowcharts, but these people can demonstrate that they understand their processes. They are now ready to tackle the next question.

Can the Company Demonstrate That Its Systems Are Under Control?

Most of the literature on total quality relative to this question works with manufacturing processes and statistical process control. At the author's company, that has been both a hindrance and a help. It has been a hindrance in the sense that defining *under control* has led to some of the train-on-the-tool-first thinking that creates so much resistance, the tool in this case being statistical process control. The author must admit that, except for the testing systems, his company cannot demonstrate that any of the core processes in the R&D lab are under statistical control, even the safety processes. This frustration leads too many people to attempt to say that this stuff does not apply to the R&D process.

Transitioning from understanding the processes to demonstrating that they are under control is the point at which a total quality culture is really implemented. To grapple with this transition, the author has attempted to establish a progress measurement scale, which is outlined in Exhibit 1. Each department is expected to make continuous progress by moving to higher levels of core-processes control.

It is probably obvious to most readers of this chapter that no department within R&D started out at even a level I state. And the benefits of just getting to level I, in terms of employee involvement, are significant. Getting to level III is the breakthrough in measurement systems, and level IV really introduces the company to substantive understanding of variability and system capability.

Exhibit 1
Progress Measurement: Core Process Plan of Control

Level I The process is recognized by all members of the department as a core process, even though process documentation does not exist.

Level II The process is documented. Each major step in the process is recorded, and cycle times are generally known.

Level III Client requirements are documented. Measurements are defined, where possible, for key requirements.

Level IV The process is flowcharted. Critical control factors relative to client requirements are identified, and a tracking process is in place.

Level V A continuous-improvement plan, relative to Level IV knowledge, is being applied to the process.

Level VI The process is completely standardized and stable. Client requirements and critical control factors are routinely measured, displayed, and improved.

The author's physical testing lab is operating at a legitimate level V state, whereas all departments are at least at level II. Thus, the company still has a long way to go, but with this type of progress measurement, the company can make incremental improvements and still feel good about it.

Can the company understand what its customers want?

The previous steps have vastly aided the company's internal capability. However, concentrating on demonstrating that the processes are under control has also greatly aided the R&D community in working with the manufacturing locations and designing experiments for investigating process improvements. The breakthrough came when the company admitted to the following realization: if the manufacturing processes on which the company is designing the experiment cannot be shown to be in statistical control, just what is the experiment going to show? Thus, the company first had to expend significant resources in working with its manufacturing locations to help achieve process stability. Once that has been completed, everybody can focus more on process improvement.

Does the Company Have a Plan for Continuous Improvement?

Knowing what level the company is operating at and having a well-thought-out set of next steps is the essence of the ability to answer this question, at least prior to reaching level V or VI.

Ultimately, the ability to answer this question says that the company knows what its customers want, understands the systems it employs for satisfying that want, demonstrates that its system is behaving reliably (even if it is not satisfying the want), knows the critical control factors for adjusting the process and has a well-documented justification for what it wants to improve next. In addition, this plan is continuously updated.

Really getting serious with this question is where the concept of benchmarking comes in. As the company moves from a reliable system to a more capable system to a world-class system, fundamental knowledge of what the best and the brightest are doing grows in importance. Too many organizations start a total quality program with a massive benchmarking effort. The author's observations tell him that this approach places benchmarking before system understanding and frustrates the majority of people who have to deal with it.

The author knows that the creation of a fully competitive organization is not this simple, but he does think that the essence of the total quality movement is this simple. The accounting department at the author's firm's R&D lab has never had any statistics training, has never been to one of the total quality seminars, and does not know any of the modern charting techniques yet. But the staff has systematically addressed the four questions in a way that has led to dramatic improvements in speed, accuracy, quality, and internal-customer satisfaction. The R&D professionals have had most of the contemporary training, and the application of the four questions is still the focus of their strategy. They are not only focusing on these questions relative to new products and processes but on the way they carry out their work.

However, executing these steps in continuous improvement is where the lessons of organizational change learned during the last few decades come in. The four questions will lend not to an intellectual breakthrough but to an intellectual failure, unless the company tunes in to what it has learned from other high-performing organizations.

THE BASICS—BELIEFS AND CONCEPTS

Most of the changes that are brought into an organization are changes in practices—that is, changes in the way things are done. To remain competitive, an organization must adopt new practices, sometimes very rapidly. A few organizations seem to be able to do it almost continually, but most cannot.

Although there are many organization design models to help explain this phenomenon, the one that is most meaningful is outlined in Exhibit 2. Exhibit 2 shows that an alignment must exist, or a balance must be achieved, among a company's basic beliefs about people, its basic concepts about management, and its practices.

The basic beliefs should make possible a sense of absolute stability. The basic concepts should be quite stable but should also evolve to help the company

Exhibit 2
Organization Design Balance

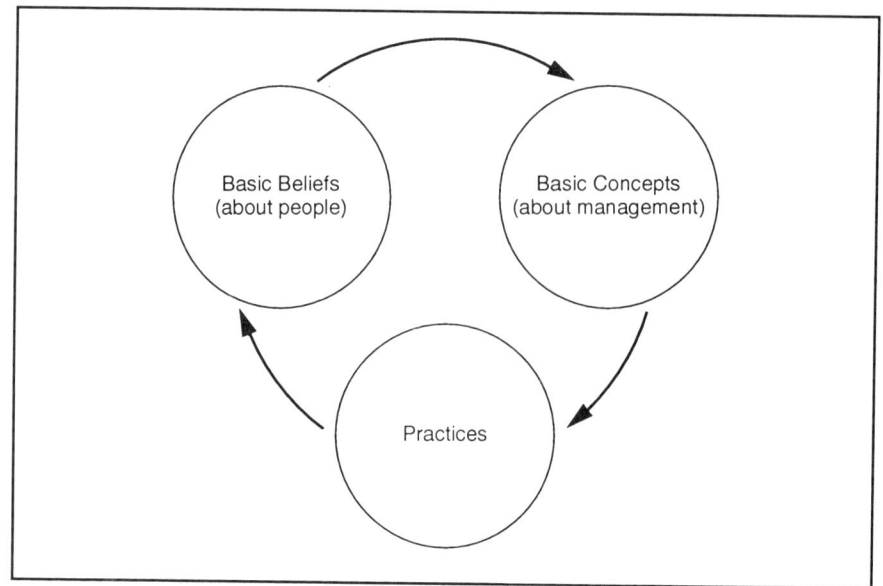

continue to translate the basic beliefs, given contemporary conditions, and support the more rapidly evolving practices. Most of the excellent companies have had the leadership stability to internalize the beliefs and the management stability to institutionalize the concepts. Therefore, the practices that fit are modified early enough, and because their consistency with the beliefs and concepts will be put to the test, no air drops should occur.

Can the company demonstrate that it understands its processes?

What are the basic beliefs? Well, total quality is the ultimate manifestation of employee involvement. Thus, the whole organization should be driven by these basic beliefs:

- People are intelligent and creative.
- People are responsible.
- People are motivated.

Most contemporary cultures still have not grappled with just what these concepts mean. Procter & Gamble's manufacturing organization fundamentally grappled with what they meant 30 years ago and left most of its competitors in the dust. A recent thought expressed by an R&D management professional reinforces these notions:

> A half-century of indisputable social science data shows that participative, person-centered management results in optimum employee performance, growth and productivity . . . . Yet far too many technical managers disregard this data and perform as impersonal autocrats.[1]

Although all managers should look in the mirror relative to that harsh judgment, there is another explanation of why managers ignore indisputable data. This reason is that, as most managers see them, participative management, employee involvement, and human relations programs fail miserably. That is where basic concepts about management come in. The basic beliefs require the right management concepts in order to work. Too many managers act out participative management as permissive management, and too many employee involvement programs involve employees in everything but what matters. Thus, it is the author's belief that the lessons of high-performing systems show that organizations must be characterized by the following concepts:

- Management by principle.
- Goal directedness.
- Extensive information sharing.
- A pervasive sense of accountability.

TOTAL QUALITY

- A skills-development orientation.
- Minimal status differences.

The organization-development literature and personal experience show that those organizations that put in place the basic concepts around the basic beliefs could then adopt the practices that were generally observed in high-performing systems. Organizations that airdropped a practice without both the beliefs and concepts in place have failed miserably. Therefore, organizations that move to total quality practices without learning these lessons will fail miserably.

Management by Principle

Individuals may be intelligent, responsible, and motivated, but they need to exist within a specific set of organizational principles in order to channel collective energy in the right direction. The principles by which excellent organizations manage are spelled out, whether they call them principles or something more specific to their culture. They are few in number, lofty, lean heavily on ethics and fair play, and are widely published.

Goal Directedness

Here is what companies are seeking to achieve. Goals in high-performing firms are always there, always few in number, and always very specific. There is a lot of measurement and feedback. This concept ought to be a natural to support total quality practices, but the only success stories the author currently has heard are from organizations that already had the concepts in place. Jumping to a lot of statistical charting of various measures of performance in and of itself does not solve the problem, particularly when the historical drivers of performance have never been identified, publicized, measured, had targets set, had targets met, had achievement celebrated, or had higher targets set.

Extensive Information Sharing

While the author was in the education and consulting professions, he interviewed thousands of people from all levels in many different organizations. The people in high-performing systems all showed a common characteristic—they knew a lot about their company. And this knowledge was not what is usually found in employee newsletters. These people were routinely informed about strategy, costs, yields, productivity, profit margins, and goal attainment.

The managers of high-performing firms do not view information as a source of power, they view it as a source of empowerment. It seems that the four questions cannot be made operational without this concept being fully operational.

Pervasive Sense of Accountability

Success is invariably a case of balance, of recognizing both what feels good and what discipline must accompany that feeling. Excellent organizations hold individuals, not just groups, accountable for results.

> Can the company demonstrate that its processes are under control?

People in high-performing organizations accept this accountability and act on it. The author's view of the organizational development world indicates that a large number of failures in employee-involvement change efforts can be explained by the fact that this concept was not only omitted from the picture but was overtly discouraged.

A pervasive sense of accountability implies that norms are in place and enforced that spell out:

- When you say that you are going to do it, do it.
- When you know that you are going to miss a deadline or commitment, mention it before the deadline arrives.
- Better to say no than to do an unsatisfactory or incomplete job.
- Follow-up is not only acceptable but expected.
- It is necessary to evaluate not just whether a task is done but how it is done.

A Skills Orientation

A skills orientation involves the following characteristics:

- Real labor power planning based on needed capability.
- Intensive recruitment.
- Selection testing.
- Extensive orientation and early-entry training.
- Growth tied to skill acquisition.
- Required periodic renewal.

These are the observable features indicating that this basic concept is in place. Most change efforts focus on the skill being required in the change, whereas a skills orientation involves much more. For example, many total quality efforts are completely focused on teaching people statistics and problem solving, whereas the improvement of a system's ability to apply what amounts to the scientific method involves all of the previously mentioned concepts. Even worse, in many change efforts, the training event becomes an end in itself; it becomes the change.

If the skills-orientation concept is not already in place, a movement to total quality practices presents a marvelous opportunity to really foul things up if someone begins with training. As noted earlier, the author encourages beginning with the four questions; the skill-acquisition component of the answers should begin to follow.

Minimal Status Differences

All individuals seek some form of status, and to the extent that it is based on performance, there is a place for some status differences in organizations. Thus, this basic concept is primarily a hierarchical issue. If an individual is going to accept accountability for a process, then that person wants to accept it with the manager, not for the manager. Whether the person is a technician, secretary, or tester, he or she wants to participate with a feeling of mutual ownership. Extremely visible status differences in an organization, whether between manager and nonmanager, or professional and technician, will destroy a movement toward group problem solving and process ownership.

These basic beliefs and concepts constitute the principal lessons learned during the previous decade. Many firms rejected the knowledge; others rejected the risk of change. To ensure success, the essential elements of stability must be put in place. And stability equates to constancy of purpose.

TO PRACTICES AND CONSTANCY OF PURPOSE

Twenty-seven years ago, the author was being interviewed for a job by one of the country's highest-performing companies. As the author was being escorted into the magnificent foyer at corporate headquarters, something seemed to be out of place. In the midst of all the marble and mahogany, hanging from the ceiling was a large, grotesque lead ball. The ball was hanging down to within three feet of the floor, was two feet in diameter, and looked as though it had been beaten repeatedly with a nine-pound hammer.

The author asked about the ball, and his escort replied that it represents corporate strategy. The escort then proceeded to demonstrate. He reached behind the reception desk and extracted a nine-pound hammer. He then told the author to take the hammer and, with as much strength as he could muster, make that ball move with one mighty swing.

The author proceeded to grab the hammer and bash the ball with every ounce of his strength. As technical people could predict, the only detectable movement was in the author's head, neck, arms, and shoulders, caused by the vibration of the hammer striking the ball. The ball never wavered. The escort then said that the big bash never gets anyone anywhere and took his little finger and began giving the ball small but repeated pressure. Within one minute, the ball was swinging across an arc of about one foot. Within a few more minutes, it had an arc of about 10 feet. The escort was still applying pressure only with his little finger. Pretty soon, the ball was swinging almost from wall to wall. Employees coming through the foyer had to take evasive action. They seemed unperturbed, however, because they had obviously seen the demonstration before.

The escort finally summarized by saying that the organization knows what it wants to do. It keeps up the pressure. It has built tremendous momentum by not deviating from its basic values. Those who do not really know what they want to do are constantly attempting the big bash.

Does the company have a plan for continuous improvement?

That experience had a profound effect on the author. To this day, he believes it to be one of the most meaningful philosophical illustrations of what management is about that he has ever encountered.

To the extent that organizations have gone through the past few decades by getting in place the beliefs and concepts necessary to sustain a high-performance culture, they can embrace new practices like total quality with relative ease; they just keep pushing on the ball. The momentum is already there. And the author believes that defining total quality through the four questions will greatly facilitate leveraging the momentum.

To the extent that organizations have not tied in the basic beliefs and concepts to operations, they have a problem. They have a doubly difficult task, but the history of change says that it is a doable task. The challenge is to get it done right by balancing the tried-and-true beliefs and concepts with these exciting new practices.

In the decade following the author's visit to that high-performing company, it began to lose its way. The observable evidence began, of course, with a decrease in profitability. They then lurched from one practice to another. With their employees, customers, unions, and suppliers, they lost credibility by constantly changing course. Sixteen years after the author's interview, he was asked back as a consultant. As he anxiously entered the foyer, sure enough, the lead ball was gone.

SUMMARY

Total quality is an intellectual windfall to the manufacturing and R&D professions. The author's organization has embraced it by defining it as the answering of four simple questions. But previous decades did offer the author's company some sound principles on which to base both its change efforts and its constancy of purpose. Total quality is simply an exciting set of practices that will ultimately be improved on. It is the ability to create stability through basic beliefs about people and basic concepts about management that will determine whether companies effectively internalize these new practices.

Note

1. Tingstad, J.E., "Why Do Technical Managers Ignore The Data?," *Research Technology Management* 35 (January–February 1992), p. 8.

NEW PRODUCT DEVELOPMENT

SECTION

4

New product development concerns almost all companies. To ensure growth, a company generally needs to develop new products. To simply maintain existing sales and profits, a company often must develop new products to compensate for the shrinking sales and profits of mature products.

In the first chapter of this section, Scott Schaefer describes how his company improved the effectiveness of its new product development by implementing an internal structure to guide concurrent product development activities. Then Peter S. Petrunich tells a less happy story. He describes an organization in which there was very little communication between the R&D department and other functional areas. Petrunich's analysis underscores the need for a proactive product development program.

David P. Sorensen, Kerry S. Nelson, and John P. Tomsyck review the most frequently used methods for evaluating US industrial R&D programs. Then, they describe the 3M Company's technical audit process and the many advantages it provides.

John L. Schlafer's case study tells a tale of product design evolution. He discusses the product design domain of a water treatment product and uses it to show how technologies must be studied and adapted for a product design strategy to evolve. Colin MacPhee's case study describes a new product development strategy of a company that emphasized risk minimization.

Timothy H. Bohrer presents the new product idea paradox: some people who consistently generate new ideas claim that they can do so only when given complete freedom and control; however, the freedom and ability to create are most highly developed in structured environments.

John J. Moran recommends the use of focus groups to capture the voice of the high-technology customer. His chapter outlines when to use these groups and provides checklists for getting started.

Thomas Fidelle examines new product development from the viewpoint of an applied researcher who must coordinate the many company functions involved in the process. Barbara Purchia tells how Applicon created a program for improving its software development process.

As the former vice-president of engineering for a manufacturer of gasoline dispensers for service stations, Phillip R. Taylor writes about his experiences with a large-scale concurrent engineering project.

ORGANIZING FOR EFFECTIVE PRODUCT DEVELOPMENT

Scott Schaefer

This case study describes how Hutchinson Technology, Inc. (HTI) improved the effectiveness of its product development by implementing an internal structure to guide concurrent product development activities. This structure includes a system of checks and balances and critical decision points to ensure the products meet market needs as well as satisfy HTI's business goals.

HTI, as a contact photo-etching company, was founded in 1965. Growth was slow until the mid-1970s when the computer industry started to expand quickly—with the increased demand for computing power came increased demand for nonvolatile memory in the form of rigid disk drives. HTI was a key subcontractor of suspensions, a critical rigid disk drive component. A suspension is the small precision spring that allows the read/write head to fly a few millionths of an inch above the rotating magnetic disk.

Today, HTI is the world's leading supplier of suspensions to the computer industry, and the majority of its product is shipped from two US manufacturing sites directly to customers in Japan and throughout the Pacific Basin for assembly into disk drives. HTI offers more than 100 different suspension styles, and it produces approximately five million suspensions per week.

In the mid 1980s, it became apparent that HTI was in a strong position not only to manufacture suspensions but also to design them for its customers. The computer industry was changing so quickly that companies had to design rigid disks in less than one year. In addition, because suspension designs and manufacture are interdependent, HTI had a vested interest in designing suspensions that could be economically manufactured in high volumes. Thus came the stimuli for HTI to develop an effective structure to design the highest quality, most cost-effective suspensions in the world.

WHAT IS PRODUCT DEVELOPMENT?

For the purpose of this case study, product development is the time spent between product conception and mature manufacturing. The product can be an extension of a base product or a brand new product requiring new technologies. It can be developed internally or with outside sources. The methods discussed in this chapter apply to all of these situations.

SCOTT SCHAEFER is vice-president of disk drive product development for Hutchinson Technology Inc.

HOW ARE PRODUCT DEVELOPMENT ACTIVITIES ORGANIZED?

Shortly after HTI decided to make the improvement of product development a corporate objective, it recognized that it needed a guiding structure to ensure product quality and development efficiency. A cross-functional team of managers from HTI's engineering, manufacturing, quality, reliability, and marketing functions was assembled to design an effective and user-friendly guideline that all product and process developers could use. The team developed a five-phase PPD guideline that addresses each critical activity in a typical development effort. PPD structures product and process development activities and ensures the integrity of each development phase by mandating ongoing reviews of technical and business issues. Exhibit 1 illustrates the five phases of PPD.

Product quality is ensured through a series of verification tests and peer reviews called product assurance reviews. Exhibit 2 is a checklist for these reviews. Business issues are addressed in a series of executive reviews known as approval board reviews. They are shown down the center of the PPD backbone in Exhibit 1. Supporting this backbone are several strategic plans, which are developed along the way. Some examples of these plans are shown on the right side of the PPD backbone. Diamond shapes represent points of strategic review. During these reviews, product development stops until the previous phase's requirements are seen to have been adequately satisfied.

Phase One

The purpose of phase one is to understand the market need and define the market problem that the developers want to solve:

- The size of the problem.
- Its importance to the customer.
- The value the customer places on a solution.
- The window of opportunity.

The development team conducts market analyses to define the problem, and it develops several possible product solutions. Several other development activities are also initiated, as indicated on the product assurance checklist. Peer reviews test and verify the market requirements of the concept through requirement verification reviews. The approval board writes a business proposal.

Phase Two

Phase Two's task is to develop the product concept by:

- Refining the concept.
- Demonstrating its technical feasibility.
- Beginning testing and modeling.

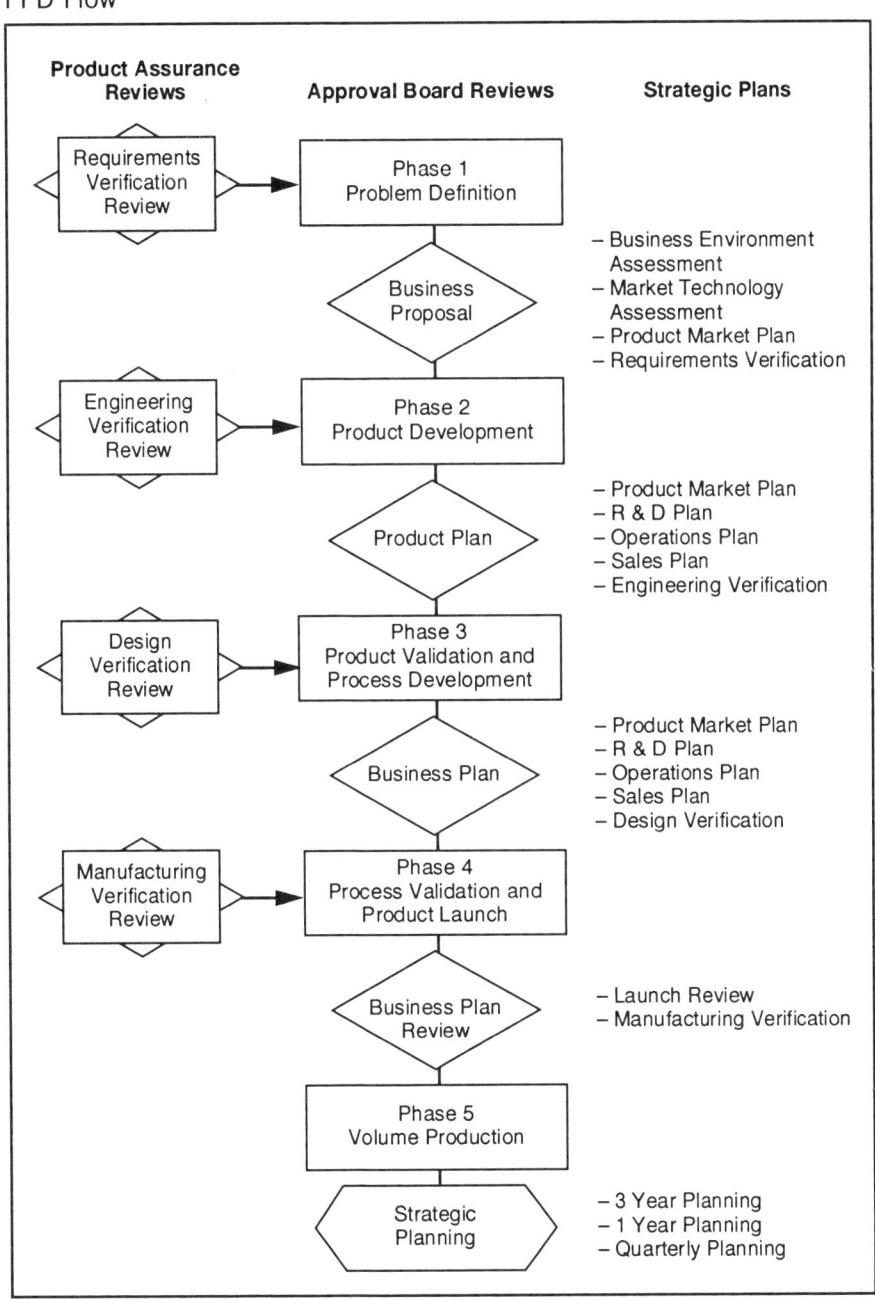

Exhibit 1
PPD Flow

Product Assurance Reviews | **Approval Board Reviews** | **Strategic Plans**

Requirements Verification Review → Phase 1 Problem Definition

Business Proposal
- Business Environment Assessment
- Market Technology Assessment
- Product Market Plan
- Requirements Verification

Engineering Verification Review → Phase 2 Product Development

Product Plan
- Product Market Plan
- R & D Plan
- Operations Plan
- Sales Plan
- Engineering Verification

Design Verification Review → Phase 3 Product Validation and Process Development

Business Plan
- Product Market Plan
- R & D Plan
- Operations Plan
- Sales Plan
- Design Verification

Manufacturing Verification Review → Phase 4 Process Validation and Product Launch

Business Plan Review
- Launch Review
- Manufacturing Verification

Phase 5 Volume Production

Strategic Planning
- 3 Year Planning
- 1 Year Planning
- Quarterly Planning

ORGANIZING FOR DEVELOPMENT

Exhibit 2
Product Assurance Checklist

		Requirements Verification Review	Engineering Verification Review	Design Verification Review	Manufacturing Verification Review
		Problem Statement	Mock-ups and Engineering Models	Prototype	Final Product
MARKET & PRODUCT TECHNOLOGY ASSESSMENT	Assessment of Competitive Products	I	F	R	
	Assessment of Competitive Technologies	I	F	R	
	Environmental Requirements	I	F		
	Intellectual Properties Strategy	I	F	R	R
DEVICE	Requirements with Conflicts Identified	F			
	Verification and Validation	I	U	F	R
	Device Specifications with Conflicts Identified		I	F	R
	Component and Material Specifications		I	F	R
	Packaging and Labeling Requirements		I	F	R
PROCESS	Suppliers Evaluated and Selected		I	F	
	Process Concepts and Alternatives	I	U	F	
	Verification and Validation		I	U	F
	Establishment of Yield, Cost and Volume Estimates	I	U	F	R
SERVICE	Serviceability, Maintainability, and Warranty Requirements	I	U	F	
RELIABILITY	Reliability Requirements and Specifications	I	F	R	R
QUALITY	Quality Requirements and Specifications		I	U	F
REGULATORY	Regulatory Requirements and Standard Practices	I	F	R	

Key:
I   Initiated
U   Updated
F   Finalized
R   Revised only as needed

The development team performs in-house testing to verify that the concept meets design requirements. Engineering models may be tested with select customers. The team uses customer feedback and in-house test results to modify or verify the concept design. Peer reviews verify the technical feasibility of the concept through engineering verification reviews. An initial business plan is written and reviewed by the approval board.

To develop a successful product, a company needs corporate freedom to stimulate innovation and corporate structure to manage the process.

Phase Three

The purpose of phase three is to complete the product design and specifications, to:

- Complete product qualification tests.
- Complete component qualification tests.
- Begin process qualification tests.

The development team makes the product in limited quantities for target beta-site testing and review. The operations personnel evaluate requirements and anticipate resources needed to respond to current prototyping needs and future volume requirements. The final product design is tested during design verification reviews (DVRs). Then, the approval board completes and approves the business plan.

Phase Four

Phase Four addresses two issues:

- The qualification of manufacturing processes, systems, and controls for high-volume manufacturing
- The promotion and market launch of the product.

The team prepares for formal product launch. A carefully composed launch plan introduces the right quantity of product to the right customers at the right time. Peer reviews test and verify the final manufacturing processes and systems through manufacturing verification reviews. The promotion and launch plan is completed and reviewed by the approval board.

Phase Five

Phase Five ensures that the product meets the customer's requirements for quality, volume, and price. The operations team (from manufacturing, engineering, quality, sales, marketing, and accounting) charts business performance against strategic goals developed through ongoing corporate planning.

WHAT IS THE VALUE OF PPD?

Developing highly successful products consistently over a long period of time is not easy. Only a few world-class corporations can claim some sustained success. It seems to require a delicate mix of corporate freedom to stimulate innovation and corporate structure to manage the process. PPD was created to provide the appropriate degree of structure for HTI's new product development objectives.

Many critical decisions are associated with developing a successful product, and it is easy to get distracted by some and forget about others. A structure such as PPD reduces the occurrence of costly oversights because following it encourages development teams to balance all aspects of the product and process development activity. During PPD, key business and technical requirements must be addressed; in many cases, these requirements must be reviewed more than once to ensure that earlier assumptions are still valid.

ROADBLOCKS IN IMPLEMENTING PPD

Because PPD is a new way of performing day-to-day business activities at HTI, it affects many of the organization's functional groups. The following sections discuss some of the issues HTI encountered during implementation of PPD.

The Fear That PPD Is Too Bureaucratic and Time-Consuming

Initially, some at HTI feared PPD would slow the development process. Formal training of the principal users helped relieve this concern. Executive management worked closely with the development teams to coach and reinforce the appropriate use of PPD. In some cases, they became temporary team members. The more PPD is used, the more proficient the teams became.

Every activity in the PPD guideline need not be started from scratch. In many cases, the bulk of the work or knowledge that the PPD process required already existed and needed only to be reverified or documented.

PPD Is Not a Substitute for Thinking

Care had to be taken to make sure PPD was not promoted as a cookbook recipe for success! It can be easy to get caught up in the details of following a process like PPD, believe that the objective is to complete the task at hand, and lose track of the larger objective, which is to create successful products. Using PPD helps organizations manage the risks associated with product development, but it does not guarantee success. Accountability and authority for successful products depends solely on the skills and abilities of development team members. They must accept that responsibility, for no guideline can substitute for clear thinking!

Adherence to PPD Guidelines in Crisis Situations

Because people naturally revert to old habits when a crisis occurs, management must constantly remind personnel of the value of using PPD. Enforcing the use of verification reviews and approval board reviews is the best way to deal with crises. Nothing should go to market without having passed all the points of strategic review. The temptation to by-pass the system is reduced as PPD is embraced by the entire organization.

Accountability and authority for successful products depends solely on the skills and abilities of development team members.

SUMMARY

Using PPD has helped HTI evolve as a company. Now, the concept is being adapted to guide the development of such internal products as business operating and software systems.

PPD should be a living guideline. It should be continuously audited and revised to meet the company's changing needs. Training should also be ongoing. Employees must be updated on added improvements; new users must be taught the methodology.

PRODUCT DEVELOPMENT EFFECTIVENESS: EXECUTION IS KEY

This chapter discusses, using a specific case study, how to make R&D effective when there is very little interaction between the R&D department and other functional areas.

Peter S. Petrunich

As both a functional R&D team member and an administrator, the author sees the issues of qualitative and quantitative measures of R&D's contribution to the overall success of a business and enhancement of that contribution as ongoing pursuits. It has been generally recognized that the contribution an individual technologist and an R&D team make to a company depends on the quality of interaction with other key functions and the consequent integration of the vision, mission, and business plan. This chapter discusses how to make R&D—or in this specific case study, product development—effective when some or all of these important interactions are missing or ineffective.

Senior management in most well-run organizations recognizes the importance of R&D; the substance of this recognition, however, ranges from something short of pure showcasing to total commitment and support. Consequently, the state of R&D within an organization is often far from ideal. Most businesses are under extreme pressure, particularly those in the US, to provide short-term results. Stockholders and investors often are romanced with the narrative of financial success through commitments to R&D and a hi-tech future. However, these two points are in direct conflict, because R&D expenditures, in the financial short term, are a profit detractor, at least as reflected in current accounting practices. Some suggestions have been made to amortize current-year expenditure over future years, consistent with the lag on return from R&D investment. None of these has been universally embraced. Although recent government discussions have supported the need to provide R&D incentives with federal tax incentives, the deficit and the recently approved federal budget belie this objective. Management often takes the short view and reduces R&D expense during hard times, rather than making a concerted effort to improve effectiveness.

PETER S. PETRUNICH has more than 15 years' experience in various aspects of industrial product development, including management, for US Fortune 500 and foreign companies. He also has held management positions in marketing and manufacturing, served in a number of professional and trade associations, and published a number of technical papers.

BACKGROUND

The current case history addresses a real experience with a reactive, as opposed to proactive, product development program in a contemporary business environment. This problem ultimately led to optimization of the product, fundamental changes in the product development effort, and proactive development of a superior product.

The quality problem that precipitated this case evolved from a previous product development project during a period of transition. The product line was not effectively supported in anticipation of divestiture of the business by the large parent corporation A. Acquisition of the business by a smaller holding corporation B then led to a reorganization, loss of much of the already limited staff support, and a subsequent period of turbulence while the new company established itself. Within the portfolio of Corporation B, this product of the newly acquired subsidiary was one key component of the business unit strategy. Because this transition occurred during the tail end of the last economic expansion, product demand was strong. The newly formed company was meeting the financial goals initially set and later guided and approved by the parent corporation B. Meeting these goals was of vital importance to ensure the perception of a successful acquisition and continued support for capital and other resources.

THE TECHNOLOGY ORGANIZATION AND FUNCTION

The technology organization was composed of a product development group, a geographically dispersed quality assurance group, and a customer technical service group. The technology manager was physically located at a headquarters site approximately 100 miles from the key primary processing facility and 600 miles from the final fabrication facility for the subject product. Agendas for the two manufacturing managers were not necessarily in concert with that of technology. Consequently, individual on-site quality assurance engineers were torn between the strong influence of their landlords and the distant direction provided by their geographical dislocated organizational superior. The marketing manager, at the same time, was putting pressure on both manufacturing and technology to provide increased quantities of high-quality product to meet growing customer needs.

PRODUCT HISTORY

Within this scenario, the subject key product had been developed and introduced just before the acquisition. It was developed to overcome an earlier composition wear deficiency that was caused by cost and environmentally driven process changes.

The product for this application was important not only because it represented nearly 30% of the sales of the business unit but because it was the single design with volume suitable for profitable automated production.

The rest of the product line was composed of a large variety of material and design combinations, usually sold in relatively low volumes.

Of the subject product sales, 25% were to customers C and D. Problems developed approximately a year after this consumable product was introduced to the larger customer C, which had converted to its exclusive use. The problems manifested themselves in the form of structural failure of the body and hardware. It was uncertain whether the cause of the problem was inferior quality of the product, other causative factors within the main component of which it was a part, or operating conditions. Nonetheless, the apparent problem was lack of toughness of the subject consumable product, which led to fracture and, in some cases, failure of a critical machine. While the failures at customer C were being investigated, a similar situation developed at customer D, the equipment and operating conditions of which were similar to those of customer C. A number of other customers still were reasonably satisfied with the performance of the same product.

The prolonged period of investigation that followed included visits to the customer's shops, quality assurance investigations, and meetings with key staff members. Although actions were taken to improve the toughness of the product and provide the customer with maintenance improvement suggestions, these effectively failed. Both companies disqualified the product, terminated purchases, and gave the business to competition. The competitive product, though not so wear resistant, was perceived as tougher and produced far fewer equipment failures. Reaction through to corporate management was, as would be expected, intense because sales losses were significant and immediate, with no final solution at hand. The company faced its first major crisis.

> **Although recent government discussions have supported the need to provide R&D incentives with federal tax incentives, the deficit and the recently approved federal budget belie this objective.**

Definition and Analysis

At this point, a major review and post-mortem was undertaken at the executive level. What became apparent was:

- The initial development of the product was flawed, and the product was released prematurely.
- The real problems had not yet been defined clearly.
- A thorough historical search had not been made, which would have provided clues to product weaknesses.
- Some key quality assurance practices were lacking.

At face value, these conclusions pointed to ineffective execution of fundamentals. Understanding why the failings occurred was crucial to preventing recurrence.

Premature Release. As noted previously, turbulence both before and after the company was acquired contributed to this shortcoming. Before divestiture, the additional tests and modifications needed before commercial release were short-circuited. Survival of the business was at stake, and a new product was needed to replace the old, which was being displaced by competition. This environment produced a high-risk mentality that permeated the entire organization. In this case, as in the others, the axiom that haste makes waste prevails. Strong leadership by the R&D team, as a partner in any business plan, is the key to minimizing the risk of failure during new product introduction.

Real Problems Versus Apparent Problems. Given that circumstances prevented optimization of product development and introduction, the simple issue of problem definition was one of singular importance. Again, although the need for effective definition of the problem can be grasped intuitively, practically it is frequently not the focus of a product development plan. Often, the mushroom concept (i.e., keep them in the dark and feed them manure) prevails, consciously or unconsciously. As unreasonable as this may seem, the author has seen it happen many times during his more than 30 years associated with R&D. In the subject case, the direction and the quality of the information given the project leader were woefully inadequate, and he did not have the first-hand exposure over a significant time span to properly define the problem himself. Problem definition in this case was by its nature elusive and required statistical definition as well as empirical insight. The type of field failure also could not be duplicated in the laboratory. The result was that many apparent problems were addressed without resolving the real problems. This experience was most disconcerting to the project leader, who was perceived by some as the reason no solution was forthcoming. Because he sensed this fact, he became confused and frustrated and experienced low morale.

Literature Search. Although fundamental to the scientific method, historical literature and personal experience specific to the toughness problems had not been researched thoroughly. Consequently, clues to needed direction had not been uncovered. Focus for this search also was not sharp because the field problem itself had not been properly defined.

Quality Assurance Practices. As was discovered during the reviews, statistical process control, particularly at the secondary and fabrication operations, also had been neglected. The importance of key process steps and execution appeared secondary to manufacturing needs, such as introduction of inventory reduction methods and output requirements. In addition, changes in the process and fabrication had been made in the interest of accommodating manufacturing needs without thorough investigation of their impact on the product and customer needs. Lag time from implementation of changes to customer use was long. Therefore, sorting the impact of these changes on the product's durability was most difficult.

The quality program was torn between the needs and priorities of the manufacturing manager and the importance and execution of statistical process control methods. Total quality management was an unknown, unpracticed concept.

> **Management often takes the short view and reduces R&D expense during hard times.**

THE REACTIVE SOLUTION

Because the premature release of the product was history, the focus became one of defining the problem. To understand the problem as the customer perceived it and to fully appreciate the interaction of this consumable with the rest of the machine as well as operation of the entire system a new approach had to be taken. Not only had customer confidence in the supplier company been lost, but an adversarial attitude existed. This confrontation developed because a solution was not forthcoming, and there were suggestions that poor equipment maintenance was a causative factor. Total lack of credibility existed at this point.

Defining Customer Problems

The principal liaison with these customers was a senior individual whose prior responsibilities had been sales management. His understanding of the culture and systems involved was from a supplier point of view. He did not display, nor did the senior applications manager have the needed depth to delve into, statistical detail of the interaction of systems, of which this product was a part. A most objective evaluation, which led to disqualification, was conducted by the manager from Customer D who had responsibility for this product and the attendant system. He was deemed a creditable candidate to define the problems and was hired specifically for that purpose and reported through the marketing organization. His knowledge of maintenance, service, and the culture of the customer's industry immediately was put to use to reestablish a creditable relationship both with customers and with those in this industry segment in general.

Historical Data Review

Requalification took some time. The other two analytical deficiencies had to be addressed in parallel with the field work to define the extent of the problems. A study team composed of members from all the involved functions was commissioned and chaired by a senior technical marketing executive. The charge to this team was to define the problems, recommend and implement proper solutions, and regain qualification as soon as possible. First priority was acquisition and study of all written information related to product toughness. This review showed:

- A number of process steps were predisposing the product to excessive mechanical stress.
- This predisposition in the subject material and design combination was most influenced by control of two primary process steps and two fabrication steps.
- Although the basic material and design combination produced optimum low wear, it was inherently sensitive to certain operating conditions that induced mechanical stress.
- A laboratory test procedure did not exist that would comparatively duplicate or simulate the failure.
- The solution was not as simple as making the material mechanically tougher.

Quality Assurance

The study team determined that a number of primary and secondary operations simply were not within established control limits. Three of these were related to toughness, as established in the historical review. Beyond that conclusion, the range of two of the control limits needed to be tightened, because their upper limits made the product more susceptible to mechanical stress during operation. Although methods to simulate field results on a comparative basis continued to be sought, they eluded the experimenters. With the time constraints to reapproval, conscious decisions were made to adjust and control these key variables, consistent with direction provided by historical experience.

In parallel, the newly hired industry service manager confirmed that a number of tolerances and conditions of operation were not within acceptable practice at both customers. He was aware that these conditions existed at company D. The economic reason for their existence was lack of an adequate maintenance budget. The fact remained that the competitive product was more tolerant of these abnormalities than was that of the author's company.

CONCLUSIONS: REACTIVE SOLUTION

The ultimate test was operation under field conditions. A decision was made to integrate the changes indicated by the literature study into two experimental material and design combinations and to statistically evaluate these improvements on the equipment of customer D. The industry manager's creditability with his ex-employer was key to getting acceptance of this proposal. Consequently, approximately three months after disqualification, tests to regain acceptance began. Because these optimized products represented consensus of the entire study team, confidence was high that they were significant improvements over the previous product. Quality assurance was mandated to assure execution of every process step. (The subsequent focus on the quality assurance effort led, two years later, to the fabrication plant achieving ISO-9002 certification.) Statistical study of both field conditions and all product failures continued during the test. Both experimental materials proved to be more resistant to failure, and the most resistant, on the basis of this field test, was requalified one year later. Two years later, after an exhaustive field test of comparative toughness with competitive products was completed, the same product was reapproved by customer C.

OVERALL ANALYSIS AND PRODUCT DEVELOPMENT

Analysis of this case from the period of acquisition provokes discussion of a number of points. The parent corporation B's vision is one of technical excellence.

The technology organization was composed of a product development group, a geographically dispersed quality assurance group, and a customer technical service group.

The initial objective of the technology organization was to fix the technical infrastructure that had deteriorated before divestiture and thereby contribute to the company's financial success. Fast return, high probability, and reactive, quality-driven projects were the focus. The subject case is an example of one of these. However, execution of some of the basic steps in problem solution was ineffective in this case. The lack of focus on problem definition was the single major cause of the loss of business. There are many other reasons for these deficiencies, as has been previously noted. The single most pervasive problem during this period and beyond was the absence of strong, cohesive teamwork, particularly between functional managers. This condition was exaggerated by the geographical separation that supported individual fiefdoms and interests. Finally, the divide-and-conquer senior management philosophy contributed to mistrust.

Interim measures were taken to improve effectiveness following the subject case. These included separating quality assurance and customer technical service from technology responsibility. An executive-level position was established for quality assurance to provide senior management focus on this function and the direction for statistical process control, ISO-9000 certification, and total quality management.

Customer technical service was reorganized into the marketing organization to complement the successful experience with the industry manager's role. Technology was left with a role of product and process development, primarily the former. The intent was to move to more proactive projects, which would support expansion of the businesses. This did in fact occur and ultimately resulted in development of a superior product that was designed to satisfy the needs of new-generation equipment and gain market share from competition. The project that led to this next generation product was initiated by the technology organization because the business plan had not defined its need.

A number of approaches have been debated to address the improved effectiveness and financial performance of the company, including R&D. Centralized versus decentralized R&D has been a key point of discussion. This debate was driven by a corporate mandate to move to a world-business-unit approach for the key product lines. The individual business unit managers generally wanted decentralization and incorporation of R&D within their individual organizations. The R&D manager favored an expanded centralized approach. Currently, rationalization and decentralization has been the restructuring of choice in the prevailing weak economy. Although this latter approach allows for greater control by the business units, it also can contribute to inefficiency and loss of flexibility within the technology organization.

How the company integrates the R&D function and restructuring, in general, are at best secondary issues. The primary issues for improved financial performance are the value contribution of individual employees and their commitment to team effectiveness. A wise, seasoned marketing manager once stated to his people after the latest reorganization to cut costs and improve efficiency, "I don't care how badly they screw it up, we'll figure out how to make it work anyway." The essence of his message was that eliminating or reshuffling boxes on an organization chart may be the means to an end but never the final solution. If the selection process is not thoughtfully and objectively implemented, restructuring the R&D function, or the company itself, will not be successful. People make an organization and company work. An ideal mix of quality people would function effectively without a formal organization.

On the other hand, structural change cannot make effective negatively predisposed individuals or groups. To the extent that instability and mistrust are enhanced by restructuring, individual and team productivity will be decreased in the mid- to long term. The effect will be opposite that perceived for the short term and generally desired. The value of individual contributions as well as intra- and interteam support and cooperation remain the keys to R&D and company effectiveness. Management at the highest levels must nurture this philosophy. To the extent that management is successful, R&D and the company also will be.

INDUSTRIAL R&D PROGRAM EVALUATION TECHNIQUES

This chapter reviews the most frequently used methods for evaluating US industrial R&D programs. Then, it describes the 3M Company's technical audit process and the many advantages it provides.

David P. Sorensen, Kerry S. Nelson, and John P. Tomsyck

US industry's evaluation techniques have always been designed to maximize the benefits that internally funded R&D can create. The most obvious of these benefits are company growth through the introduction of new products and the protection of current business through improvement of existing products. The attainment of these benefits by many companies has been the driving force for continuous efforts to update R&D program evaluation techniques.

Before World War II, the amount of R&D in US companies was small but growing in importance. From 1921 to 1938, the number of US companies with research staffs of more than 50 persons grew from 15 to 120. Technical programs were usually conducted by individuals or groups of two or three. The director of research had personal knowledge of every program and typically was the dominant force in determining what work was performed. Formal evaluation techniques were not required because the size of organizations naturally limited the number and scope of R&D programs. Lines of communication were short, and the entire management team understood its company's markets.

World War II had a profound effect on the evolution of US industrial R&D. During that period, the US became a global leader in science and technology, and many extraordinary wartime innovations were made (e.g., the Manhattan Project, radar, and synthetic rubber). These breakthroughs resulted from coordinating the efforts of large teams of scientists and engineers that shared a sense of urgency with plentiful resources to increase the speed of the innovation process. US industrial firms have been applying this strategy ever since.

DAVID P. SORENSEN was named director of technology analysis in 3M's Corporate Technical Planning and Coordination Department in 1981, and, in 1989, he was appointed executive director. In addition to the technical audit system, his current responsibilities include 3M's technology studies, technical computation, information services, and technical development functions.

KERRY S. NELSON has worked as both a product development scientist and a research and development manager in laboratories aligned with 3M's memory products and imaging systems businesses. In 1989, he became director of technology analysis.

JOHN P. TOMSYCK held a variety of technical and management positions in 3M laboratories during his first fifteen years with the company. His current title is senior technical studies specialist in the Corporate Technical Planning and Coordination Department.

The war also showed that investing in science and technology can provide competitive benefits that far outweigh their cost. Thus, after World War II, total spending on R&D in the US increased at about 10% per year.

In addition to increasing R&D funding levels after World War II, US industry also made two operational changes that are keys to understanding the use of R&D program evaluation techniques. First, R&D was organized as a specialized technical function. Second, laboratories began to become more isolated from companies in a physical sense. In the 1950s, industry began building laboratories in suburban settings, away from manufacturing plants and even corporate headquarters. The increase in the number of technically trained employees (especially those with advanced degrees), the building of new laboratories in isolated settings, and the organizational autonomy that R&D had achieved led US industrial research efforts toward an academic style. Because many US management teams did not understand the scientific process, the communication link to R&D began to weaken. To a large degree, this problem was ignored in the 1960s and 1970s because company growth was almost guaranteed by the continual growth in the US economy. The rise of global competition in the 1980s, combined with a slowing US economic growth rate, changed this attitude of noninterference. Whereas relatively little was being published about R&D program management and evaluation before the 1980s, the technical manager now has access to a wealth of information. During the past decade, outside R&D program management and evaluation-consulting services also emerged.

EXAMPLES OF R&D PROGRAM EVALUATION IN US INDUSTRY

Constantly rising costs (in this case, R&D funding) and slowing return (i.e., decreasing revenue growth) has aroused the attention of chief executive officers across corporate America. More and more they are asking whether they are getting a good return on investment in R&D. Those who are experienced in industrial R&D know that this is a complicated question. They view industrial R&D as a process of creating information about a process or product. Such information is seldom created in the straight line that suggests to most managers an efficient use of resources. Competitive pressures have also forced the need for improved R&D productivity. These situations, coupled with the fact that many of a company's leaders may not have been technically trained, create a potential communication gap. Most companies are bridging this gap through the use of R&D program evaluation techniques.

Numerous analytical techniques have been proposed as R&D program evaluation tools. The most widely used fall into the general categories of scoring models, financial appraisal methods, and risk analyses. All three of these techniques share four principles. They seek to maximize the return a company receives from its R&D funding, express that return as a function of the benefit that R&D generates, create a probability for the likelihood of a return being achieved, and predict the cost in R&D funding required to achieve the return.

A simplified equation of the preceding principles can be expressed this way:

> An industrial R&D program's index of value = [(probability of commercial success) • (net return)]/(cost of R&D).[1]

Most managers regard applying overly elaborate and rigorous techniques to industrial R&D as inappropriate.

Most of the analytical techniques used for R&D program evaluation focus on ranking existing programs by priority, selecting new opportunities to work on, or arriving at a go/no-go decision on funding.

Scoring models are perhaps the most widely used formal techniques for analyzing industrial R&D projects. They can be used effectively to assess both research and development programs. Scoring models involve the use of a list of criteria to evaluate the value of an R&D project. The list of criteria usually includes assessments of the attractiveness of the opportunity, its probability and risk for success, its match with company capabilities and business objectives, an estimate of the R&D investment required, and an assessment of competitive position. Weights are usually established for each criteria. The weighting and scoring may be linear or nonlinear, and aggregate values can be additive or multiplicative. Scoring models are attractive to US industry because they are flexible, have simple data requirements, and increase organizational communication because their use involves people from throughout the company.

The results established when using scoring models create only relative values for a set of R&D programs. However, the technique usually does not impose an undue burden on management. Its relatively simple nature appears to be compatible with management's beliefs about the complexity and uncertainty of the R&D process. Most managers regard applying overly elaborate and rigorous techniques to industrial R&D as inappropriate.

Robert Cooper is a consultant who applies scoring model techniques to R&D program evaluation. He has developed a scoring model system named NewProd to provide project diagnostics—identification of project strengths, weaknesses, critical information needs, and key uncertainties—and to help improve project selection decisions. Project selection decisions are a key element in improving R&D productivity because roughly only one industrial R&D project in four that enters the development stage eventually succeeds in the marketplace. Cooper maintains that correct use of the NewProd system does a good job of predicting successes and failures before product development even begins. He points to predictive ability validation results ranging from 73% to 84%.[2]

Financial appraisal methods are the second general category of industrial R&D program evaluation techniques. They are usually applied to product development programs. Healthy industrial companies typically set and achieve both financial and strategic goals. However, in a hierarchical sense, financial performance must come first because it enables the pursuit of strategic goals. Recognition throughout the organization that financial performance is crucial helps explain why these techniques are being used by some US industrial companies to evaluate their R&D programs.

Financial analyses of R&D programs include the investmentlike character of R&D expenditures and apply capital budgeting techniques to their evaluation. They treat expected costs as a stream of future expenditures and subsequent returns as a stream of future income. Using standard discounting techniques to establish equivalent present values, they can compare the attractiveness of a set of R&D programs. A variety of methods are now in use.

One of the key responsibilities of the R&D program manager is to assess the potential rewards of success against the likelihood of failure. Because the majority of industrial R&D programs fail, risk is involved in R&D program funding. Inevitably, there are also always more R&D program ideas than funds to support them. Decision analysis is a technique for choosing R&D programs. It is a combination of statistical analysis and decision theory. The essence of decision analysis is the acknowledgment that all decisions involve uncertainty, and this uncertainty puts R&D investment at risk.

The decision analysis method helps rank a set of R&D programs in terms that management understands, namely profits. It typically does not, however, consider the breadth of an R&D program as most scoring models do (e.g., intellectual property considerations). The four main elements used in the technique are cost of development, expected profits over the life of the product, probability of technical success, and probability of commercial success.

3M'S TECHNICAL AUDIT PROCESS

Since 1965, 3M has used its technical audit system to conduct internal peer reviews of its R&D activities. Although the informational content within the technical audit has grown since the inception of the system, its basic objectives have remained the same. It conducts periodic, technically oriented reviews of the operational activities as well as important individual R&D programs within each 3M laboratory. Unlike many other R&D program evaluation methods, it is focused on identifying the resources and actions required to help each R&D program succeed and strives to balance the feedback along these lines.

Each of 3M's more than 100 worldwide laboratories is reviewed on a three-year cycle. Information about the organization's technical or business plans is reviewed, and the audit procedure provides feedback that may lead to improvements in such areas as R&D productivity and prioritization; corporatewide technology and information sharing; new product development success rates; and coordination of global, environmental, and intellectual property activities. The audit process thus provides 3M a systematic method of examining technologies and programs and estimating probabilities of technical and commercial success. The audit system offers a consistent overview of all technical activities for management and elucidates critical issues and opportunities for each laboratory. It also provides a source of specialized internal consultation on technical problems.

The technical audit is an in-depth peer review process of all aspects of key new product and technology-building activities within 3M. A typical team of auditors includes invited members from 3M's laboratory management and senior scientist ranks (for a total of six or seven auditors who are familiar with the area of technology but who are not a part of the laboratory being audited), the director of the technology analysis department, and four to six management and technical personnel from within the unit being audited. The makeup of the team of auditors changes for each audit.

Each of 3M's laboratories is reviewed on a three-year cycle.

The technical audit is one of the responsibilities of the technology analysis department, which is part of the corporate technical planning and coordination organization. The administration of each audit, including the analysis and feedback of the information obtained, is accomplished using a three-person team composed of an audit manager, a technical specialist, and a data processing or keyboarding staff member. In addition, an audit coordinator is responsible for arranging auditors, scheduling audits, and managing the information in the technical audit data base. The technology analysis director, managers, and specialists have all had broad exposure to 3M R&D activities before their assignments in the technology analysis department. At the present time, two three-person audit teams are being used.

About six weeks before each technical audit, an audit manager meets with the associated 3M laboratory head to discuss overall laboratory issues and audit procedures and to select the programs to be audited. To shelter embryonic efforts, small programs and those in the concept phase are not audited. The technical audit is conducted as an all-day (up to three days) meeting at which members of the laboratory being reviewed present information describing their R&D programs. These oral descriptions are supplemented by detailed written materials that the laboratory provides before the audit.

The 3M technical audit uses a scoring model that has been evolving through almost thirty years of application as the primary element of its R&D program evaluation method. The model continues to be refined as the company learns more about which rating factors have proved most relevant to the success of R&D programs. Each auditor uses scoring sheets, which are filled out on the audit day, to provide the audit team with both numerical ratings and comments on each audited program and overall laboratory operation. Further evaluation information is gathered in an informal breakfast meeting attended only by the auditors not associated with the audited laboratory, the laboratory head being reviewed, and several corporate technical planning and coordination staff a day or two following the audit.

3M's long experience with its scoring model has indicated that a number of factors are important predictors of the future commercial success for its R&D programs. Key technical factor ratings include overall technology strength (breadth, patentability, competitiveness), R&D personnel (number, skills), competitive factors (knowledge of competition, targeted 3M product performance), remaining R&D investment (vs. time to complete the program), manufacturing implementation (feasibility, cost, protectability), and an overall rating of the probability of technical success. Key business factor ratings include financial potential (sales, profit), 3M competitive position (marketing channels, product value), and an overall rating of the probability of marketing success. Laboratory factors that are rated include organization and planning (strategy, focus, clarity of goals), staffing (number, skills), program balance (product maintenance vs. new product efforts), and coordination and interaction (with marketing, manufacturing, and other 3M laboratories).

After all of the information-gathering steps have been completed, the technical audit staff analyzes the quantitative and qualitative data. Items that are numerically scored are compared with the profile that other similar programs have created in the audit data base. The main conclusions and results from each audit are summarized in a document that details program strengths and concerns and makes program recommendations. The laboratory head, its general manager, and a senior R&D vice-president review this information in a special meeting. After this meeting, a follow-up review of the results is conducted for the members of the audited laboratory. A comprehensive written report of the audit results is also issued. These reports are used with a variety of other assessments to constantly produce a picture of 3M's present technical position and to consider future allocation of R&D funds.

Companies can also use technical audits to help solve the communication and prioritization problems that large organizations inevitably face. All levels of the company obtain benefits. Exhibit 1 summarizes some of them.

TRACKING 3M R&D PRODUCTIVITY

Another important part of 3M's program evaluation effort—technical studies—is performed by the technology analysis department. Like members of the technical audit teams, members of this function obtain a broad background in a number of 3M laboratories before filling this assignment. The fact that technical studies personnel work closely with 3M's R&D, controller, and marketing functions ensures a particularly powerful access to both quantitative and qualitative information.

The general objective of this work is to support 3M management's continuous efforts to make informed decisions on R&D matters. The technical studies function records the performance of past programs to help predict the success of future ones.

One of the responsibilities of the technical studies function is to provide support to technical audits by conducting follow-up analysis of audited programs and by developing updated or new means of evaluating the performance of laboratory programs. Having a data base of audit-derived parameters that extends back at least twenty-five years is especially useful in this regard. For example, it is possible to selectively define the criteria for

Exhibit 1
Benefits of 3M's Technical Audit Process

Level	Benefit Obtained
Senior management	Consistent, comparative overview of laboratory R&D. Independent assessment of R&D programs and needs.
Laboratory directors	A method to share best R&D practices. Tool for gaining additional support or management understanding.
Laboratory employees	Learning through participation as auditors. Technology advances through contact with peer scientists.

success of product development programs and then analyze how success rates of such programs are changing as a function of time. Exhibit 2 illustrates the result of this type of evaluation by depicting the success rates for three categories of product development programs during the 1970s and 1980s. During the 1980s, 3M improved already highly successful programs directed at existing business maintenance and made relatively larger improvements in the success rates of programs directed at related new business (i.e., new products related to existing product lines and businesses) and unrelated new business (i.e., new products for relatively new types of businesses). Although a number of things effected the improvement seen in the product program success rates, the technical audit results were a key contributor.

The results of technical audits can also be used to quantify degrees of risk associated with specific programs and to decide which key influencing factors for a program need improvement. 3M has evolved from looking at the impact of highly influential single variables, to the use of tables that position products in relative regions of risk, to the use of statistical analysis of a series of variables that is valuable for numerically estimating both the degree and the type of risk.

An example of the tabular analysis is shown in Exhibit 3, wherein the success rates in the various sections are derived from follow-up analysis of the related new product programs audited during the 1980s for which there can be a definite success or failure designation. The boundaries between the zones of probability of success and the zones of sales impact were developed using natural as well as statistically significant break points. The data shows that related new product programs that receive low audit probability of success ratings rarely achieve success. For products with a midrange audit probability of success, the success rate varies from 47% for those aimed at low sales targets to 61% for those aimed at high sales targets. The combination of high audit probability of success ratings and high estimated sales potential has almost guaranteed success (i.e., 96%) for related new product R&D programs.

Exhibit 2
R&D Product Program Success Rates

■ During the 1970s
□ During the 1980s

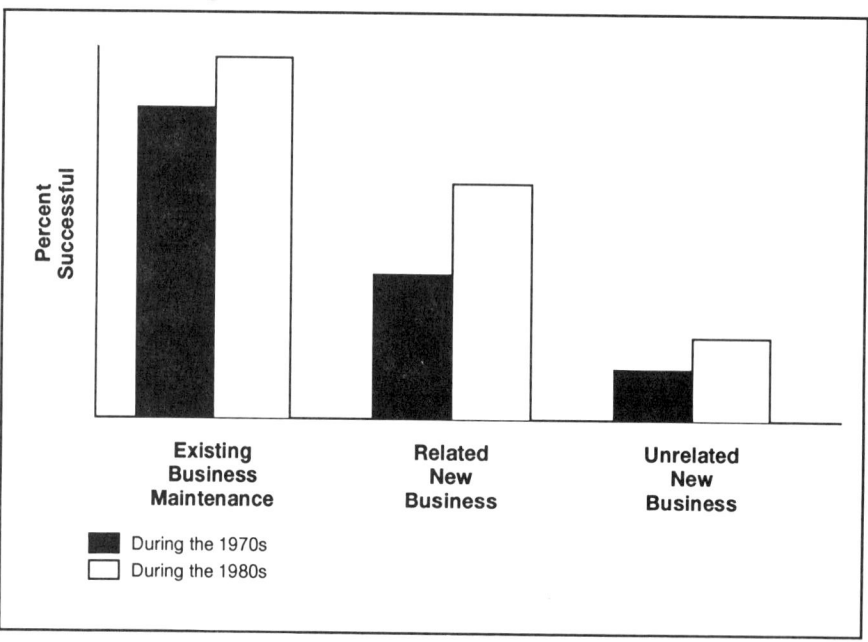

Exhibit 3
Related New Product R&D Programs Audited During the 1980s

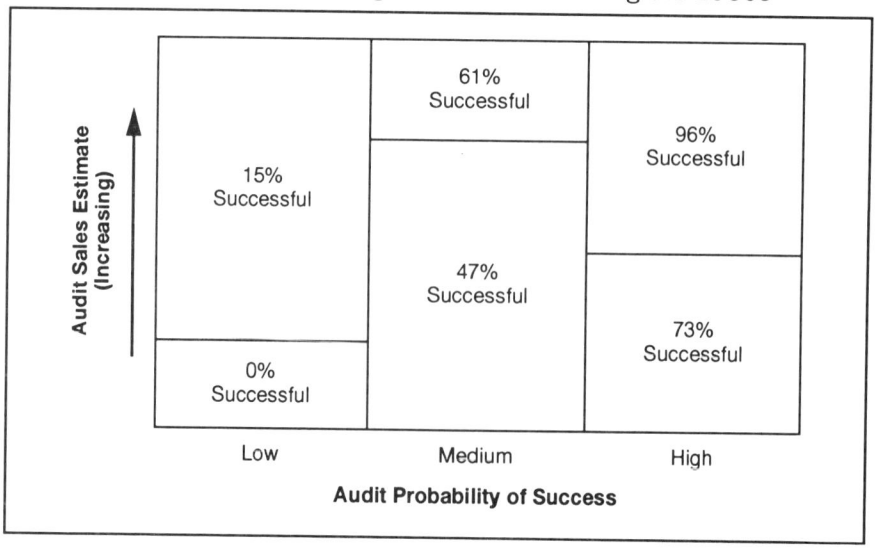

Audit Sales Estimate (Increasing) ↑

	Low	Medium	High
	15% Successful	61% Successful	96% Successful
		47% Successful	
	0% Successful		73% Successful

Audit Probability of Success

NEW PRODUCT DEVELOPMENT

Exhibit 4
Product Program Health Index

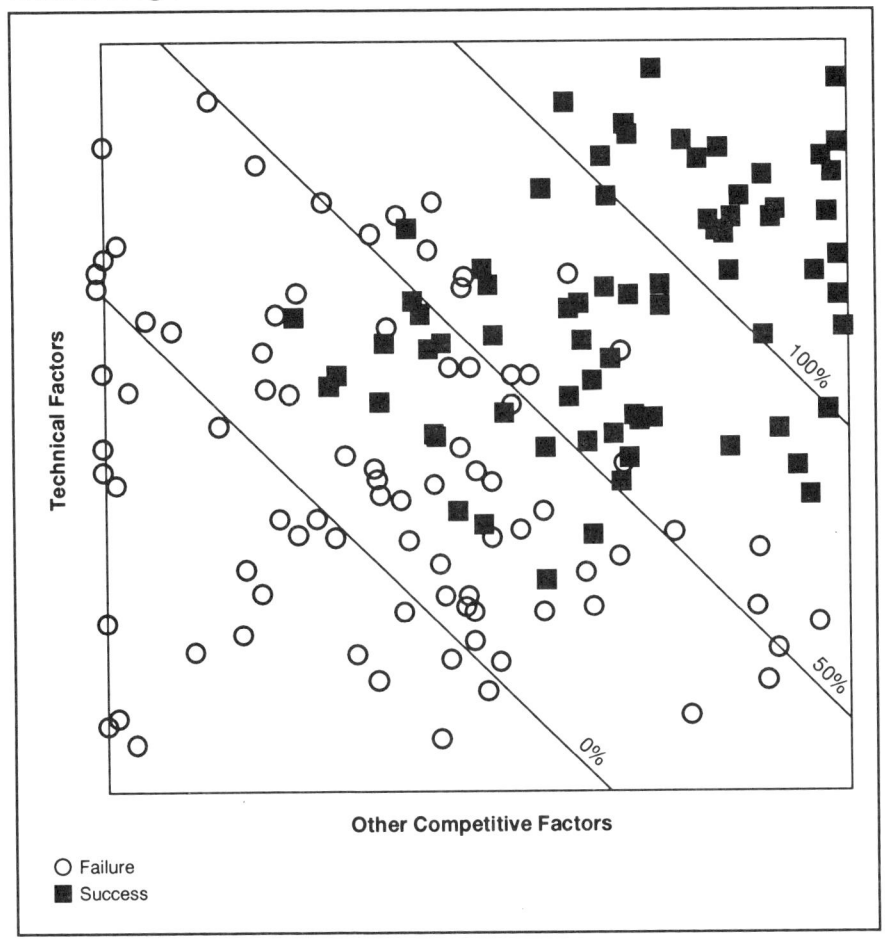

[Scatter plot with "Technical Factors" on y-axis and "Other Competitive Factors" on x-axis, showing diagonal lines at 0%, 50%, and 100%. Legend: ○ Failure, ■ Success]

The results from a more rigorous statistical analysis of how factors derived from technical audit evaluation related to predictability of program success or failure are depicted in Exhibit 4. This two-dimensional plot is the result of follow-up analysis on a large number of related and unrelated new product programs in which there were equal quantities of successes and failures. Although the influence of many factors was measured by means of appropriate ratings, only those factors most statistically significant were used to generate an overall mathematical predictive model, which was subsequently divided into two parts, a segment associated with technical factors and a segment associated with other competitive factors, to create the health index chart in Exhibit 4. Each reference point, therefore, results from mathematical positioning of a product program in the data field; the three diagonal lines designate zones of risk (e.g., all products below 0% failed, all products above 100% succeeded). Although positioning a product program on the health index chart is useful in showing the degree to which the perceived risk is associated with technical factors versus the other competitive factors, the overall distribution of data clearly indicates the need for an equal balance of good ratings for both types of factors to be positioned among the 100% successful programs.

Perhaps the most fundamental measure of laboratory output that the technical studies function monitors is sales achieved from new product introductions. All new products are tracked during their first five years in the marketplace. The accumulation and analysis of this information over many years has resulted in a variety of trend graphs and predictive growth models. It has been found that both first-year and new product sales over five years are good indicators of future company growth.

Given a competitive environment that necessitates the need for productivity improvement in industry, R&D program evaluation models will become increasingly widespread and sophisticated. When used effectively, the information these tools provide translates into competitive advantages.

Notes

1. L. W. Steele, "Selecting R&D Programs and Objectives," *Managing Research and Development What We've Learned in the Past 50 Years*, ed. M. F. Wolff (New York: Industrial Research Institute).
2. R. G. Cooper, "The NewProd System: The Industry Experience," *Journal of Product Innovation Management*, 9:113-27.

A STRATEGY FOR PRODUCT DESIGN EVOLUTION

John L. Schlafer

This chapter discusses how water treatment products found in homes and light commercial establishments can be designed and adapted to meet varied customer requirements and use patterns.

The water treatment products found in homes and light commercial establishments condition water and thereby improve its quality. The conditioning process improves aesthetic quality by removing hardness, which causes a film to form on plumbing fixtures or on laundry and dishes. The product may also improve the health effects of the water by removing unacceptable levels of lead or nitrates from the water. The conditioning appliance is usually placed at the point of entry of the water into the home or can be selectively placed at the end point of use, such as under a kitchen sink or in a bathroom.

The point-of-entry appliance conditions all water that enters the residence and is usually located in the basement or utility room. Water softeners are a popular appliance in this classification, as the owner usually does not have to attend to the appliance for weeks or months at a time. When the owner does, it involves only a simple replacement of a filter element or replenishment of a consumed chemical. The design requires the function and maintenance to be as simple and intuitive as possible. Product quality and reliability are also key requirements of the product design, because failure to perform is readily noticed by the user in most situations.

The point-of-use appliance has similar requirements, except that the customer frequently interacts with the product at a special faucet. If this product is being used to improve a health concern, which the user cannot perceive, special monitoring requirements are imposed on the user or must be included in the design to alert the user of a possible failure or service need. This design alert can consist of a water quality meter or chemical test kit for the owner to use.

Water conditioning involves chemical and physical processes, such as ion exchange, reverse osmosis, and oxidation, and customers cannot be expected to understand them. Consequently, water treatment products must have an install-and-forget customer objective as a design goal. All customer interactions must be eliminated or, if required, must be minimal and intuitively simple.

THE PRODUCT DESIGN DOMAIN IS EXTENSIVE AND INTENSIVE

The major portion of water treatment products contain ion exchange, resin, or filtration media in a pressure tank. The incoming water is channeled by

---

JOHN L. SCHLAFER, PhD, PE, is manager of product engineering at EcoWater Systems, Inc., Woodbury MN.

a control valve through the resin or media in which the actual treatment occurs. The control valve then channels the treated water into the plumbing system in which it is continuously available for consumption. Because filtration media require back-flushing to remove the accumulated substances, the filter media or resin must be reconditioned after a period of use. An ion exchange mediun and several special resins require back-flushing, plus regeneration by a chemical, to restore them to their original condition. Water softeners, for example, require back-flushing and a slow brining with a salt solution followed by a rinse to recondition the resin.

Regeneration of the media requires a control valve with some means by which to sequence and time the steps. Unfortunately, when the regeneration is in process, any water used in the home is bypassed and is not filtered. To minimize this adverse condition, a softener or filter is set to regenerate late at night, frequently at 2:00 AM. The chemicals used during this regeneration are salt for softeners or an oxidizer (e.g., air, permanganate, or chlorine) for filters. The sequencing of the control valve and the timing of the steps require a control system.

Historically, the control system consisted of a timer motor driving a series of gears or cams with switches or solenoids to time and sequence the four to six steps of the regeneration process. These designs have been used sucessfully by the industry with success and are still used today by many manufacturers.

Demographics of consumers have been changing and will continue to change. Quality, reliability, and maintainability are becoming more important as are concerns about cost. Ultimately, the consumer wants conditioned water without undesirable tasks or time spent after the appliance has been installed. Customers do not have the time or the desire to take care of an appliance; the appliance must serve the customer, not vice versa.

Regulation

Another prime influence on the industry is regulation by various government agencies. The use of chemicals to regenerate filter media has a net effect of adding to the natural environment. The regeneration process uses extra water to back-flush the media, which, in turn, is dumped to drain, and this is counter to the water conservation goals of any community. If the appliance is intended to resolve a health problem in the water, a monitor and a warning indicator are required by many regulators, and this requirement has had a major effect on design and installation. Regulations are escalating to the point at which meeting the requirements is now a major goal in product development.

Water quality and conditions vary widely throughout the US as well as in foreign countries. One well can produce water with markedly different characteristics from that from a neighboring well, depending on depth, geology, and pumping rate, and water treatment appliances must be adaptable to this wide range of conditions. The design of the valve and control system must have wide ranges in settings, timing, and frequencies and must also have many combinations of gears, cams, or switches in order to adapt to a wide variety of geographical applications. Because many parts or adjustments are possible for each product installation, trained and experienced field personnel are required.

> **Water treatment products must have an install-and-forget customer objective as a design goal.**

The individual characteristics of the user must also be considered. The users' habits and situations affect the performance of the appliance. Vacations, change of season, and added guests change the demand on the water conditioner, requiring designs and performance with added flexibility. Water use can double during hot summers, quadruple with guests, or drop abruptly to zero during vacations away from home, and these situations require wide flexibility in the design.

For purposes of this chapter, two terms are defined as follows:

- Adaptability is the degree to which a water treatment appliance can chemically treat the water, fit into the plumbing system, meet the local codes, and conform to state and local regulations.
- Flexibility is defined as the degree to which an appliance can flex with the users' changing needs.

The key element in defining these terms is focus. Adaptability focuses on the physical location in which product is installed, whereas flexibility focuses on the customer in a personal sense. Adaptability determines whether an appliance design can be installed, and flexibility determines whether that appliance, after installation, can continue to perform to customers' satisfaction. In terms of attributes, adaptability encompasses:

- Condition of water.
- Plumbing arrangement.
- State and local codes.
- Permits and laws.
- Ease of installation.

Flexibility encompasses:

- Customer's use habits.
- Vacations and guests.
- Ease of customer service.
- Literature and diagnostics.
- Ease of maintenance.

An analogy to the previous definitions is software used on a personal computer. Adaptability is the degree to which software can be adequately installed on a particular computer, whereas flexibility is the degree to which the software can meet a particular user's changing needs. These two terms are used in the following sections.

ALL TECHNOLOGIES MUST BE STUDIED AND ADAPTED

A water treatment appliance usually involves two basic technologies: chemical and mechanical. Until recently, the electrical content was minimal, and software was rarely used. The water treatment process is a chemical process, and the valve and control logic, until recently, was mechanical. The changing requirements presented previously are becoming very difficult and expensive to meet using only mechanical and chemical technologies.

Administratively, the product design had evolved almost by chance. If a problem was found in the field or if a competitor developed a new feature, EcoWater responded with something. If a supplier came forth with a cost or quality improvement, it was considered. There was no large-scale strategy to assess technologies or the customer in the broad perspective and drive the design to best advantage. In a sense, the design strategy was addressing special causes and not common causes for design improvement. Whenever a special or unique opportunity came along, EcoWater tried to take advantage of it. EcoWater did not seek out the broad, common trends of the engineering and industrial community and adapt those to the customer's benefit. One symptom of the special cause strategy is the feeling that the same problems are confronted over and over, one to three years apart. The design seems to waffle with the passage of time even though quality and cost improvement goals seem to be achieved year after year.

To break out of this pattern and evolve a strategy requires looking at the technical fundamentals and the customer needs as broadly as possible. Replacing mechanical control systems with electrical controls and sensors adds considerable adaptability to the product design. Software for control logic adds simple adaptability for specific water conditions. It adds flexibility when combined with a water flow sensor to monitor and then predict users' water patterns. Sensors and software, together, permit identification of reminders and faults as well as the ability to interpret situations and present routine maintenance reminders or failure diagnostic codes to the users. Bringing the electrical and software technologies into full use greatly enhances the adaptability and flexibility of a design, and it has potential for reducing the number of parts in the product and lowering costs.

Software capabilities in a consumer product create broad opportunities for adapting a product to wide applications and to changing customer needs. These applications are almost unlimited, but development of electronic controls and sensors to their fullest in a product line is necessary before the software potential can be used.

During a development project, EcoWater Systems' engineers, who possess strengths in each of the basic technologies, are brought in early on, so that scenarios can be identified. Time after time, software surfaces as the most adaptable and flexible technology when the wide diversity of applications and customers is studied for long-term strategy of product development. Within 10 years, software may well be the highest value-added technology in most consumer products. To bring this about, changes from mechanical to electronic systems are required. Thus careful,

If the appliance is intended to resolve a health problem in the water, a monitor and a warning indicator are required by many regulators.

knowledgeable orchestration of all the technologies through an evolutionary spiral of a product is required for success. This requires time, focused effort, responsible risk taking, communication, and continuous study of the technologies and the customer.

All functions of an organization must exemplify these characteristics, because the product development evolution also requires evolution of the organization to support it. If a product contains a new technology, all functions within the organization or company must adapt to and nourish that technology to effectively deliver its benefits to a customer. Turtles do not give birth to rabbits. They create in their own image. Thus, the organization within which a product design is evolving must also be evolving.

PRODUCT DESIGN STRATEGY—A CONTINUAL EVOLVING MEDIUM

The customer is the end point for which product design changes are intended. Change is the means to improve product and services for the customer's preference, which in turn, provides the energy to enhance the organization through market share or profits. A product development strategy must, therefore, have the customer needs and perceived preferences in mind. It must also contain flexibility and adaptability to anticipate and correct changes for the customer with passing time. The resulting spiral or helix is frequently used as a graphic way of illustrating the strategy.

The product development cycle is illustrated in Exhibit 1. A product design may go through many cycles of improvement during its lifetime. The amount of, or number of, changes in the product with each cycle is not as important as the company's ability to move through the evolutionary cycles rapidly to keep up with the customers and remain ahead of the competition. Sometimes a major redesign is necessary, whereas other times a small increment of change satisfies the immediate need for survival of a product. Regardless, a long-term, evolutionary strategy is vital for long-term survival of a product.

Exhibit 2 illustrates the time evolution of a product design; it implies the need to know the customers and their environment. Product design development consists of continual improvement cycles appropriately integrating new technologies (internal influences) to bring about and meet evolving product expectations and requirements (external influences). Product leadership measures how well an organization can meet and lead the external product influences

Exhibit 1
Product Design Development Cycle for Continual Improvement

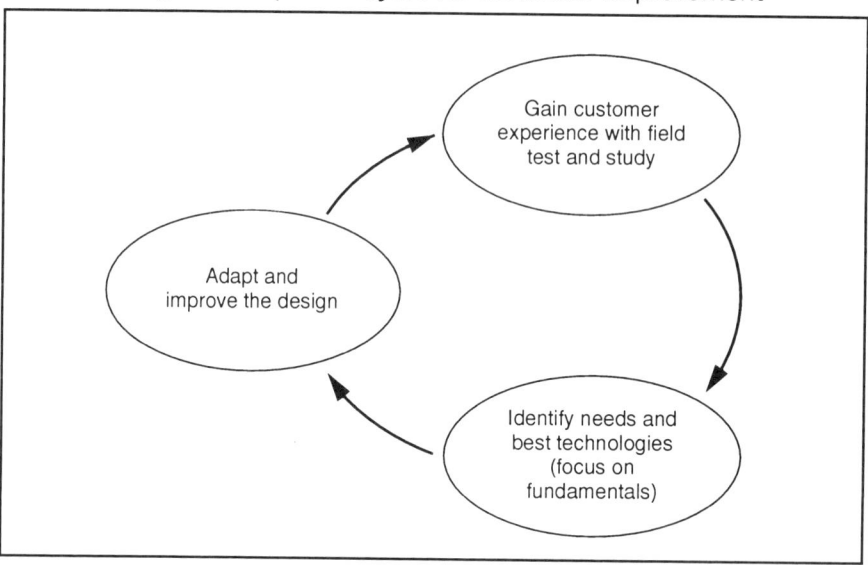

- Gain customer experience with field test and study
- Identify needs and best technologies (focus on fundamentals)
- Adapt and improve the design

Exhibit 2
Product Design Development and Product Leadership

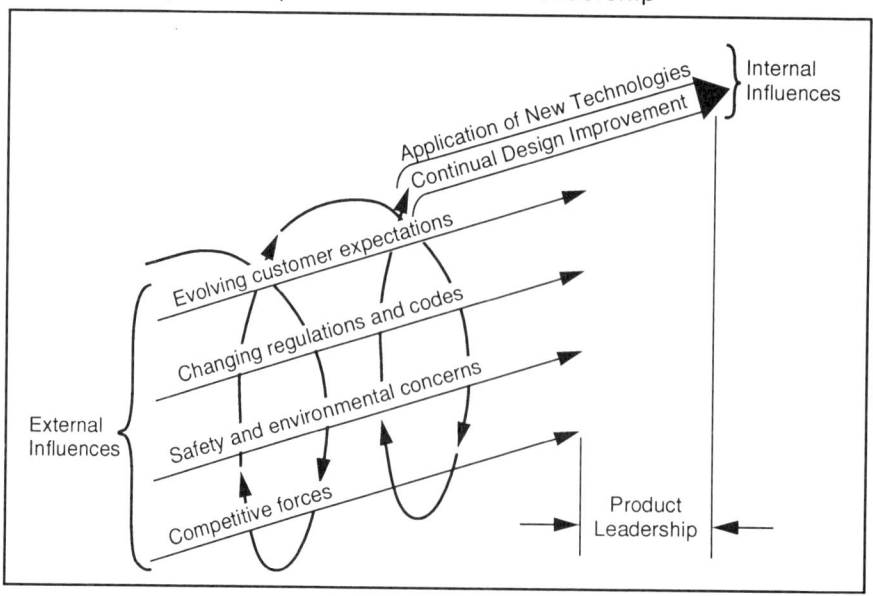

Internal Influences:
- Application of New Technologies
- Continual Design Improvement

External Influences:
- Evolving customer expectations
- Changing regulations and codes
- Safety and environmental concerns
- Competitive forces

Product Leadership

NEW PRODUCT DEVELOPMENT

with its designs. Deming points out the need for a broad perspective that includes integration of technologies or knowledge as well as the customers' needs.[1] Historically, no customer asked for the transistor, facsimile, or even an automobile. These products came from sufficiently developed technologies and foresight applied to a product. In addition, customer expectations had to be changed to be compatible with new products. Customers who shouted "Get a horse" had to be enlightened before the auto became a major product, and preferably, the continual improvement cycles shown in Exhibit 2 should lead and influence customer expectations. To evolve from designing buggies to designing automobiles required this leadership. Influence of customer expectations is vital for product leadership and important to survival in a competitive environment.

Regulations are escalating to the point at which meeting the requirements is now the major goal in product development.

During the last 10 years, the water treatment industry has experienced many new external influences threatening the survival of its products and services. Droughts, floods, population growth, increased knowledge concerning the health effects of water contaminants, improved measurement techniques, awareness of environmental effects, regulations, conservation initiatives, and numerous other factors are inducing rapid and wide-ranging demands on water treatment products. These demands have the potential for initiating dramatic change in product design.

EcoWater Systems has responded to these changes through internal efforts and acquisition. Ten years ago, products contained basically chemical and mechanical technologies, and the introduction and use of electrical and software technologies was slow. Customer acceptance and product performance seemed to meet the needs, but recently external influences previously cited and the experience gained from initial electrical and software use formed the basis for a rapid transformation of design. Six years ago, EcoWater Systems initiated major efforts to apply all appropriate technologies to its products. The goal was to have customer-preferred products that were superior in quality and reliability, cost competitive, and positioned for rapid evolution of customer expectations.

Major, rapid changes of products raised many concerns, including providing service for older designs, minimizing the number of part numbers and resulting costs, maintaining robustness of design to changes, and responding quickly to any unanticipated field problems. A foundation of electrical and software application to the products existed, and 10% to 20% of the models were electronically controlled. Customer acceptance was weak, primarily because the design offered little advantage over the older, more reliable mechanical designs.

Peter Senge points out the importance of "seeing the forest and the trees".[2] The application of new technology to replace old without broader perspectives on the part of the developers is rarely successful. Using electronics to replace mechanical gears and cams is insufficient, as the older, seasoned design will too frequently remain successful.

Electronics have a dramatic effect, however, if properly applied. They open the opportunity for using software that has the potential for great flexibility and adaptability. In addition, electronics can be very cost-effective and reliable. Successful application of these technologies to products requires both focus on detail and broad perspective.

EcoWater Systems focused on improving reliability and quality and broadened its customer interface with the electronics. This focus required development of a matrix display, simple input keys for the novice, and software with menu selection of options. The goal was to design the customer interface for simplicity, error tolerance, and rapid customer confidence building. The software even has the feature that allows confused customers to walk away, and after four minutes the control returns to the point at which the customers started. Customers know they can try again if they make a mistake.

The demands for the development of improved products led EcoWater Systems to focus on highly adaptive, flexible software and electronic technologies. Gears and cams that control the regeneration sequence and timing were replaced with electronics and operator-selectable choices dependent on customer needs. The electronics were minimized in size and protected from the environment and easily replaced standalone parts. The software includes diagnostics that inform the service person or customer what to do should an error be sensed. If the failure is catastrophic, the single, complete electronic board is easily replaced. The design goal was to assume that the service person or customer had little or no electronics knowledge. The programming label on the product or the owner's manual gives the novice the explicit steps to correct a problem. Tactile, audio, and visual feedback to the user enhance customer confidence.

It was felt that high reliability and simplicity of field service were necessary for customer acceptance. The customer had to be confident to take the risk of changing from tried-and-true mechanical controls to electronic and software technologies. The development people learned that new technologies carry added burdens associated with lack of confidence on the part of customers and service people. Customers do not want to be bothered with inadequate application of new technologies, regardless of the potential. People are intolerant of nonproductive second guessing when it consumes their time.

Making a major improvement in an organization's performance requires overcoming problems. EcoWater performed considerable training of the employees as needs were recognized. In addition, several engineers with specific capabilities were hired and the organization modified. Product development was changed, in general, from a person leading a task to teams leading programs. The team approach resulted in members learning from one another the new technologies being applied. A person who champions an idea does not have the strength, generally, to rapidly and effectively evolve a product design through an organization; that is done better by a multifunctional, multitechnical team that has been given responsibility and authority to lead a program. Thus, teams were used to develop new products or components and provide training of personnel.

Another problem to overcome was the special-cause approach of solving individual problems one at a time as they arose. The product design had to be looked at as the customers perceive the product or, preferably, the benefits they receive from the product. To change the special-cause culture at EcoWater, training on the customer relationship was done. This included off-site training and development of in-house trainers with the goal of raising customer awareness.

Vacations, change of season, and added guests change the demand on the water conditioner, requiring designs and performance with added flexibility.

Extensive field testing was started to further overcome the lack of knowledge of the customer and the product function. One program involved more than 200 field tests before full production was started to ensure that the product would be well received by the customer. This was time-consuming and costly, but the rewards were high. As full production started, EcoWater was confident that the new demand-regenerated softener would be successful and that the company had found the elements of a strategy to administer future product development.

To be positioned to meet present as well as future customer needs and expectations, EcoWater has accepted the product development strategy shown in Exhibit 2 for many of its products.

WHERE THE PRODUCT DESIGN IS TODAY

A review of an EcoWater Systems' demand-regenerated water softener is an example of the results to date concerning the previously mentioned strategy. The mechanical valve is positioned by a low-voltage electrical motor and position switch. The flow meter is a turbine in the water stream with magnets on its periphery. The rudimentary elements are the only mechanical parts in the control system besides the main water valve. Basically, the electronics must read the meter, energize the electric motor as necessary, and provide a clock-timing function. All other logic and control is in the software, stored in masked, nonvolatile memory.

This design approach pushes as much of the control and logic into the software as possible. It minimizes parts and makes future product modifications and improvements convenient and easy to implement. Software can be written with a personality, and as more is learned about the customer, the software personality can be revised to match that particular customer. The control system monitors the customers' daily water use, identifies the use patterns, and uses that information in a control algorithm to conserve on regeneration salt or back-flushing water. It automatically identifies a user who is on vacation or has house guests and responds accordingly. The software has diagnostics that provide error codes, which are clearly explained in the manual or on a product decal, so customers have the option to fix any problem before they call a service person.

The benefits of this approach are adaptability and flexibility. The new product is installed and set up to meet the demands of the water conditions with simple input of software, menu-driven responses. Then, as the owner uses water, the product monitors the quantity and timing of water used. From this information, the algorithm identifies patterns and irregularities and determines salt-regeneration and back-flushing needs that minimize wasted water and eliminate excess salt use. Customers do not need to be aware of any of the details, because the product responds to their needs without their being involved.

WHERE THE PRODUCT DEVELOPMENT IS HEADING

EcoWater Systems' product design is positioned for rapid and effective adaptation to future changes in customer needs. For example, future regulation on effluent disposal can be addressed by using various software-driven control schemes, depending on local requirements. Many of these anticipated schemes are programmable with software changes only. In addition, several water contaminants not previously treated can now be treated with adaptation through the software, by programming special control schemes. Projecting further into the future, by expanding on the number of motors and water meters, the design can be positioned for evaluation for use in multiple systems for special installations, such as commercial establishments or difficult-to-treat waters.

If a customer requests a special or unique model, a minor software change makes many variations available. This capability puts this type of product near the desirable goal of building a design-to-order product without the typical prohibitively high development costs or the long time needed to make and stock special parts. Where build-to-order is becoming commonplace in consumer products today, design-to-order is the next evolutionary step toward customer satisfaction.

There are more proprietary features that this design strategy makes available to rapidly increase customer satisfaction. The strategy of maximizing adaptability and flexibility in the design provides a strong, competitive advantage and opportunity for product leadership.

CONCLUSIONS AND RESULTS

The first demand-regenerated water softeners with this improved design were introduced into the market almost four years ago with very good results. Two years ago, additional models with greater flexibility in the software were introduced. In the last two years, the electronic-demand EcoWater Systems' models have increased from less than 20% of total production to more than

80%. This is a very rapid shift in design for a consumer product such as this, and it is only a matter of time before mechanical control systems are terminated and all products have new control systems.

The lessons learned during the past several years of product development have been numerous. But there are several conclusions worth reviewing:

> **Adaptability is the degree to which a water treatment appliance is able to meet the local codes and conform to state and local regulations.**

- New technologies and new products require a learning organization.
- Satisfactory delivery of a new product meeting customers' expectations and perceived needs requires knowing the customers and their perceptions of the product.
- Successful evolution of a product in an organization requires multifunctional, multitechnical teams to drive the development through and instill the learning required within that organization.
- The perspective needs to include looking at the forest and the trees with an astute balance of both.
- A design should include as much flexibility and adaptability as possible for the customer's benefit.

There have been many corrections, enhancements, and special features added to the early electronics and software packages. These changes, as anticipated, were implemented rapidly and easily into the design, and this same strategy will enable future product adaptations and flexibilities to meet rising customer needs.

Acknowledgment

Sir Isaac Newton is quoted "If I have been able to see farther than others, it was because I stood on the shoulders of giants." Similarly, I want to thank the people of EcoWater Systems who persevered through these programs. It is a privilege to step back from all their efforts and report on their successes.

Notes

1. W.E. Deming, *The New Economics for Industry, Government, Education* (Cambridge MA: MIT Press, 1993), p. 7, 134–137.
2. P.M. Senge, *The Fifth Discipline, The Art and Practice of the Learning Organization* (New York: Doubleday Currency, 1990), pp. 127–135.

REDUCING NEW PRODUCT DEVELOPMENT RISKS

Colin MacPhee

Managing technology to meet the conflicting needs of the marketing, sales, manufacturing, quality control, and engineering functions and reconciling those needs with the essential demands of senior management is an art, not a science.

The term *new product* has different meanings for different functions throughout a company. To the sales department, a new product is one that overcomes customers' problems with existing products. To the marketing department, a new product is one that differs demonstrably from preceding ones and therefore is easy to promote, though it need not be radically new in concept.

In contrast, a new product provides the typical engineer with an opportunity to explore a new design concept or to redesign an existing product to incorporate new technology and improve performance. The manufacturing department usually views a new product as a redesigned product with reduced component costs and assembly labor. To senior corporate management, a new product is an opportunity to improve the profitability of the company; if this profitability is not forthcoming, the new product will probably not be around for long.

Managing technology to satisfy all these conflicting views is an art, not a science. This case study details one company's attempt to formulate a more disciplined approach to new product development than it had previously used.

The need for the particular new product had been established on the basis of several factors, one of which was that the company knew that its competitors had introduced or were planning to introduce similar products. Consequently, to improve its competitive position, the company felt the need to be perceived as state of the art. In addition, some problems were associated with the existing product—specifically, high production costs and difficulties in field service—that could be overcome by a new product.

RECOGNIZING THE NEED FOR CHANGE: THE EVOLUTION OF A PRODUCT

The product discussed in this case study is the overcurrent relay component of a molded-case circuit breaker typically used in industrial and commercial electric power distribution systems. The power system is designed to meet users' needs. Still, because it is connected to a public utility that is capable of supplying excess power, the system must be protected from currents that are greater than the capacity for which it is designed (i.e., it must be protected from overcurrent). Damage can be caused by persistent small overloads or very

---

COLIN MACPHEE is the principal of the technical consulting firm Albion Enterprises. Previously he spent 22 years in R&D and engineering management for Federal Pioneer Ltd., Toronto, Canada.

high instantaneous overcurrents; in such cases, the results can be catastrophic, leading to system breakdowns (e.g., insulation failure).

The relay component monitors the current flow in the circuit and opens the circuit breaker if a fault occurs, thus isolating the section experiencing a fault. Because power distribution systems break down power into progressively smaller blocks as it flows from the utility to the user, there are often several overcurrent relay components in the system. Therefore, the more precise the relay components are and the more accurately they can be set, the better the system can be controlled.

Correcting the Problems of the Existing Product

Molded-case circuit breakers were invented during the 1920s to provide a low-cost alternative to the air-break circuit breaker. The overcurrent relay component used exclusively since that time makes use of thermal effects for overcurrent control and electromagnetic effects for catastrophe control. Because the action is mechanical, the relay component combines thermal, mechanical, and electromechanical elements. All of the components in the relay component are subject to various types of tolerance errors, so both the precision and the setting accuracy of the relay component became problems as electric power use escalated rapidly during the past 20 years. No satisfactory, low-cost solution had been found to compensate for the errors, and the problem was compounded by the high currents that had to be used during product testing and adjustment.

As is the case with many product innovations, the company in question anticipated that there might be strong resistance to new products or components because the industrial users had already developed methods for coping with the deficiencies of existing products and might be reluctant to exchange the problems they understood for a new set of potential problems. To overcome such resistance, sales personnel had to assure customers that their complaints concerning the existing product had been considered in the design of the new product, and the marketing department dealt realistically with those complaints in its product introduction program.

Introducing the New, Improved Product

Improvements that the company sought for the new product included ease of manufacture, ease of installation, and field testability as well as use of a minimum number of close-tolerance components. It was also vital to maintain the product's performance requirements. In addition, it was believed that the new relay component should be calibrated before installation in the breaker, should be easily adjustable for different breaker ratings, and should not require high-power testing during production. Finally, it was critical to produce a new product that was cost-competitive with existing relay components.

This last point is often a subject of acrimonious debate; some parties assume that a factory-cost comparison is sufficient. In fact, the lifetime product cost is more meaningful because it considers such critical factors as:

- The cost of overdesigning the breaker and the power system to compensate for the inaccuracies of the relay component.
- Warranty costs caused by lack of precision in the relay component and the difficulty of performing overcurrent tests.
- The need to remove the old products from service for routine testing.

To overcome market resistance, sales personnel assured customers that their grievances with the existing product were considered in the design of the new product.

SPECIFYING THE NEW PRODUCT

Preparing a commercial specification that accurately defines the desired characteristics of a new product is the single most important step in the development process. As such, it is central to technology management, and it is therefore essential to take whatever time is necessary to prepare a fully detailed specification before the development program is started. This conviction is fully supported by studies on the causes of failure in new product introductions.

Writing a new product specification is too important a task to be assigned to any single function (e.g., marketing or engineering). It must be a truly cooperative effort involving all interested parties—from the customer (represented by the sales department) to the senior managers of the company. Unfortunately, this is precisely the area in which technology management is weakest in most companies.

The commercial specification for a new product should consider only what is required, never how it is to be achieved. If consideration is paid to how to achieve the specification, there is a significant risk that the design will be constrained by existing technology and that the designers will be unable to give their imaginations free rein.

The specification process begins by laying down general guidelines based on the desired characteristics of the new product. The details are developed through dialogues between the product developer and the potential buyers until the document is fully detailed. Only then should the development process be allowed to start. The specification process is time-consuming and often regarded by the participants as a tedious and unnecessary impediment to getting on with the job. Without it, however, the risk of failure is extremely high. If the product is allowed to evolve through continuous fine-tuning in response to the perceived needs of the customer after the specification stage, development costs will usually be higher than necessary.

Incorporating New Technology

The commercial specification for the new relay component dictated that it should use solid-state devices. Therefore, it was determined that the relay component should be composed of:

- A current transformer for each power conductor to monitor current and provide power to operate the relay component.
- A solid-state circuit to carry out the logic and control functions.
- A tripping device to unlatch the breaker when needed.
- A frame to hold these parts in proper relationship for easy assembly, calibration, and insertion into the circuit breaker.

In addition, the low power required by the relay component suggested that a battery-powered, truly portable test set could be designed to permit in-circuit field testing of the relay component and its associated circuit breaker. The testing equipment was added to the list of requirements.

CONTROLLING THE DEVELOPMENT PROCESS

It is characteristic of a properly executed new product development process that the level of uncertainty over the result is at its greatest when the project commences and that this level diminishes as the project progresses. It is also true that the costs increase rapidly as the project nears completion; accordingly, the risk increases at the same rate.

The development process must be designed to control risk at all stages. Therefore, it is vital to begin with a fully detailed commercial specification and to manage the development process so that the specification requirements are met at each stage of the process before authorization is given for the next stage.

The Phases of Product Development

To ensure that the commercial specification requirements were met in the case at hand, the development process was broken down into discrete phases, as described in the following sections.

Phase 1: Reducing Uncertainty. During this phase, the first steps were taken to reduce uncertainty. The development team studied the commercial specification, decided on a preliminary approach to solving the technical problems, and scoured technical journals and other sources for information about competitive products and patents. After the team conducted a feasibility study, the project was ready for the first design audit.

In general, the purpose of the design audit is to review the work of the development team, examine its conclusions, and resolve any conflicts, always keeping in mind the requirements of the commercial specification. Only when each member of the audit team is satisfied is the next phase authorized.

The design audit team includes the engineering area managers (usually designated as chairpersons) and representatives from field service, manufacturing, manufacturing engineering, marketing, purchasing, quality control, and test equipment engineering. Each departmental representative is responsible for ensuring that its concerns are addressed and resolved at each design audit.

At the first design audit of this product, the project engineer presented the development team's report on feasibility, drawing attention to possible conflicts with patents or competitive products and stating the team's opinion on the probability of meeting all the commercial specification requirements. Problems that arose were discussed by the audit team and resolved. In some cases, resolving a problem required further work during phase 1, so authorization to begin phase 2 was withheld and the design audit was rescheduled accordingly. Thus, the audits were designated as first-, second-, third-, and fourth-phase audits, because each phase included more than one.

> **The prototype specifications were reviewed with particular attention to test requirements, which significantly affect manufacturing and capital costs.**

Phase 2: Constructing and Testing the Working Model. Phase 2 began with the construction of a working model of the product, followed by detailed testing to ensure that the performance requirements were met by the model. The second-phase design audit considered the results of these tests and paid particular attention to the proposed method of construction to determine the product's ease of assembly and ease of service. With the consent of the appropriate members of the audit team, it was agreed that some minor changes would be made in the construction to simplify assembly but that these changes would be incorporated in the prototype design scheduled for phase 3 in order not to delay the project by repeating part of phase 2.

Phase 3: Specifying the Prototype and Choosing Components. Phase 3 opened with the specification of the prototype and the task of choosing components for reliability. Preliminary bills of materials and engineering drawings were prepared and detailed cost estimates were made. Test requirements were established, test points were incorporated into the design, and test specifications were prepared. During the entire process, the various departments represented on the audit team were kept fully informed of developments affecting them, and their input was sought to minimize conflicts and delays during the audits.

During the third-phase design audit, the prototype specifications were reviewed with particular attention to test requirements because they significantly affect manufacturing cost and capital costs for test equipment. Before authorizing phase 4, the audit team scrutinized the development team's ability to meet commercial specification requirements up to this point—an indication of the probability of success in future phases. It was crucial to proceed only if this indication was favorable because the development costs—and hence the risk—would escalate very rapidly in the following phase with the purchase of tooling and sizable quantities of components for prototype sample manufacture, along with the development of production test equipment.

Phase 4: Releasing the Preliminary Design.

During phase 4, the preliminary design was released to the manufacturing, purchasing, test equipment engineering, and quality control departments to permit those departments to gear up for production. The tasks these departments had to perform at this stage included:

- Designing jigs and fixtures.
- Modifying the circuit-breaker tooling.
- Developing new labeling for the circuit breakers.
- Determining labor and space requirements.
- Sourcing components.
- Discussing quality control requirements with subcontractors.
- Designing in-house test equipment.

Several prototype samples were made and tested exhaustively for compliance with the commercial specification as well as additional manufacturing and quality control specifications. Subsequently, certification testing was carried out as the samples were distributed to other departments represented on the audit team to determine how well their needs were met. The results of certification testing were reviewed during the fourth-phase design audit, with each department reporting on any difficulties it encountered in preparation for production that might affect the design. Because these departments were extensively involved in the development process, no problems requiring design modifications were reported, and phase 5 was authorized. The release date for the product was set for February 1988 to coincide with the Electrex 1988 conference, the biennial electrical engineering showcase held in England.

Although the risk to the company was very high at this point, it was offset substantially by the certainty that the product would perform as required, given satisfactory manufacturing procedures. The possibility remained that some unforeseen factor would emerge to affect the performance of the product in the field, but this risk was adjudged minimal because of the stringent procedures followed in the development process.

Phase 5: Completing the Engineering Drawings and Bills of Materials.

In phase 5, the engineering drawings were completed and the final bills of materials were drawn up. This information package was released to the marketing, manufacturing, purchasing, quality control, and test equipment engineering departments to allow them to complete their preparations for the pilot run and for full production. The stakes increased rapidly with the purchase of tools, equipment, and subcomponents necessary for the pilot run.

The development team had not yet completed its job at this point, because it was charged with overseeing the pilot run and evaluating the relay components produced. It was also responsible for determining the failure rate of the subcomponents used and the efficiency of the production testing method both in detecting failed subcomponents and in minimizing testing labor. After these tasks had been completed, the team documented the technology used in developing the product and prepared generalized procedures for troubleshooting failed products and for transferring technology to production.

> **Improving technology management became more important as the solid-state products became more complex, causing more difficulties in transferring technology from engineering to manufacturing.**

Phase 6: Performing the Pilot Run. The purpose of the pilot run was twofold: to evaluate the product as made in a manufacturing environment (which differs markedly from that of the development laboratory producing prototype samples) and to evaluate the production process to permit fine-tuning before actually commencing production. Phase 6 was designed to control the pilot run and to ensure that these evaluations were carried out.

The manufacturing process proved satisfactory, and the product met the full requirements of the commercial specification. However, problems arose with the supplier of the current transformers, whose interpretation of the component specification differed from what was intended. It was therefore necessary to design and manufacture a simple test unit to inspect incoming components until more satisfactory supply arrangements could be made.

Systematic analysis of failed subcomponents in products fabricated during the pilot run suggested that these subcomponents should be purchased only from suppliers prepared to guarantee their quality. The engineering department was directed to modify the bills of materials to list only approved suppliers in order to minimize subcomponent failure. After the pilot run evaluations had been completed and the modifications described had been made, a final design audit was held and the product was released to the manufacturing department.

ANALYZING THE RESULTS

The new product development case described here represents a concerted effort to improve technology management, with the objectives of reducing product development failures to a minimum, controlling risk at all times, and facilitating the transfer of technology from the development group to manufacturing. The need to improve technology management became more important as the solid-state products developed by the company became more complex. Greater product complexity initially caused considerable difficulties in transferring technology from the engineering function to the manufacturing function, which in turn caused significant delays in scheduled product release dates. These delays led to a loss of confidence on the part of the sales force and disappointing results from new product introductions.

The company's senior management faced the challenge of trying to obtain the active participation of all the major departments in its efforts to improve the management of technology. In the past, other departments had regarded new product development as the engineering department's job, and they had their own jobs to do that kept them fully occupied. As a result of these ingrained attitudes, not all members of the audit team attended meetings or carried out the necessary detailed reviews at every stage of the process. Consequently, there was some overlap between the phases. Nevertheless, the process was successful in controlling risks by not permitting the product to proceed to the next phase without the appropriate audit.

The technology management improvement effort was successful in most respects: a successful product was introduced, the risks were controlled during the development process, and the product developers were able to transfer to the manufacturing department detailed information on the product technology as well as the results and the pitfalls to be expected during the production process. Success could be attributed to the unremitting efforts of the engineering department to make the process work and the recognition by other departments represented on the audit team that this process (or a similar one) was the way of the future. All participants came to realize that the traditional method of developing products was unworkable in the current environment of short product lifetimes and unceasing pressures to reduce manufacturing costs. Eventually, all the appointed departmental representatives participated actively in the team.

As anticipated, the most difficult part of the entire process was to establish the detailed product specification quickly enough to meet the intended launch date. At the Electrex 1988 conference, the centerpiece of the company's display was to be a new line of molded-case circuit breakers incorporating the solid-state overcurrent relay component. Despite the pressure to succeed that this expectation exerted, difficulties were experienced in completing the specification before the project began. Immediate everyday operating problems sidetracked some project team members. As a result, several minor details of the specification were not made available until they had become absolutely essential. No solution was found other than persistent efforts by the engineering department to pressure the delinquent departmental representatives to provide the necessary input.

The product was first shown to the general electrical engineering community at its planned release at Electrex 1988, where it was judged the best new product at the show. And as planned, the relay component was subsequently incorporated into higher-rated circuit breakers; although an unforeseen technical problem was detected in this application, it was subsequently eliminated with a minor change in the product.

IMPLEMENTING THE GAME PLAN

During the new product development program described, the senior management of the company was changed and a major effort was begun to reduce overhead costs by increasing the efficiency of all departments, with particular reference to those associated directly with product development. This effort

> **The most difficult part of the process was to establish the detailed product specification quickly enough to meet the intended launch date.**

has taken precedence over all others; as a result, the benefits arising from the attempt to improve technology management described in this chapter have been deferred.

This case history has bearing on any product development effort. Any small or medium-sized company (i.e., up to $250 million in annual sales) with an extensive product line in an established industry can use the process outlined as a formal mechanism for developing and screening new ideas. Without such formalized procedures, any new idea—regardless of its merit—will probably be developed only if it has some appeal to management. In such precarious circumstances, the only way to improve the chances of launching a successful new product is to manage the risks actively, using the approach described here or a comparable methodology. Otherwise, the risk of failure is great indeed.

THE NEW PRODUCT IDEA PARADOX

Timothy H. Bohrer

Some people who consistently generate new product ideas often claim that they practice their craft only when given complete freedom and control of the process. The paradox, however, is that the freedom and ability to create is most highly developed in a structured environment.

Research, development, and new product managers have been intrigued for years by the differences in creativity that are displayed in new products. Some so-called new products are simply minor changes to existing products, whereas others represent real breakthroughs in satisfying user needs. Companies and individuals have the opportunity to improve their new-product success rate and the impact of those new products by improving the management of creativity.

THE STRUCTURE CREATIVITY PARADOX

A concept that has proven effective in generating new product ideas is structuring creativity. Structured creativity often is seen as a paradox, or threat, by many creative people. Some people who consistently generate new product ideas often claim they practice their craft only when given complete control of the process. In fact, they will claim that they need absolute freedom from any interference (or input) from management or customers. The paradox is that, in reality, the freedom and ability to create is most highly developed in a structured environment.

WHY NEW PRODUCTS SUCCEED

To effectively understand why a structured environment encourages successful innovation, it is useful to review why products succeed. Stated simply, new products succeed when they meet the needs, both functional and emotional, of customers more completely than other available choices. Researchers often deal more easily with the functional needs of the customer; these are hard, measurable performance attributes such as size, strength, or availability. But researchers must also understand and consider the customer's emotional needs and how those needs can best be met.

Typical emotional needs include security, the need to belong, or a desire for status. On a purely objective or functional basis, there is limited justification for spending two to five times the price of an average full-sized automobile to acquire a status symbol. Nonetheless, status and other emotional needs have

---

TIMOTHY H. BOHRER is vice-president of technology at the James River Corp. Packaging Business in Cincinnati OH.

created a strong market for these vehicles. In addition to top-speed capabilities that cannot legally be used, this high-end market looks for the newest technology and innovations. Reliability is an issue for some people, whereas others are willing to risk the problems of owning a new model in order to be among the first to drive it. This balance of emotional and functional needs must be understood before researchers can target creative forces effectively.

As there is a difference between functional and emotional needs, there is a difference between features and benefits. The features of a product are what it is and does. The benefits, on the other hand, are what the user experiences. Features are generally functional in nature, whereas benefits can be functional, emotional, or both. For example, in snack food packaging, a feature is the protective gas barrier provided by flexible lamination; the functional benefit to the user is a fresh product. The resulting emotional benefit is security, the knowledge that when the package is opened a snack of predictable quality will be found.

Unsurprisingly, features which do not cause the user to experience or perceive a benefit are not valued by that user, and features that are not valued by users are seen as superfluous. This can lead customers to conclude that goods or services are overpriced.

NEW PRODUCT PROCESS

There are many valid ways to break down the new product process, depending on priorities. The following is a seven-step model that emphasizes structured creativity:

1. Understanding and discovering customer needs.
2. Evaluating current offerings.
3. Seeing performance gaps as problems to be solved.
4. Creating alternatives.
5. Choosing promising ideas.
6. Developing the product.
7. Making and delivering the product.

In the following sections, the first five steps will be examined in detail.

Understanding and Discovering Needs

Effective creativity is the result of discovering and understanding customer needs. For any enterprise, products that do not correspond to real needs of real users are a waste of resources and can put the organization at a competitive disadvantage. Understanding customers and discovering their true needs is the key to successful new product development.

The User's Perspective. The first and the most important effort in product development should be to assume the user's perspective. Having new-product people talk directly to users is a powerful way to develop an in-depth understanding of how they choose and use products. Identifying and working with those users who tend to buy first is extremely valuable, because these lead users offer the most leverage in the marketplace, and they are prepared to take new-product risks to gain the advantage of being first. Lead users also tend to be the most willing to work jointly with suppliers to create new products and are helpful in introducing them successfully.

> A concept that has proven effective in generating new product ideas is structured creativity.

Industrial lead users are often most accessible, and they are willing to spend time evaluating new products and concepts. These industrial customers are eager to participate as beta site tests of new products and often dedicate resources that can increase the probability of success. When Federal Express Co. adopted hand-held computers for use by their delivery agents, achieving a breakthrough in package-tracing capability, it came as a result of the company's willingness to experiment with service-enhancing technology.

Finding lead users for new consumer products is more difficult, but they can often be reached through distribution channels. Specialty mail order catalogs provide an effective means to reach consumers whose prior purchase records indicate a willingness to try new products in a particular category.

When attempting to gain a user's perspective, it is important to dig beneath initial statements, because few people are good at describing their short- and long-term needs clearly. Digging deeper through active questioning can uncover unspoken or even unthought needs. Too often, the supplier is expected only to provide answers rather than ask questions. Technical people must suspend their expert persona from time to time to become investigators who pursue a deeper understanding through thoughtful questioning and careful listening.

Wishing as a Tool. A third concept is useful in helping customers express their deepest needs: the concept of wishing. Too often, people think and speak narrowly about needs for products. They are reluctant to express needs that might seem absurd or unmeetable. In contrast, it is much safer to say "I wish." Wishing stretches creativity. Using wishing as a tool can encourage the customer to express needs that could lead to ultimate satisfaction. In a creativity session to discuss tamper-evident packaging concepts, one participant wished packages would bleed when tampering had occurred. This wish led to a range of concepts that used color-change technologies to display evidence of tampering.

As mentioned earlier, it is important for technical people to consider both the emotional and functional aspects of needs in order to create successful new products. Consideration of emotional needs cannot be left soley to the marketing department. These emotional needs must be addressed in the early stages of design. Creating a product-needs form that encourages potential users to address features and benefits in both functional and emotional terms can be very useful.

How product developers perceive customer needs can create a clear framework for creativity in the process of developing new product ideas. Successful products must meet real needs; helping the user discover new needs leads to innovative new products.

Evaluating Current Offerings

The next step in the new product process is to evaluate current offerings. First, do real choices exist? If not, why not? In some instances, users essentially are limited to a single choice. As an example, prior to deregulation of long-distance telephone service, no real choices existed for US residents, and the range of product and service offerings was limited. When new competitors entered the market, users suddenly could choose from a range of new products. Researchers must understand whether real choice exists and what hurdles may be blocking new alternatives.

How Are Users Choosing? Researchers should next evaluate the full range of options, considering cost and performance, on the basis of users' definitions. Next, it is necessary to determine how people are choosing among the options and ask what can be inferred about the product and their needs from their choices. Much has been written about consumer reaction to The Coca Cola Company's introduction of a new version of its cola, and most who have studied the New Coke phenomenon have concluded that a better understanding of consumers could have mitigated, if not eliminated, the debacle.

The final activity in evaluating options is to assess how close to ultimate satisfaction the available options come and how a new product could come closer to the ideal by eliminating performance gaps.

Seeing Performance Gaps as Problems to Be Solved

Having evaluated current offerings, the researcher is in a position to treat performance gaps as problems to be solved. Researchers should attempt to determine whether users view these gaps as critical and how much interest they have in solutions. The real opportunity for a new product can be understood by determining how big the gaps are between existing offerings and ultimate satisfaction and how important these gaps are.

The most important result of closing a need gap with a new product or with improvements to existing products is the additional value that is delivered to the user or customer. From a well-developed understanding of user needs, researchers can estimate how the user would value the benefits from the new product. Attempts should also be made to predict how competitors will react to a new product and what can be done to mitigate any negative effects resulting from their reaction. Finally, a company must gauge the costs of introduction—

including development, capital, marketing, and operations—against the rewards—in volume, margin, and market share—of achieving different levels of performance improvement.

Effective creativity is the result of discovering and understanding customer needs.

At this point, it is also prudent to begin to assess how a specific opportunity compares with others a company may be considering. By ranking opportunities, companies can determine how to use their resources most effectively.

Creating Alternatives

Three levels for new products can be considered:

- Modest improvements
- Major enhancements.
- Totally new ways to deliver features and benefits.

Product developers should understand clearly how much latitude they have in this area. A customer, user, or market that will accept only small changes in products will not be well served by an effort to create totally new systems. Conversely, users who are demanding breakthroughs in product function and performance are not likely to be satisfied by creative efforts aimed at only modest improvements. Clarifying the goal is an important prerequisite to focusing creativity.

Applying Creativity Effectively. Creativity can be best applied with careful and conscious preparation. The creative process is far too important to leave to random occurrence and must be deliberately managed using proven techniques.

The key to creating marketable product alternatives is to provide strong leadership to help innovators channel their creative energy. Assuming the right people are in place, they must be allowed to work in an environment that encourages creativity. At the same time they are granted creative freedom, the environment must be structured in a way that encourages them to keep sight of the goal. Ideas must be considered in the context of real goals and objectives.

Unfortunately, creative skills often are not used effectively. Few of the many new products introduced each year represent real breakthroughs, and many fail to provide ultimate satisfaction.

Self-Censoring. One significant barrier to the creative process can be described as the self-censor. Most of us learn to suppress subconscious thoughts that may not be acceptable to ourselves or others. Unfortunately, many creative ideas are lost this way. Young children say what they think and mean what they say. Their bluntness is refreshsing and exasperating, their creativity unbounded. Adults and society teach them as they grow older that not all statements or actions are acceptable; they learn to censor their thoughts to prevent certain ones from surfacing as socially unacceptable words or actions. Although this learning may be essential for surviving and succeeding in society, it also can result in real and significant barriers to exercising creativity.

Making connections is the key to creativity. Creativity is exercised by making connection between seemingly unrelated ideas, objects, or concepts. To the extent that a person's subconscious continues to create these connections but the conscious mind refuses to consider them, that person may lead a safe but relatively noncreative existence. To improve the exercise of creativity, it is necessary to help people learn to circumnavigate their self-censor, particularly during the early stages of the new idea process. Their capacity to live with ambiguous and possibly nebulous ideas in the early stages of development must be encouraged.

Unfortunately, what individuals and society tend to do is systematically undermine the honing and development of creative skills. All too often, people measure and focus on wrongness, not rightness. Ask children how they did on a spelling test, and they will probably reply, "I got two wrong," instead of "I got 28 right." Similarly, most people recognize the golfers who, on the day they shoot their best round ever, focus on the bad shots rather than the many good ones. Self-talk is typically concerned with wrongness, not rightness; people catigate themselves for errors but rarely congratulate themselves for things done well. An outcome of measuring wrongness instead of rightness is a perceived need to be 100% right the first time. Speculative or risk ideas and thinking out loud are rarely encouraged in the board room or on the shop floor.

The bottom line is that for most people creativity is just not worth the risk, and the result is a suppression of potentially great ideas. In the early stages of development, it is difficult to know how good an idea is. It must be explored, shaped, and improved. If self- and group-censorship are strong, and new ideas that are not completely right are ridiculed, potentially great connections are suppressed. Creativity killers include comments like, "We did that before, and boy, was it a disaster," or "That's interesting but not very relevant, is it?" Facial expressions and body language can convey enough disapproval to squelch new ideas. The result of these behaviors is safe ideas that do not yield big rewards. A cartoon makes the point: it shows a man at the end of a boardroom table looking sternly at the executives around him, saying, "What we need is a brand new idea that has been thoroughly tested." Leaders and managers need to examine their behavior to determine whether they are sending that message and inhibiting creativity.

How Creative People Work. The contrast between the paralysis of rigorous self-censoring and the flexibility of creative people is dramatic. Creative people start by believing that they are creative. This is critical, given the extent to which self-image determines capabilities. Second, these people are able to conrol their self-censors. They survive in society by using tact and discretion when required, but when it is time to be creative they are able to let down this barrier to creativity.

They are willing to be partially right at the early stages of the creative process and then find ways to improve on initial concepts. Creative people play with ideas, build on the ideas of others, and are pleased when others build on theirs. They are willing to explore the impossible, the unreasonable, and the outrageous, because they understand that ranging far afield often yields the most creative solutions. They may roam widely, but they do not lose sight of the true objectives.

> **Creating a product-needs form that encourages potential users to address features and benefits in both functional and emotional terms can be very useful.**

Most organizations employ, or have tried to employ, brainstorming as a technique for generating ideas. Brainstorming, however, is an unstructured process that frequently lacks discipline. Resulting frustrations include sessions being dominated by one person, loss of creative momentum, and no mechanisms for selecting and improving promising ideas. There are structured approaches to creativity that can be much more productive.

Approaches to Structuring Creativity. Two specific approaches to structuring creativity warrant detailed exploration because they are relatively easy to adopt and are valuable additions to an organization's arsenal of creativity tools.

The first approach is mind mapping, a concept based on the understanding that the mind operates in a nonlinear fashion. Therefore, linear processes for organizing thoughts, like typical outlining techniques, are seen as counterproductive to creative thinking. The object of mind mapping is to collect thoughts on paper in a manner that is similar to the way the brain works. This technique allows people to think in a free form by making it easy to add and expand on ideas. Thoughts can be collected as they occur, rather than forced into the rigid, sequential ordering required by an outline format. Mind mapping also works well with groups.

A second approach to structuring creativity is practiced and taught by an organization named Synectics, Inc., which grew out of new product creativity endeavors at Arthur D. Little. The people involved in this venture began to study differences in the creative performance of groups and came to some important conclusions. From these conclusions has been derived a structured approach to creating and refining new ideas and products.

The Synectics approach requires members of a group to first diverge through creative thinking patterns and then to converge in the creation of useful and valuable ideas for development. Four specifically defined roles are a key part of the Synectics process:

- The client defines the problem and makes decisions regarding the direction the creative process takes.
- Experts as well as resources bring knowledge relevant to the subject area and help test during the converging process.
- The facilitator manages the creative environment, steers interactions in ways that minimize self-censoring and censoring of others, and encourages new ways of looking at problems.

People can learn skills used in the Synectics process, but participants need to be selected carefully. Clients have one of the hardest roles because they must be able to describe the problem in a clear but somewhat open-ended way, participate in creative activity, and make decisions about the direction a session should take and which ideas should be developed further. In addition, the client must demonstrate that diversity is valued and encourage far-ranging activity, while keeping track of ideas and deciding which deserve further attention.

Experts who are too expert can stifle the process, particularly by categorizing ideas as infeasible on the basis of past experience. If there are too many experts in a group, they may dominate other members or become trapped by old paradigms. The difficulty in selecting experts may be one of finding people who have a high level of knowledge on a subject but who are open enough to value the ideas of others. Although they are valuable as individual contributors, overzealous experts have limited potential for working on joint projects.

Resources can add variety and contribute new ways of looking at problems, as long as they have the self-confidence to state their ideas. Good resources often can make unusual connections that become the basis for highly innovative solutions.

The facilitator's role is to evoke the best creative efforts from the group. This requires helping people seek greatness in their ideas, capturing all ideas, ensuring that everyone participates, quickly (but diplomatically) stopping any behavior that could harm the creative environment, making sure the client is satisfied with the progress, and rejuvenating the group when its creative energy ebbs. Effective facilitation is hard work often requiring specialized training and practice.

A typical group will lose energy, and creativity will decline after spending a certain amount of time on a problem. The group may also become stuck in a creative rut, repeatedly coming up with similar ideas. Facilitators are trained to use various techniques, including excursions and triggers, to revitalize the group and stimulate new thinking.

An excursion is the mental equivalent of taking a break by walking around the block or taking a shower. Participants are asked to think about seemingly irrelevant subjects and remember some of their thoughts. They are then asked to force-fit some aspect of those thoughts as potential solutions to the problem.

For example, a facilitator might have each person in the group name his or her favorite movie character and describe that character in a few words. After the group hears from all the members, they are asked to describe distinctive ways these movie characters might tackle the problem at hand. A group working on developing tamper-evident packaging was asked to name its favorite fictional detective or spy. The qualities the group attributed to these characters were then used to trigger ideas for methods to detect package tampering.

> **A significant barrier to the creative process is the self-censor.**

Triggers involve combining divergent thoughts to create new connections, again leading to different ways of seeing problems. An example would be to think about crossbreeds of plants, animals, and other things and use them as the basis for a short story. The story can then be used to trigger new ideas about the problem.

Many companies have used the Synectics process internally to generate new product ideas, and it has also been used successfully as a collaborative effort between suppliers and their customers. MindLink software, based on Synectics approaches, offers a basis for personal creativity sessions. Frequent use of this structured approach to creativity can help those involved learn more effective ways to manage their personal creativity and become more proficient in using the skills on a day-to-day basis.

To review the Synectics process, a skilled facilitator and a decisive client lead the group through a process that begins with a discussion of many disparate ideas. In the convergence process, the client chooses promising and intriguing ideas for refinement and development, after which the group works on ways to overcome barriers to implementation. This is where structured approaches are more powerful than looser brainstorming techniques, because structured approaches drive toward a limited number of concrete options for which strengths and weaknesses have been identified and approaches to overcoming hurdles proposed.

Choosing Promising Ideas

As the client chooses the most promising ideas, he or she must guard against a natural tendency to abandon the best ideas in favor of known, safe approaches. It takes energy and commitment to turn initial ideas into workable new products, but the rewards of breakthrough solutions are worth the investment. The opportunity in choosing promising ideas, and the challenge, is to balance intuition with hard data. This requires an ability to combine instinct with science.

Testing Ideas with Users. An invaluable tool for new product developers is testing ideas with users. As soon as it is possible to describe the idea and its features and potential benefits in an understandable fashion, it is time to get some feedback from the user. In the case of industrial products and customers, face-to-face conversations about product ideas are probably most fruitful. Customers need to be asked what they like about the ideas and in which areas they would like to see improvements, rather than being asked for a standard product critique outlining what is wrong with the ideas. The objective is also to get the customer thinking in a more creative fashion, one that parallels the process used to create the product ideas. Prototypes or mock-ups help users understand the product and describe their likes and dislikes more clearly.

For consumer products, focus-group sessions offer structured opportunities to test ideas with users. Storyboards that describe the use of the product are a good tool to get consumers talking about the ideas. By encouraging consumers to speak openly about the concept, not just saying what they think the leader wants to hear, valuable feedback can be obtained. It is even possible, by listening carefully, to hear creative improvement suggestions in consumer comments. It is almost never too early to get users involved in assessing new ideas, provided that it is handled in a constructive fashion.

Assessing Hurdles. It is also important to assess potential hurdles to commercialization. What are the technical challenges? Are there capital requirements? What other considerations come into play? For example, does successful implementation require that another company develop a new raw material or piece of hardware? Are there regulatory hurdles imposed by the government that must be overcome? These are the kinds of issues that need to be assessed in preparation for a risk and reward-based decision and before a commitment to move forward. This is a conscious, deliberate decision that says the potential benefits outweigh the risks, and it provides the basis for an initial commitment. Because investment at this point is limited, this is the time to walk away from an idea that does not seem to have enough potential.

Developing, Making, and Delivering the Product

This discussion has focused on the first five steps in a seven-step product development model. The last two steps, developing the product and making and delivering it, must be accomplished with the same high level of energy and discipline. Moving on to development and introduction requires major resource commitments and should not be undertaken casually. Structured creativity can be applied in these stages to overcome problems that arise.

CONCLUSION

Creativity is a key company resource. Sadly, systems and practices generally inhibit its effective use. It is such an important resource and tool that it cannot be permitted to occur randomly. Conscious choices need to be made about how to manage creativity and how to apply it to the toughest problems. The opportunity is to turn the paradox of structured creativity into a new paradigm—a paradigm that leads to new and better products.

Acknowledgments

MindLink is a trademark of MindLink, Inc.

Mindmapping is a registered trademark belonging to Tony Buzan.

Synectics is a registered trademark belonging to Synectics, Inc.

The Synectics approach requires group members to diverge through creative thinking patterns and then converge in the creation of useful and valuable ideas for development.

USING FOCUS GROUPS TO CAPTURE THE VOICE OF THE HIGH-TECHNOLOGY CUSTOMER

Although focus groups show that customers do not fit a neat analytical framework, their input is critical for a successful, profitable product. This chapter describes how to use focus groups as one method of capturing the voice of the customer.

John J. Moran

Today, many firms are trying to increase the quality of their new products and services by improving the R&D process. Invariably, their efforts include capturing the voice of the customer and integrating it with R&D efforts. This chapter describes how to use focus groups as one method of capturing the voice of the customer. It outlines when to use focus groups and provides checklists on how to get started. Each of the elements of the successful focus group are examined along with helpful insights and examples showing how customer input was used to improve the new product development process.

Focus groups show that frequently customers do not fit a neat analytical framework, yet their input is critical for a successful, profitable product. The candid, unfiltered feedback of what potential customers think of an idea, how it will fit into their operations, and their struggle with all the changes associated with a new product or service is a fast, powerful, hard-to-deny perspective that in the final analysis measures the quality of the R&D effort.

Because of their personable, conversational style, focus groups are a very palatable form of market research. Most market research is quantitative, based on a large number of objective observations of selected respondents representing a particular group or market segment. Quantitative approaches have meaning in the aggregate; they provide a precise but not necessarily accurate picture of the market. After all, markets are made up of individual customers, not numbers. Focus groups address this deficiency, providing a microcosm of the marketplace, with all its diversity, idiosyncrasies, and inconsistency. They can be particularly valuable to the technology company in which the emphasis on the technical capabilities of the product can often assume disproportionate importance.

A partial list of projects that focus groups are ideally suited for includes:

JOHN J. MORAN is a certified management consultant and a principal with J.J. Moran & Associates in Santa Monica CA.

- Evaluating the acceptance and application fit for a new product concept.
- Increasing the awareness of the customers, workflows, processes, and operating environment.
- Looking at general product areas to identify areas of opportunity or to generate new ideas.
- Testing the market impact of a competitive product or promotion.

A few examples of focus groups that the author has run will illustrate how focus groups have been effectively used by clients.

A manufacturer was contemplating a new version of a product that would require some reconfiguration of the typical data center. The focus groups uncovered a strong opposition on the part of the data center manager to any remodeling. Not only was remodeling impossible in many instances, but more important, data center managers were strongly opposed to the idea of modifying their data center to accommodate an outside product. There was a subtle but strong feeling that the unique product requirements should be handled by the manufacturer and not be passed on to the customer. The manufacturer decided to design the product to eliminate the necessity for any data center construction and avoided a potential major problem with the new product.

A firm was considering a new line of products to facilitate the filing of documents in offices. Because filing is such a generic activity and the product involved advanced scanning and storage technologies, the manufacturer decided to brainstorm applications with various segments of the targeted market. Focus groups were held with IS and functional department managers and secretaries. Surprisingly, each of these groups had dramatically different views of the product. IS managers were very concerned with the size and unwieldy nature of image-based files. The functional managers were concerned with the products' technical capabilities and cost. Secretaries immediately took to the concept and generated multiple ways it could be used to improve their personal productivity and effectiveness. The manufacturer had adequate time to modify certain product capabilities and features in positioning the product for each segment.

A manufacturer was about to implement a new packaging and pricing program for product support. Support services such as field maintenance, telephone hot line, operator training, and product documentation were grouped into packages that would combine and price the services differently. Focus groups run prior to introduction uncovered major weaknesses in the existing support program. The manufacturer was able to first address these inadequacies and avoid a potential disaster with his existing customer base.

HOW TO ORGANIZE A FOCUS GROUP

The flexibility and ease of implementation of focus groups belie the level of discipline necessary to realize the true benefits. Without specific objectives, careful screening and recruitment of participants, and professional moderation, a focus group can easily become entertainment at best or misleading information at worst. The opportunities for mismanagement of this powerful tool are great.

There are four basic elements to the typical focus group project:

Because of their personable, conversational style, focus groups are a very palatable form of market research.

- Study design.
- Recruitment of respondents and facility arrangements.
- Moderation of the actual focus group session.
- Analysis and reporting of results.

Study Design

The overall study design has two basic elements:

- Whom to talk to.
- What to find out from them.

It is important that the information requested relate directly to alternatives or decisions to be made by the focus group sponsor. Working with a moderator who can bring an overall business perspective as well as a knowledge of both the technology and the market to the process is particularly important for technology-based products. Focus groups are flexible and must be molded to meet the unique information and business needs of the sponsor. Exhibits 1 and 2 present a checklist of the cost and design elements that should be defined early in the focus group design process. Focus groups can run between $5,000 and $10,000 per group depending on variables that include, travel, incentives, number of groups, video, and report requirements.

Exhibit 1
Focus Group Cost Elements

- Facility rental
- Participant recruiting
- Participant incentive
- Moderator fees
- Meals (participants and observers)
- Videotaping, tape editing, tape reproduction
- Audiotape transcription
- Travel
- Miscellaneous (mail, telephone, report)

Exhibit 2
Focus Group Design Elements

- Respondent identification and recruiting
- Market geography and number of sessions
- Ordered information needs
- Time allotted for the session
- Written questionnaire; purpose and content
- Presentations or props used during the session
- Client identified or blind
- Final report requirements

It should be noted that focus groups are qualitative in nature. They provide indicative rather than conclusive information. Most often, they provide in-depth information from relatively few respondents who are typically leading edge or innovative users. Usually there is a trade-off between getting both the in-depth qualitative information and quantitative statistics representative of a target population. However, a large project that the author managed for a client addressed this limitation. They held a quantitative study (conjoint analysis of product attributes) with a group of randomly selected statistically representative respondents. They then held a series of focus groups with these same respondents to better understand their applications and preferences. The very successful but expensive project met the clients' objectives, though many of the groups necessarily included several respondents with limited interest in the new product concept.

Recruitment of Respondents and Facility Arrangements

An important element of the design process is to reach early agreement on the targeted respondents for the groups. Telephone recruiting of respondents is frequently done from a client-supplied list, lists purchased from an outside supplier, or a list developed by telephone screening of a larger population. The importance of recruiting the appropriate respondents is critical. Groups have little value if the wrong respondents are recruited. It is important to note that the nature of the product can also dictate geographic restraints. The author has run groups on expensive capital equipment for which an adequate concentration of respondents existed only in major cities.

Moderation of the Actual Focus Group Session

The unique requirements of high-technology companies require levels of objectivity and technical competence that can best be met by a professional moderator. The role of the technical focus group moderator in this environment is rapidly changing from group psychologist to marketing consultant. Gone are the days when the moderator shows up two hours before the session for a short briefing and mails in a summary report on the session. Clients are demanding more of the moderator to ensure they get their money's worth.

A client checklist to evaluate a potential moderator for a technology-oriented group could include the following:

> **Focus groups provide a microcosm of the marketplace, with all its diversity, idiosyncrasies, and inconsistency.**

- *Technical competence.* Good moderators must understand the technology. They must be able to speak the language to be able to work with the group and the sponsor.
- *Marketing experience.* The moderator should bring a good sense of both marketing and business to assist and guide the client in relating the overall study design to specific product or marketing decisions.
- *Moderating skills.* The moderator must also possess the necessary mix of interpersonal skills to effectively carry off the group. Often, seemingly minor techniques can make a big difference with the high-technology group. For example, one useful approach the author has learned is to include a short questionnaire for the respondents to complete at the start of the group. This allows slower respondents to think through their position prior to the group discussion.
- *Motivation and objectivity.* The moderator should have an interest in the subject yet not be predisposed toward any viewpoint, strategy, or product.

The moderator must ensure that the information flows from the group. This is not a sales effort or an opportunity to educate misinformed respondents. Often the moderator must profess knowledgeable ignorance concerning a topic in order to flesh out important customer opinions and beliefs.

Analysis and Reporting of Results

Generally, the moderator will prepare a report or presentation that synthesizes important messages from the group to support findings, conclusions, and recommendations for the sponsoring organization. Qualitative information is difficult to analyze, and the hands-on familiarity with the process and unique perspective based on direct conversation with the participants makes the moderator's report an important deliverable. Typed transcripts and audio- and videotape recordings of the session are also typically provided. The audience for the report may also require the production of a composite videotape summarizing several groups. This can be an effective multimedia message for management tying key respondent comments to important findings or recommendations.

HOW TO GET THE MOST OUT OF FOCUS GROUPS

As the popularity of focus groups increases and organizations learn how to use the technique, their demands for quality are increasing. The following suggestions can help increase the overall value to the sponsor.

Preparation

Because of the flexibility of the process, it is important that appropriate expectations be set. It should be kept in mind that focus groups are a qualitative form of market research. They are best used for developing insights and direction rather than obtaining conclusive findings. Like the markets they reflect, the reactions of individual focus groups often can differ considerably. With this in mind, it is highly recommended that more than a single focus group be held to address the needs of a particular project.

Setting Priorities

The necessary homework should be done to identify and prioritize the issues to be addressed by the focus group. There is often a tendency to pile on interesting questions or to turn the group into a quantitative question-and-answer period, both of which should be resisted. In most cases, a moderator's guide should be prepared as an agreed-on roadmap prior to the initial group. This guide typically outlines the planned sequence, timing, and discussion topics. A reasonable list of prioritized topics enables the moderator both to cover all critical issues and to give the group its head for in-depth discussion on important subjects. Exhibit 3 provides a sample moderator's guide for a two-and-one-half-hour group. Note that the written questionnaire is given early in the session. The guide also indicates the planned order of topics and frequent changes to keep the session moving. If appropriate, the author has found it valuable to include sponsor personnel (e.g., product R&D personnel) for the summary wrap-up discussion with the participants.

Exhibit 3
Sample Moderator's Guide

1. Orientation 0:00–0:10
   —Introduction and Welcome
   —Agenda
   —Ground Rules
2. Respondent Introductions 0:10–0:30
   —Name, Company, Responsibility
   —Use of Product, Application
3. Written Questionnaire 0:30–0:45
   —Break
4. Discussion 0:45–1:30
   —Topic Area 1
   —Topic Area 2
5. Presentation 1:30–1:45
6. Discussion 1:45–2:10
   —Break
7. Summary/Wrap-Up Discussion 2:10–2:30

Recruiting Carefully

As mentioned previously, a very important element is the screening and recruitment of participants. A careful balance of product awareness, market segmentation, and session topics is required to target the appropriate respondents. Screening criteria typically include either the use or familiarity with a particular product or service or certain management, administrative, or budget responsibility. The client and moderator should also agree on the right incentive and presession follow up to ensure attendance. It is generally a good idea to also recruit two or three extra respondents to compensate for no-shows.

The flexibility and ease of implementation of focus groups belie the level of discipline necessary to realize the true benefits.

Preparing Observers

An often overlooked but important element in the process is setting the tone for observers from the sponsoring organization. Viewers should observe the proceedings with the overall project objectives in mind. There is often a tendency toward selective listening to a loud or dominant respondent or a respondent who shares the observer's point of view. Observers should keep an open mind during the session, using naive listening to draw in as much information as possible; the uninformed or ignorant respondent may be saying a lot about the market or product. A follow-up debriefing group attended by the moderator and observers immediately after the focus group can be very helpful in gaining perspective from others concerning important points and messages they heard during the group.

FOCUS GROUPS AND THE R&D MANAGER

This chapter has summarized the important elements of the typical focus group and how it can be designed and executed to provide value from an overall company perspective. Another important aspect to explore is specifically how the focus group can add value for the functional manager, specifically the R&D or engineering manager.

The value of focus groups for the marketing manager is obvious: the information can be used directly for advertising, marketing, collateral, and other product-positioning activities. The value to the R&D manager is more subtle but potentially even greater. In the hands of an aggressive, far-sighted R&D manager, focus groups can provide new perspectives on the R&D process itself. It is this opportunity to think outside the box and address the current R&D product development paradigm that can yield considerable value.

Focus groups represent an opportunity for R&D managers to get closer to the customer. This can make those managers not only better managers but also more valuable members of the product development team.

Specifically, customer input can help R&D managers better address the following elements of their jobs:

- *Getting creative.* R&D personnel are natural problem solvers. Left to their own devices, however, they frequently focus on familiar technical problems. Often, a customer comment or problem can result in a new creative approach resulting in a vastly improved product.
- *Prioritizing.* It is important to factor in the voice of the customer into the priority equation. Addressing only schedule or technical priorities can miss important elements that should enter into the setting of priorities.
- *Generating new horizons.* The variety and extent of customer needs will often start the R&D manager thinking about a continuum or family of products rather than trying to develop the initial perfect product or design.

The variation in customers and the concept of market segmentation are important aspects for the R&D manager to understand. Very often, innovator or leading-edge customers will make up an attractive and vocal segment of the market. Unfortunately, they frequently represent less than 20% of the potential for a new product or service.

The following case makes this point clear. In a focus group the author was moderating, 2 of 10 respondents were very technical and got into the bits and bytes at every opportunity, leaving the other 8 respondents relatively quiet. After getting the less technical respondents into the discussion, it became clear that their issues were packaging, user interface, and support, not the technical performance issues of the minority. At the conclusion of the group, it was obvious that the client would be wiser to spend its limited resources on telephone hot-line support and an improved graphical user interface rather than increasing the raw processing speed of the product.

Focus groups and the palatable way they package the voice of the customer offer the R&D manager the opportunity to become a better product development team member by becoming a better businessperson. Focus groups give the R&D manager a chance to address the bigger picture, a chance to get a feeling for the customer's perspective and mind-set. Only by addressing these considerations will companies be able to develop products that are more responsive to the market and ultimately more profitable for the company.

SUMMARY

The use of focus groups as a marketing tool is in a period of transition. The tool, long accepted in consumer marketing, is increasingly being used in the industrial area, particularly in technology-based companies. Comanies are beginning to see the value focus groups have in presenting market information in a true multimedia format. More important, they recognize the impact this conversation with the market can play in capturing the voice of the customer and introducing a marketing perspective throughout their organizations.

Focus groups can also play an important role in expanding the vision of the R&D manager. Only by listening to the voice of the customer can the forward-thinking R&D manager effectively link products and technologies to market needs.

THE ROLE OF APPLIED RESEARCH IN CORPORATE DEVELOPMENT

Thomas P. Fidelle

Applied research is governed by the same general principles that govern less practice-oriented forms of research. However, the methods that work best in applied research, the way goals are set and accomplishments are measured, the skills that applied scientists need, and the resources that applied scientists must cultivate and use effectively are different.

Applied research is usually associated with industry and basic research with academia; however, this is an oversimplification. Much excellent applied research is conducted in universities as well as in industry. Conversely, relatively little basic research is conducted in industry, and that's probably how it should be. Industry can draw on academia for the scientific foundation necessary to support its applied research programs, and academic institutions can draw on industry for financial backing and guidelines to ensure the relevance of their work.

The case studies discussed here are based on industrial applied research simply because my experience is centered almost exclusively in the industrial arena. The business issue as well as the technical ones are covered because, quite bluntly, the ultimate goal of industrial research should be to make money for the company.

It comes as a surprise to some observers that most companies make products or offer services that the consumer never sees as finished products. These products or services are called intermediates and will pass through one or more stages of development before reaching the consumer. This is especially true among chemical and petroleum companies, which make up a substantial portion of the capitalization in the US. For example, Dow Chemical realizes close to $15 billion in annual revenue, yet less than $2 billion of sales are directly related to consumer products. It behooves a company that supplies mostly intermediates to understand as much as possible about its customers' technologies, given the limitations that may be imposed by secrecy. That is the principal responsibility of applied research in industry. As a supplier of intermediates, the company cannot value its products, especially new products, unless it can establish the value of those products as the customer sees them.

THOMAS P. FIDELLE, PhD, is the manager of applications research at Great Lakes Chemical Corp., West Lafayette IN. Previously, he worked for Celanese Corp., where, among other responsibilities, he served as the applications engineering product manager for two of the PBT-based product development projects: carpet fiber and stretch denim.

Sometimes this evaluation can be achieved through rigorous specifications. More often than not, however, specifications alone cannot predict downstream performance, especially when unexpected and unusual developments come about (as is frequently the case with substantially different new products). It may be necessary for the applied research department to be familiar with a broad variety of complex technologies if the products they sell are to enter several markets.

HOW APPLIED RESEARCH FITS

Because of the uniqueness of its role, the applied research group may be an integral part of the R&D organization or a separate group outside R&D or, in some companies, part of the marketing department. For example, DuPont coined the term *end-use research* for its applied science department and had several specialized groups occupying several buildings at the Wilmington Experimental Station. Dow Chemical still uses the term *Technical Service & Development* for its applied technology area. Before its acquisition by Hoechst AG (Frankfurt, Germany), Celanese Fibers Group had a marketing and technical department that reported directly to the president of the Fibers Marketing Co. As the name suggests, the group performed marketing as well as technical services.

Exhibit 1 depicts the spectrum of research and development in a specialty chemical company. This medium-sized company annually spends between 5% and 6% of gross revenue on R&D. Some synthesis among the various research areas is accomplished, which is aimed primarily at new products. Because of limited resources, the synthesis effort must be directed at technologies at least somewhat familiar to the company. A major share of R&D resources is dedicated to process development that is aimed at cost reduction, efficiencies, environmental requirements, quality, or product development.

There is room for knowledge building when the need exists, both from a fundamental and an applied perspective. When new products or processes emerge, they must first be demonstrated at the pilot level, then on the production-scale level. The R&D department is likely to be involved in all phases of scaleup.

The freedom to innovate must be present throughout the entire R&D structure; the organization is feckless without it. Frequent interaction among R&D staff members and other functions outside R&D is also essential for success.

The R&D applications group commands a unique position in the organization. It is the key link between R&D and sales and marketing and—directly or indirectly—with the customer base. Ideas may originate from within the R&D organization or from an outside source. Several potential sources are shown at both sides of the R&D spectrum in Exhibit 1.

Exhibit 1
The R&D Spectrum

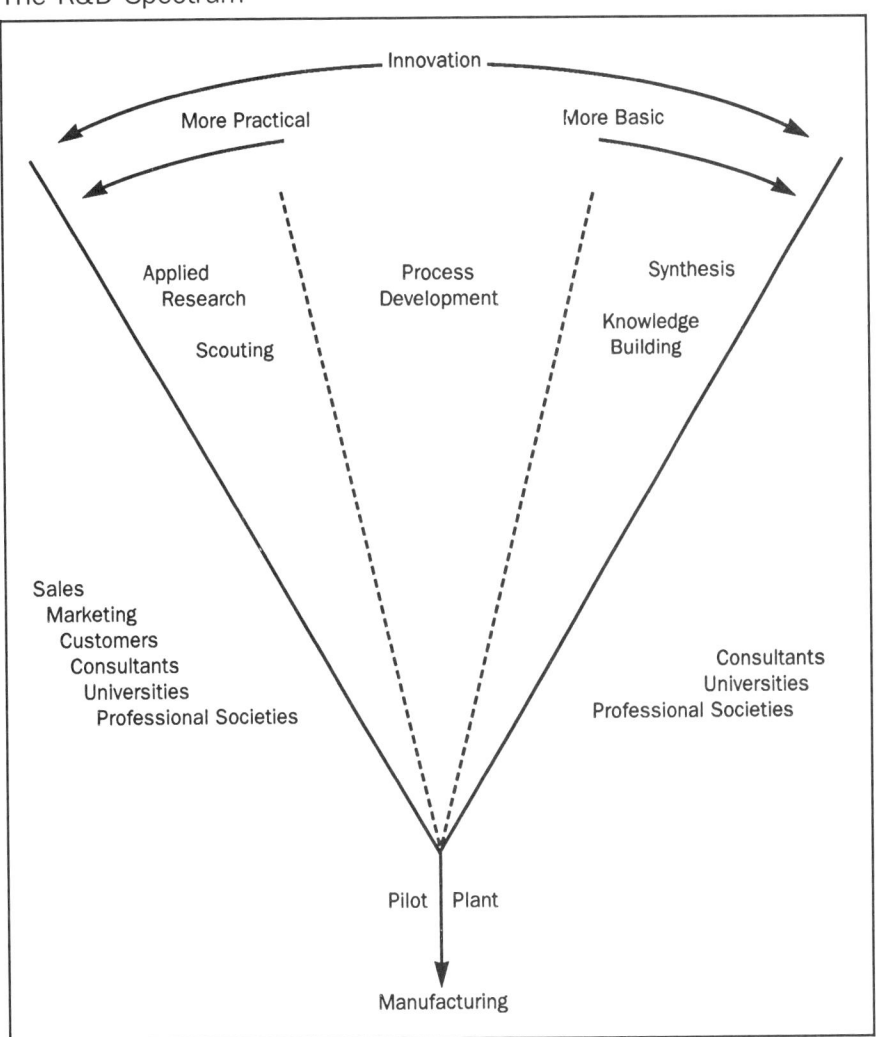

```
                         ─── Innovation ───
              More Practical              More Basic

          Applied            Process         Synthesis
          Research         Development
                                            Knowledge
           Scouting                          Building


  Sales
  Marketing
  Customers
  Consultants                                   Consultants
  Universities                                  Universities
  Professional Societies              Professional Societies



                          Pilot │ Plant
                             ↓
                        Manufacturing
```

Sources of Invention

It is important for R&D managers to note that R&D departments do not have a monopoly on ideas. Many bright ideas emanate from other sources, and it is the responsibility of R&D, led by the applications group, to tap those outside resources effectively. Nothing can isolate and ultimately destroy an R&D organization as readily as a narrow development philosophy that fails to recognize the rich outside resources available for obtaining useful ideas.

THE ROLE OF APPLIED RESEARCH

In responding to either opportunities or problems, the applications group has a relatively short-term focus—that is, between two years and five years for developing significant new products. If there is no separate technical service department, the applications group is likely to be the first line of technical support for the sales force. In such cases, however, they must constantly strive to avoid getting bogged down in daily battles at the expense of longer-term objectives.

The Tasks of Applied R&D

A typical distribution of effort for the applications group is:

- Routine technical service (30% of effort).
- New product development and introduction (40% of effort).
- Finding new applications for existing products and exploring new technologies (30% of effort).

Routine technical service consists of troubleshooting, particularly regarding quality problems, and specifying the use of products to customers. Troubleshooting typically may require some detective work or laboratory evaluation. Advising customers on product use may be accomplished with existing information or it may be necessary to go to the laboratory or pilot plant to derive the data.

In new product development, the applications role is to take the new product offering at the early stages of development, to determine the value of the product in the customers' eyes, and to participate in all the subsequent stages of development, from introduction through commercialization. This includes preliminary evaluation, defining performance criteria and specifications in the interest of the customer, reviewing cost and selling-price scenarios, making product presentations to prospective customers, and sometimes conducting field trials on location with the customer.

The tasks of the applications function are the most difficult to define and perform. Finding new applications for existing products sounds easier than it is, especially for mature product lines. However, it is the responsibility of applications research to constantly seek new markets for existing products as well as new technologies for the company to grow into. If the applied research department is large enough, it makes sense to organize the department around functional product lines.

An Organizational Example. One applied research department was divided into three groups, each consisting of six to eight scientists and a group leader. Two of the groups were aligned with the core businesses of the corporation. These were well-defined markets that had not only some mature products but also some opportunities and the need for new products. The objectives for these two groups were reasonably well defined, and the projects generally had development time frames of between one year and three years. There was a respectable amount of routine technical service provided by the two groups.

The third group in the organization had lofty goals, much more poorly defined objectives than the other groups, and considerably more freedom to diversify outside the company's core business area. Little technical service was provided, and only a limited contact with the traditional customer base was required.

This group's resources for new product ideas included available literature, technical conferences, university contacts, start-up companies, consultants, entrepreneurs, and others involved in state-of-the-art technologies. On average, projects took from 5 years to 10 years to complete, from idea to commercialization. It is often difficult to justify such expansive and risky undertakings in R&D, given the bottom-line mentality that currently besets many US corporations; however, technology-based companies that fail to make sacrifices today for long-term growth are doomed to atrophy as their product lines mature and are replaced by new products with new technologies.

> **It is the continuous responsibility of applied research to look for new markets for existing products and for technology that the company can grow into.**

THE ROLE OF THE APPLIED SCIENTIST

The applications researcher, or applied scientist, functions as the focal point for new developments and must possess certain skills in order to be successful. Over the years, I have interviewed hundreds of people for applied research jobs and hired many of them. In describing the position to the applicant, I often use the diagram in Exhibit 2, which depicts the four key elements of a business and the way they must interact to be effective. Most large companies are organized along the lines of Exhibit 2. The larger the company, the greater the premium placed on communication and cooperation.

For the business to be successful, communication among the business functions must be excellent, particularly if new product development is meant to play a significant role in that success. A key link in this chain of communication is that between R&D and manufacturing. R&D must be able to scale up new products in the plant, and plant personnel frequently must rely on the R&D department to solve problems.

In my experience, the most difficult communication bridge to build and maintain is that between the R&D and the sales and marketing functions (as indicated by the dashed line in Exhibit 2). Weak links in the communication chain tend to be caused by mutual misunderstanding of the other's professional goals and point of view. A central figure is needed who understands the diverse perspectives of each party and can speak their languages.

The applications researcher is equipped to play such a role, coordinating the overall effort in introducing a new product. As mentioned, the idea for the new product may originate in R&D or come from any number of outside sources. The first decision on whether or not to proceed with the idea is likely

Exhibit 2
Functional Interaction

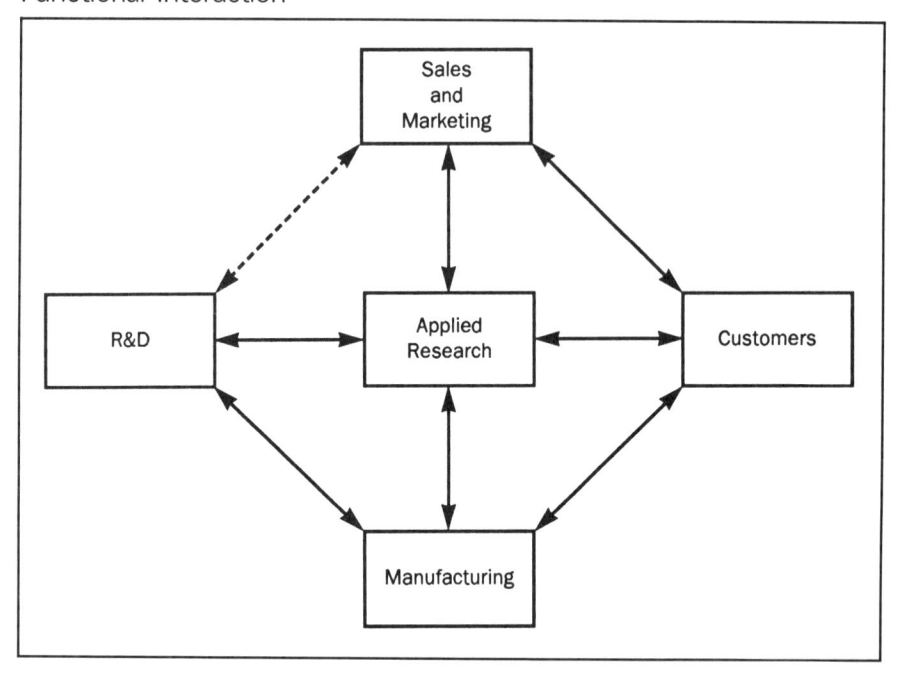

Diagram showing Sales and Marketing (top), R&D (left), Applied Research (center), Customers (right), and Manufacturing (bottom), with bidirectional arrows connecting Applied Research to each of the other four nodes, and additional arrows between Sales and Marketing, R&D, Customers, and Manufacturing.

to be made by the applications researcher. If the idea originates in R&D, the decision will be based on the idea's commercial feasibility; if the idea comes from an outside source, the decision is likely to be based on its technical feasibility and estimated cost.

Likewise, the research applications department is usually the first gate for a new product to pass through when it is synthesized or formulated on a laboratory (i.e., prototype) scale. At this stage, the product is evaluated in terms of its ability to perform its ultimate end use. As the project unfolds, the applications group evaluates several versions of the product as it is scaled up and begins to formulate specifications on the basis of its process capability and market needs. At this stage, the sales and marketing function identifies potential customers, projects volumes and prices, and plans initial field trials.

If significant start-up capital is needed, engineering and manufacturing are brought in at this stage to aid in planning and specifications. Patents need to be filed before any public disclosure is made. Environmental and toxicological issues may need to be addressed early on. An applications researcher is likely to go along on the initial calls to prospective customers to promote the merits of the product and discuss any special handling requirements. If the customer allows, the applications researcher may be heavily involved during initial stages of field trials.

During the scale up to manufacturing, an applications researcher is responsible for setting specifications according to customer requirements. If any major changes become necessary, some or all of the foregoing steps may need to be retraced to obtain continual feedback as the results of such changes. In short, the applications department serves as the focal point for new product introductions, and the applications researcher is well suited to act as the project manager.

> **Between R&D and sales and marketing, a central figure is needed who understands the diverse perspectives of each party and can speak their languages.**

A PROFILE OF THE IDEAL APPLICATIONS RESEARCHER

When interviewing candidates for applied research, look for the following characteristics:

- Social skills—Does the applicant speak clearly and concisely and express thoughts logically? Does the candidate maintain eye contact? I expect eye-to-eye contact for at least 80% of an interview, which reassures both parties that they are having a meaningful, honest exchange. Does the candidate make a favorable first impression, dress acceptably, and exhibit a commanding presence? Is this person likely to be a leader?
- Technical qualifications—At a minimum, a bachelor of science degree is required, in a subject ideally but not necessary related to the company's products.

Advanced degrees are welcome but not required; in my view, an advanced degree can be a handicap if the applicant is too narrowly focused. I'd rather have a person with average formal technical training who is innovative and has initiative rather than a technical genius who can't carry ideas to fruition and is inflexible in the subject area. Experience and a successful track record in applied research are highly desirable, especially in related fields.

- Innovative ability and hands-on style—People who can generate their own ideas and build on the ideas of others are well suited to applied research. In my experience, fewer than 1 out of 20 technical people are naturally innovative; the rest of us have to work at it constantly. In addition, the job requires a hands-on person—that is, someone who enjoys actual lab work and interacting with the physical product and with colleagues.
- Winners and achievers—The candidate should exhibit a few significant accomplishments as a professional and as a student and should also exhibit a positive attitude when faced with challenges.
- Multifaceted—Can the applicant effectively handle two to three projects simultaneously? In general, candidates who would rather work on several short-term projects that last from 1 year to 3 years than on one long-term project lasting from 5 years to 10 years make better applied researchers.

☐ Goal orientation—The candidate's career goals may reveal a general attitude. For example, does the candidate intend to remain in research or to move into management, sales and marketing, or another function? Because of the variety of their experience, people who have worked in applied R&D for a few years often make excellent candidates for sales and marketing or general management.

HOW APPLIED R&D PERFORMS ITS JOB

The following two case studies illustrate different aspects of the applied R&D department's function.

Grappling with the Difficulties of Corporate Growth

During the early 1970s, Calloway Chemicals was a small specialty chemical company. A public company traded on the American Stock Exchange, in 1973 it reported sales of $30 million and an 11% after-tax profit. Calloway started as a regional supplier of dyestuffs and finishing chemicals for the textile industry in the southeastern US. As an alternative to the large multinational companies, Calloway offered a broad range of products, attractive prices, and excellent service. None of its products was proprietary or unique; the company simply followed its larger competitors with me-too products. The R&D department's spending was less than $1 million (less than 3% of gross sales).

Throughout the 1970s, the company grew at a compound rate of 22%, and by 1980 it was generating $120 million in sales annually with an R&D budget of $2 million. Most of the R&D budget was being spent on process development aimed at cost reduction, because at this point there was increased competition in the industry. In addition, the company began to experience slower growth and loss of market share to smaller, more flexible companies that were doing exactly what Calloway had done to become successful: copy its large competitors.

A new, three-pronged corporate strategy for the 1980s was designed to solve this problem:

☐ In R&D—Increasing the R&D budget to 5% of sales, implementing more new product development (synthesis), and finding new opportunities outside the core business (i.e., textiles).
☐ In corporate management—Aggressively pursuing acquisitions and joint ventures to facilitate growth.
☐ In sales—Establishing a global sales posture by branching out to other US regions and to Europe and the Far East with novel and proprietary products.

The first leg of the strategy was to beef up the R&D function. Previously, the company had never had an applied research group. Dominated by textile chemists, the R&D department had a strong internal focus, with most of the outside influence coming from direct sales and customer service. The company hired an R&D director with a strong background in applied research and made him accountable to the corporate president. (The R&D department had previously reported through operations and did not have strong support or contact at the senior management level.)

The new director formulated a five-year plan with the following goals:

> The new director highlighted new product and business opportunities that would result in resumed growth for the company and would recapture market share.

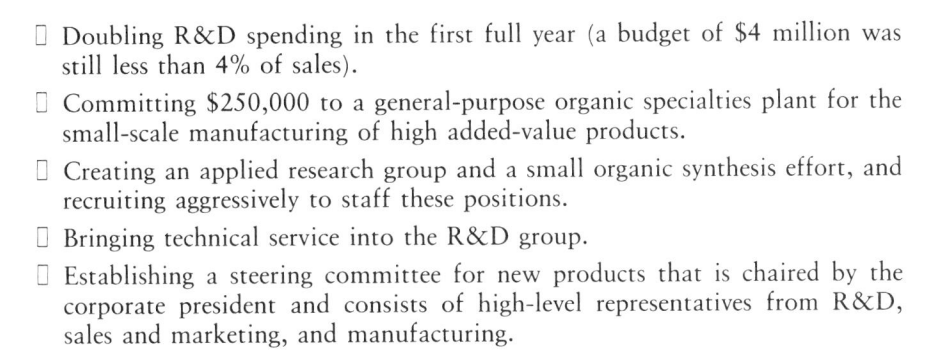

- Doubling R&D spending in the first full year (a budget of $4 million was still less than 4% of sales).
- Committing $250,000 to a general-purpose organic specialties plant for the small-scale manufacturing of high added-value products.
- Creating an applied research group and a small organic synthesis effort, and recruiting aggressively to staff these positions.
- Bringing technical service into the R&D group.
- Establishing a steering committee for new products that is chaired by the corporate president and consists of high-level representatives from R&D, sales and marketing, and manufacturing.

Creation of the steering committee was a key element, as the new director could not make sweeping changes without broad-based support from senior management. First, the president had to be sold on the plan; the new director then had to actively solicit support at key levels throughout the organization. As might be expected, the most resistance came from the sales and marketing staff members, who expected their power and influence to erode after the changes in the R&D department went into effect. Eventually, the new director was able to convince them otherwise by painting a vista of new products and worldwide business opportunities that would result in resumed growth for the company and enable it to recapture a significant market share.

The turnaround in organizational positions was time-consuming. It took a year to reorganize R&D and to staff the new positions with high-caliber people. It took another year before the group began to produce workable ideas and new product road maps began to emerge. Three years from that stage, new products were finally established in the marketplace and began to earn respectable revenues. Thus, the constant investment of time and resources paid off.

By 1988, the corporation had grown to just under $500 million in annual sales, putting it back on the 20% annual growth curve. More than $100 million of the growth resulted from products that were developed in the 1980s. By 1987, 10 new product patents were being issued to the company every year, and new technology had resulted in two joint ventures totaling $50 million annually. Nearly 25% of R&D costs were offset by annual licensing fees paid for by technology developed after the reorganization.

By 1987, the R&D directorship had been elevated to the corporate vice-president level and the R&D department had close to 100 scientists. Twenty-five percent of R&D personnel were involved in applied research. Much of the success of this corporation was attributed to the emphasis on applied research and the leadership exercised by the applications researchers who acted as project leaders in developing new products.

A Technology Looking for a Use

During the 1950s, Celanese Corp, in a joint venture with UK-based Imperial Chemicals, launched a company called Fiber Industries Inc. Imperial Chemicals brought polyester and nylon fiber technology to the relationship, and Celanese brought expertise on marketing fibers in the US. Eventually, the joint venture became one of the largest polyester fiber producers worldwide, with sales in excess of £1.5 billion annually. In chemical terms, the polyester of choice for synthetic fibers was (and still is) polyethylene terephthalate, or PET, the simple condensation polymer of terephthalic acid and ethylene glycol. This fiber offers the right combination of chemical and physical properties for many apparel, home furnishing, and industrial end uses, ranging from T-shirts to tires.

Another simple condensation polymer, based on terephthalic acid and butanediol and called polybutylene terephthalate, or PBT, could be made easily with the equipment used to make PET. Although PBT has unique physical properties compared to PET and is used extensively in injection-molded plastics, the material never seemed to have any value as a fiber during the early years of applied development of polyester.

Celanese prided itself on fiber technology and employed some of the best applications researchers in the industry, who doggedly tried to find a use for this new fiber. They first looked at hosiery yarn. Nylon was then the miracle fiber for hosiery. DuPont had chosen hosiery to introduce nylon because it was cost-effective; the new polyester polymers were quite expensive in small volumes, however, so this use seemed impractical though possible. In retrospect, fabricating a knockoff of nylon for this end use would have been a Herculean task.

PBT had a few minor advantages (e.g., it produced less static electricity) but one major deficiency relative to nylon that eventually killed the hosiery application. Although PBT had significantly better resiliency than PET, it was still inferior to nylon in that category. After many wearings and washings, PBT pantyhose tended to sag in the knees worse than nylon hose, resulting in a shorter life for the PBT product. Several hundred thousands of dollars and several years of work went down the drain with this failure. The project was well on the way to commercialization before it was killed.

The applications researchers were not dismayed, however. Even before the demise of the hosiery product, they found that PBT made a superior carpet yarn compared to PET. Nylon was the ideal carpet fiber, as it is today, and this was the second major use discovered for nylon polymer. PET polyester had been commercially introduced as a carpet fiber during the late 1960s, and because it was less expensive than nylon, it was rapidly embraced by the tufted carpet manufacturers. After a couple of years in use, however, the polyester carpet fibers became crushed and matted, which led to the eventual failure of the product. Polyester carpets also soiled more readily than nylon carpets and were more difficult to clean.

The performance limitations in each case could be determined only after hundreds of goods were manufactured and tested.

Celanese introduced PBT-based carpet fiber in 1970, and at least two commercial programs were under way with appreciable retail sales before it was discovered that the PBT fiber was more flammable than PET. Many rolls of carpet were rejected by the mills because they were unable to pass the mandatory flammability tests before shipping. This was another expensive lesson for Celanese.

One could argue that both of these failures were examples of untenable applied research because the deficiencies should have been recognized before the products got so far down the road; of course, these problems are much clearer with hindsight than they were during the development process. The performance limitations in each case could be determined statistically only after many hundreds of goods were manufactured and tested. Such is the nature of new product development and applied research.

Fortunately, there were a few brave souls at Celanese who were ready to come forward with the next potential application for PBT fiber. During the late 1970s, an opportunity presented itself. Stretch garments were becoming increasingly popular in the US. Stretch wovens were in demand in men's and women's trousers, and the denim manufacturers were interested in making fashionable stretch denims. The applications researchers at Celanese found that PBT had unique viscoelastic properties that made it an excellent filling yarn for stretch denims. They did their homework thoroughly this time, and the first successful stretch denim jeans were introduced in the early 1980s. The fabrics were produced by Burlington Industries, and the denims were marketed under the Fortrel ESP trademark name. That material continues to be a successful product.

This case illustrates the challenges of applied research, some of the potential pitfalls, and, in this instance, the protracted time line from the origin of an idea to its ultimate commercial success.

A CRITICAL LINK

Applied research can be challenging and time-consuming, and it can result in eventual commercial successes or failures. This chapter examines the methods that work most effectively, the way goals should be set and accomplishments measured, and what resources the applications researcher should cultivate and use to ensure ultimate success.

As stated, the freedom to innovate must be present throughout the entire R&D structure; without it, the organization can become weakened and ineffective. Frequent interaction among R&D personnel and other functions in the organization is critical to ensure a company's success. The applications research group commands a unique position: it is the key link between R&D and sales and marketing and—directly or indirectly—the customer.

TRANSFORMING THE SOFTWARE ENVIRONMENT AT APPLICON

Applicon, Inc., has a program for improving its software development process by raising the quality of the software it produces and the productivity of the development process. This chapter details how Applicon developed the program and how the program is doing.

Barbara Purchia

Almost four years ago, Applicon, Inc., decided to improve the quality of its products and services. It embarked on a program to ensure that the entire company was trained, and actively participating in the quality process. Engineering, the first organization to pursue the quality initiative, was particularly aggressive in implementing it. The vice-president of engineering stated, "Engineering is where our products originate and a logical place to start focusing our quality improvement efforts." All of engineering, starting with senior management, was trained in basic quality improvement.

An engineering quality management team (EQMT) was formed, consisting of senior engineering management and representatives from all engineering departments. Its goal was to provide the leadership to implement and sustain a comprehensive quality improvement process throughout the engineering organization. In addition to quality process training, training sessions were held in several key quality areas, including quality function deployment (QFD). QFD is a planning tool for defining user requirements, rating their relative importance, estimating how far products have to be advanced to fulfill these requirements, and determining technical solutions to address the users's needs. The outcome of a QFD planning session is an ordered list of actions necessary to satisfy customer requirements. Engineering has successfully used QFD to help determine process improvements and to rank them by priority. The EQMT also sponsored two volunteer committees to address two major areas of quality improvements: software process and measurements.

The software process improvement team focused on improving the development process. It conducted several engineering surveys and tailored QFD sessions to identify the biggest problems and potential solutions within the engineering organization. Number one on the list was the need for a standard software process methodology. After presenting the results to the EQMT and getting approval, the team developed a software process model that would be the cornerstone for all engineering efforts. This software process model covers the development cycle and provides guidelines for transforming the initial

---

BARBARA PURCHIA is senior manager of the development process improvement group at Lotus Development Corp.

product or feature requirements into a software solution. The model is now the basis for meeting schedule and quality objectives.

The second committee, the quality measurement committee, recommended and defined criteria and processes for measuring quality and productivity improvements. The company could not determine whether quality and productivity were improving without measuring them. The committee also held QFD sessions to determine and rank in priority areas for measurements and to determine the methods and tools needed to address each of them. The results were presented to the EQMT and the quality measurement committee then began developing tools for gathering the information.

Although progress was being made on software process improvements through these voluntary committees, more emphasis and full-time commitment was needed. In September, 1990, Applicon formed the software planning and process improvement department (SPPID) to improve all facets of the software development process for all engineering groups within Applicon. This was a major step. An entire group now focused on a variety of process improvement areas to ensure that Applicon was developing products effectively and efficiently according to a defined software engineering process.

ACTIVITIES AND ACCOMPLISHMENTS

SPPID developed plans to advance the maturity of the software development process by using the Software Engineering Institute (SEI) software maturity scale as a framework. The department focused on seven areas of improvement for the engineering organization:

- Process.
- Planning.
- Software process assurance.
- Training.
- Tools.
- Measurements.
- Communication.

Process

SPPID assumed ownership and responsibility for the software process model, which was approved by the EQMT in January, 1991. It contains standards and guidelines for developing software and describes how to use the model, including phase completion reviews and responsibility and accountability information. The model was introduced to engineering through a series of meetings. Although several groups began following the model immediately, the majority waited until their current development cycle was completed. Today, use of the model is required on all projects undergoing regular, active development. Overall reaction to the model has been positive and constructive.

After the model had been used for one development cycle, opportunities for improvement were identified and an effort begun to streamline the process. As use of the model increased, areas of redundancy, missing and unnecessary areas, and areas that were not used effectively became evident. Document deliverables were reviewed, and in some cases, several documents were combined and some were eliminated. For example, individual test plans were originally used when all the information could be covered in an overall project test plan. The contents of templates were reviewed and modified. Reviewers looked for process bottlenecks. The list of people necessary to approve deliverables was revised; for example, the signatory requirements of the vice-president were reduced. The use of phase completion checklists and reviews was revised so that phases that were document intensive, such as the requirements phase, no longer required a formal meeting to indicate that phase requirements were satisfied. The project managers could exit certain phases by submitting a phase completion checklist. Checks and balances were maintained by an audit mechanism that verified that each checklist item was completed satisfactorily. Each item had therefore been inspected, approved, and placed under configuration management control.

Part of this streamlining project involved measuring the success of process improvement efforts. SPPID distributed a survey to engineering management and technical contributors to establish a baseline and provide early validation of the areas targeted for improvement. This high-level model did not describe process details. Phase output and activities were only briefly described. During 1991, SPPID also elaborated on the deliverables of the model, defining and documenting standards and guidelines for each model deliverable. It produced eight standard and guideline documents. For example, a process document entitled "User Requirements Standards and Guidelines" was produced for the user requirements document deliverable. These associated documents typically contain the following:

> **Engineering's number one problem was to standardize software process methodology.**

- Standards and guidelines relevant to producing the deliverable. This information relates to the purpose of the deliverable and what is necessary to produce it.
- Tools and techniques for preparing the deliverable.
- A template for the deliverable. The template contains a description of the required and optional information for each of its sections, as well as descriptions of the expected contents for each section.
- A checklist for reviewing and ensuring completeness of each deliverable.
- A sample deliverable.

The templates and checklists are kept online so developers can readily access them.

These standards and guidelines served several purposes. First, they further defined the software development process. This was an important step toward a long-term objective of having a clearly and precisely defined software development process. Second, these documents were welcomed by the development groups because they provided examples of what each of the model deliverables was intended to represent. Third, the templates and checklists were easily accessible. Finally, the standards documents required that certain activities be undertaken to ensure a repeatable software development process.

One of the engineering directors was responsible for a complex, multiyear project requiring more than 100 people. His department has been using the model since the project's inception. He states, "We needed a way to effectively manage our project, coordinate the efforts of this large project team, ensure consistency in these efforts, and recognize dependencies between subproject groups and between the project and the rest of engineering. The software process model has allowed us to accomplish this." Because the project is so large, he has needed to modify parts of the model to accommodate the project's special needs. "The flexibility of the process model has been key to its acceptance and success in managing our project," he stated.

Planning

Although Applicon had a planning, reviewing, and tracking process, it was inadequate. Semiannual engineering planning meetings had been held since 1989; but their format, scope, and schedule was not as tightly controlled or formalized as necessary. Meetings were organized by one of the development directors to foster communications and coordination. They lasted 4 to 6 hours, used a loose format, and were presented by the engineering managers. Meetings were open to all areas of the company. Issues were raised, assigned, and reviewed.

However, these forums were used to communicate not only plans but also changes to plans, sometimes before senior management had been informed of them. In addition, senior management support and accountability for action items was uneven, and several action items remained unaddressed even though some groups deemed resolution crucial. The vision was also very short term, usually the contents and schedule for the release in progress. In addition, the organization did not use a common and consistent project planning mechanism. This caused communication and coordination problems that undermined trust in the presented dates and functions.

As a result, the position of an engineering planning coordinator was created to ensure the development of consistent, comprehensive, accurate, and integrated plans, as well as to implement and coordinate the engineering planning process. The engineering planning process model describes planning activities, including required documents and the planning cycle. The schedule for planning and review meetings is announced in advance. Each meeting is documented, including any resulting action items, and distributed to senior engineering management.

Today, all plans are developed and presented using common templates and terminology. Major plan milestones are based on the phases of the software process model. All affected parties review and approve the plans. Although this process may seem inconsequential, the results have been substantial. The scope of the plans has expanded from six months to three years. These plans incorporate risk identification as well as quality objectives and key milestones, and they include dependencies from groups external to the development team and describe expectations of these groups. The use of common, consistent, and regularly scheduled meetings for planning and reviewing has improved the planning process significantly and improved intergroup and interdepartmental communication and coordination.

> **The software process model is used with all projects undergoing regular, active development.**

The changes have been noticeable at all levels of management. In early 1991, the vice-president of engineering stated, "We suffer from a lack of long-range planning. We have to change plans because we are too focused on the short term." By January 1993, he said, "The strength of the process is the focus it gives us up front in planning and specification and the stability that results from it."

An engineering five-year plan describes the long-term goals and overall engineering priorities, as well as a high-level implementation plan, product evolution plan, and schedule. Producing a five-year plan is difficult, and this one takes a wide view, tracking market trends and competitive factors, and detailing the company's differentiators and core competencies. The five-year plan was translated into an annual plan, which defines the product development deliverables and schedules for the year and describes quality assurance and software process objectives. Both documents are distributed to marketing, sales, and all other functional organizations. According to the vice-president of engineering, "We must constantly strive to improve our time to market. We cannot improve if we do not plan ahead. If we emulate our competition, we will be behind them. The planning process has provided a framework to allow us to focus on up-front planning and develop ways to validate the quality of our products."

In addition to the planning cycle, project management training was provided and standards and guidelines were developed for individual project plans. These standards and guidelines contain the project plan template; project tracking and reporting requirements; project review activities; and tutorials on project scheduling, estimating, risk identification, and management. Project tracking information consists of project tracking and risk forms that managers must submit monthly. Risk forms consist of a description of a risk, its priority, the number of months that it has been an active risk, a backup plan, and a date by which an action must occur if the risk is not resolved. A monthly dependency tracking mechanism helps with coordination and communication among interdependent groups with Applicon. On project completion, project managers must submit project release information consisting of key project measurements.

Software Process Assurance

SPPID monitors process use, helping with process implementation and identifying opportunities for process improvements. A software process assurance (SPA) function was created to accomplish these objectives. SPA helps define standards for engineering, teaches engineering personnel about the role of SPA, and monitors and facilitates compliance with engineering standards, recommended practices and procedures, and quality plans. SPA receives copies of software process model deliverables from each product and reviews them to ensure that the requirements of the documented standards and guidelines are understood and followed. These activities also help identify improvement areas within the process. SPA occasionally participates in document inspections and in phase completion review meetings. If there are any problems, SPA discusses them with the project manager. As a last resort, SPA reports nonconformance to senior management. However, SPA is not responsible for the quality of products or for product testing.

Training

Education and training have been critical components of many of Applicon's improvement activities. It trains all of engineering on the software process model. Using a combination of outside and inside resources to provide training in both management and technical areas, the company has provided training on:

- The software process model.
- Technical review techniques.
- The inspection moderator.
- Software project management.
- Quality function deployment.

The following sections highlight two of these courses.

Inspection Moderator Training. During 1991, SPPID began a program to educate engineering personnel about software inspection techniques and to encourage their use. Software inspections are a formal technical review technique. Reviewers first inspect the material on their own and then participate in a defect logging meeting facilitated by a trained moderator.

Inspections can increase software quality. They also increase productivity by eliminating some rework time. In addition, detecting defects early in the development cycle lowers the cost of defects and improves productivity.

Inspections force people to concentrate exclusively on an item, whether source code or documentation, and hearing comments on that item at one time, rather than through several iterations, saves time and improves the end result. Inspections also ensure traceability. They tie pieces together, including parent documents or code and applicable standards and guidelines. Additional benefits include improved teamwork and communication, as well as the provision of an educational mechanism for the project team.

> Although Applicon had a planning, reviewing, and tracking process, it was inadequate.

Inspections efforts at Applicon began with a three-day seminar on inspection techniques. Training material was prepared for an in-house training program. It included one-half a day of overview training on inspection techniques and one day of training geared specifically for inspection moderators. The program caught on fairly quickly, and today Applicon has more than 30 certified moderators. Nearly all groups in engineering have sent a member to the moderator training course. In addition, a moderator's toolkit is available online, and many of the moderators' first inspections have been observed. The program seems nearly ideal for grass-roots adoption. Interested engineers can frequently try it in their groups with only minimal management involvement.

Inspections are now mandatory for all requirements documents. Each time an inspection is conducted, an inspection summary form is completed and sent to SPPID. The form, which includes such information on the inspection as the number of hours spent in preparation, number of defects found, and severity of each defect, is used in computing statistics on the effectiveness of inspection techniques. Inspections of documentation and code saved Applicon more than $2M in 1992.

Software Process Model Training. During 1991, SPPID helped develop a course on the software process model for a major part of Schlumberger, its parent company. This course covers not only information about the software process model but also tools and activities that occur during the entire process, such as project management, configuration management, and verification and validation. During 1992, this course was taught to all Applicon engineering and all engineers within Schlumberger Technologies in the United States and Europe. SPPID helped provide the training for more than half the courses taught. Participants and management responded positively, and the training program is provided for all new engineers. In addition, a tailored version of this course was taught to all Applicon marketing product managers.

Tools

The company is also trying to standardize its management and technical tools. SPPID has been a focal point for tool evaluation and use, trying to minimize redundant work within engineering. During the last two years, SPPID has evaluated and recommended several tools and, when necessary, produced tutorial documents describing recommended tools. The following sections describe two tools that were evaluated.

Test Coverage. One of the company's intermediate goals was to base its software testing process on quantitative objectives. As a first step in this direction, SPPID evaluated its test coverage methods. These methods involve setting numeric goals for the percentage of the lines or basic blocks of the program executed during a test session.

The VAX/VMS layered product PCA includes test coverage analysis functionality. Although the performance analysis capabilities of this product had been used by developers, its test coverage capabilities had not. SPPID began evaluating PCA on a moderately large application. This was successful even though SPPID found the PCA manual large and poorly organized. The reader had to sort through irrelevant material to find needed information. As a result, a brief document was produced, summarizing how to use PCA specifically for test coverage, and it was distributed to all engineering personnel.

Several groups were asked to use PCA for test coverage analysis on CAD/CAM applications. SPPID was available to assist with questions or problems concerning the product. It is now being used as the test coverage analysis tools for many projects.

Electronic Bulletin Board. Because the engineering function at Applicon is distributed between two locations, a mechanism was needed to aid communications. The tool selected was Digital Equipment Corp.'s VAX Notes for VAX/VMS systems. After this tool was purchased, a tutorial document was produced and distributed to all of engineering. The tool became a communication channel between SPPID and engineering. Minutes from committee meetings were posted on the bulletin board, and a conference was established for feedback on and discussion of process documents under review. SPPID provided encouragement and advice to help establish VAX Notes as a valuable communication mechanism. Intersite groups were the first to use it heavily. Today, there are more than 40 conferences containing over 2500 topics.

Measurements

Measurement is fundamental to Applicon's long-term goal of using statistical process control to view software development and set objectives.

The quality measurement committee developed a set of measurements to be used within engineering and began developing tools for obtaining these measurements. Some of the information it planned to gather included defects per KLOC (thousand lines of code), KLOC per person month, scheduling accuracy, and time spent on engineering activities. This was another area in which it was necessary to apply a dedicated resource to the effort, and the committee was dissolved. A more robust measurement plan was developed, and semi-annual reports on quality and productivity were provided. Each manager submits information to SPPID on a monthly basis and also after the project is released. This information provides tracking data on planned progress and on defects.

> Education and training have been critical components of many of Applicon's improvement activities.

However, this area is being revised. Measurements are being improved, and the collection of measurement data is being consolidated and automated. A lot of measurement information had to be gathered manually, and work began on developing tools and a data base structure to collect information on the relationships among the data. Certain areas were chosen for examination, such as scheduling effectiveness and efficiency, quality, and productivity.

Communication

Close communication with the engineering development groups has been an important part of virtually all of Applicon's process improvement efforts. Improving communication among the engineering development groups has also been a high priority. As described previously, the use of an electronic bulletin board has helped improve communication, especially between engineering sites and intersite groups. This section describes additional communications techniques used.

Documentation. Each member of the engineering organization has a software engineering notebook that contains all approved standards, guidelines, templates, and checklists, as well as information needed to perform an engineering task. A project manager's guide contains information on project management at Applicon. It includes training requirements for both managers and the engineering staff, as well as monthly reporting forms and requirements.

Newsletter. A semi-annual newsletter edited and produced by SPPID is distributed to engineering. Originally, contributing authors were exclusively from SPPID; however, a recent issue contains articles by engineering managers describing their experiences. The newsletter also contains information about various process improvement areas, process tips and techniques, and occasional interviews with senior engineering management.

EVALUATION

Applicon has been successful at implementing process improvements for various reasons. A software process model is being used for all active development projects. Project managers recognize its usefulness, especially with the management of interdependent projects. Expectations are clear because a common terminology is used, and the contents of project deliverables are clearly understood. That the model is currently being streamlined is another indication of its applicability, usefulness, and institutionalization. A set of engineeringwide quality objectives are being established for all active engineering projects. The setting of quantifiable and verifiable requirements for the entire engineering organization is crucial to the company's improvement efforts.

The company's ability to plan and meet schedules has shown significant improvements, especially its ability to develop and meet long-range planning objectives. Applicon is in the third year of a five-year plan. For the last three years, it has met or exceeded all planned, annual objectives. According to the vice-president of engineering, "Our latest release of software, Bravo Version 4.0, is our most significant accomplishment so far. In almost every area, we have exceeded our commitments. This is aided by our disciplined approach to software. We now have the sense of control needed to let people think creatively. This has helped us improve our efficiency to the point that we have covered ground in just one year that some of our competitors will take several years to accomplish."

On-time delivery of products has increased dramatically. In 1990, 53% of all products were released on schedule, 31% were one quarter late, and 16% were two quarters late. In 1991, 89% of all products were released on schedule, with the remaining products one quarter late. In 1992, 95% of all products were released on schedule, with the remaining 5% one quarter late.

Responsiveness to customer reported problems has been improved also. Today, more than 75% of all in-coming high-priority problems are resolved in fewer than 10 days. In addition, the company's software maturity has shown steady improvement. Progress has been measured through SEI assessments and self-audits, and Applicon has developed and is implementing action plans based on the results of these measurements.

Several elements have been critical to the successes achieved. First is senior engineering management's clear commitment to the improvement process. The vice-president of engineering solidly supports improvement efforts and has conveyed this to his staff. Senior engineering management is also involved in continuous process improvements, providing feedback and suggestions.

One of the clearest and strongest examples of this commitment has been engineering management's willingness to commit full-time staff to the improvement process. The pace of the process improvement efforts greatly accelerated when the SPPID was created. Initially, SPPID worked hard to define and document a process that is optimal for the organization and to educate engineering. Over the last year, SPPID staff has reduced staff levels because the process has been accepted throughout the organization and because each manager recognizes his or her responsibility to the process. One of the engineering directors stated, "Today, each manager is a process person."

Another essential element of Applicon's efforts has been to set measurable process improvement and quality objectives, which has encouraged managers to participate in many of SPPID's efforts to evaluate or deploy new methods and tools. As part of the company's overall quality initiative, each development group is required at senior engineering management project review meetings to describe, commit to, and prove it is progressing toward quality objectives. The requirement for explicit quality objectives and an emphasis on trying to make them measurable have given the development groups additional incentive to use such new methods as test coverage analysis.

The SEI maturity scale, and more recently, the capability maturity model for software have been very useful as frameworks for implementing plans to improve process and maturity. They have also helped Applicon move from a word-of-mouth culture to a documented and defined culture. Senior management focus on the maturity levels has facilitated the obtaining of commitment and process usage by the engineering organization. Improving process was not just an objective for SPPID. Engineering groups also had objectives related to maturity.

In many instances, SPPID has tried a tool or technique first to test its usefulness and applicability. For example, it used inspections heavily, reviewing many of the process standards and guidelines and modules from the software process model course. This practice had several advantages. SPPID became increasingly proficient in the technique and tailored it for the company. Local experience that was readily available helped groups to adopt and implement a tool and technique.

Some improvement efforts have not worked as well. Using committees to drive efforts has met with mixed results. The pace of their efforts has been slow because committee members have other responsibilities that typically receive higher priority than committee activities. Although committee activities are considered during employee performance appraisals, they affect an appraisal far less than primary job responsibilities do.

Committees work slowly, but they can be useful if they have clear objectives and senior management says it expects concrete results. Either lack of focus or of clear expectations from senior management cripples a committee. Committees are good for buy-in, but to be effective, they must strive for technical consensus, as opposed to compromise.

Another difficulty has been establishing an historical measurements data base. When it started, SPPID was not willing and, in some cases, not able to gather accurate historical data. Consequently, SPPID has taken longer to establish its data base than it would have liked because of the length of its development cycle. This fact has affected its ability to use this data for estimating and planning future projects.

The company's ability to plan and meet schedules has shown significant improvements, especially its ability to develop and meet long-range planning objectives.

FUTURE DIRECTIONS

Applicon intends to continue to improve its software development process so that it is repeatable, fully defined, and well-managed. It wants to plan and deliver products efficiently and effectively. This includes estimating and scheduling with a high degree of confidence and assuredness. One of the key elements in process improvements is the ability to measure progress. Applicon still needs to improve, streamline, and automate data collection and then improve its analysis of the data. Another area Applicon will address is reuse, how to maximize use of existing software, algorithms, and designs.

As Applicon's engineering organization grows and evolves, its level of maturity will also evolve. The software process model and supporting deliverables will change and improve, a healthy indication that the process is being followed. More tools will be used to develop quality software, including more state-of-the-art CASE tools.

The key to the future is continued management commitment. Applicon has implemented an organization and a development framework that will be followed regardless of organizational changes, such as turn-over or reorganization.

Process improvement has become part of the engineering culture and part of the corporate culture. Applicon must ensure that all areas of the company are aware of SPPID's progress and processes, and that everyone works together as a team.

MAKING CONCURRENT ENGINEERING HAPPEN

Philip R. Taylor

As vice-president of engineering for a manufacturer of gasoline dispensers for service stations, the author experienced firsthand the trials and triumphs of a large-scale concurrent engineering project. This experience illustrates some of the problems encountered—and their solutions—in implementing a team approach to product development.

In its simplest form, concurrent engineering is a team approach to product design, involving marketing, engineering, manufacturing, and other departments at all stages. This approach is in contrast to the sequential approach, still the most commonly used by US manufacturers, in which each department in turn performs its specialized steps and then hands off its work to the next department in the chain. Basic diagrams of the sequential and concurrent approaches are shown in Exhibits 1 and 2, respectively.

DISADVANTAGES OF SEQUENTIAL ENGINEERING

In Exhibit 1, the right-pointing arrows indicate a design information handoff to the next department. Optimistically, this represents forward progress, though in many companies, this transfer is cynically referred to as throwing the design over the wall. The left-pointing arrows indicate feedback of errors, incomplete data, and misunderstandings, which have been discovered by the next department in the chain. This error feedback process is an inevitable consequence of the sequential approach and often leads to a prolonged cycle of redesign at each stage of development.

A second consequence of the sequential approach is a general lack of accountability for any project schedule or for design errors. Each department can and therefore does blame the preceding department for incomplete or erroneous data. In addition, each department can and often does blame the next department in the chain for the extra time needed to receive feedback and correct errors.

A common third consequence of the sequential approach is a final product that does not meet customer requirements. This happens because:

☐ Marketing is not involved except at the very early stages, before most of the important trade-offs are made.

PHILIP R. TAYLOR is a business systems consultant for leading manufacturing and service firms.

Exhibit 1
Sequential Product Design

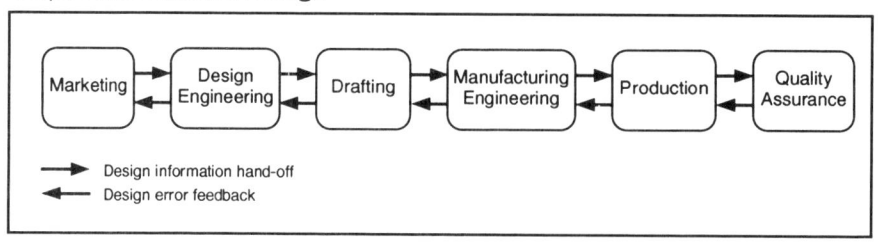

Marketing → Design Engineering → Drafting → Manufacturing Engineering → Production → Quality Assurance

→ Design information hand-off
← Design error feedback

Exhibit 2
Concurrent Product Design

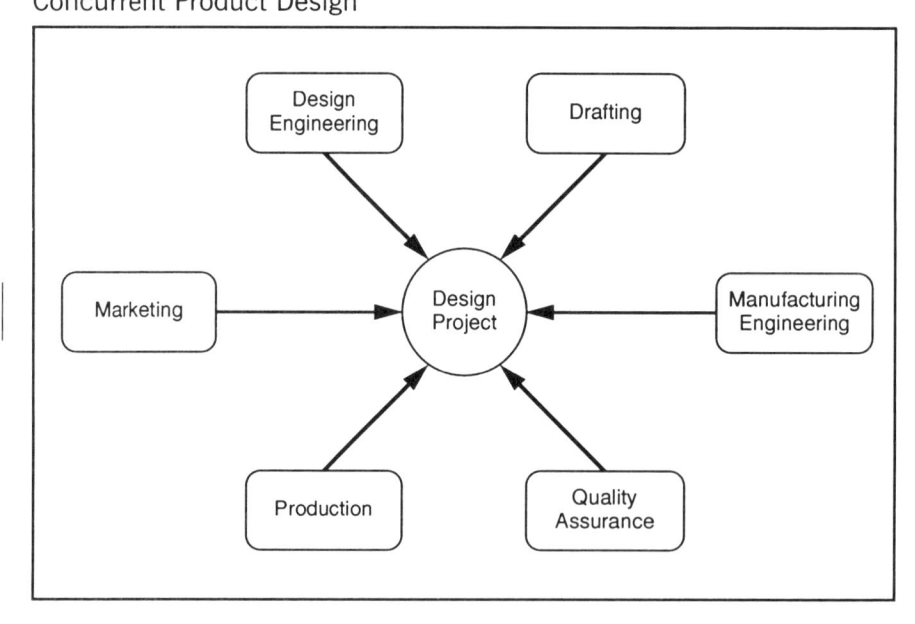

Design Engineering, Drafting, Marketing, Manufacturing Engineering, Production, Quality Assurance → Design Project

☐ As the schedule and budget run out, a compromised or even unfinished design is hurriedly thrown over the last wall, to manufacturing and quality assurance.

For these reasons, the traditional (i.e., sequential) approach to product design is slow and expensive, provides little accountability for results, and generally leads to products that are late to market and do not meet customer needs. Over time, the cumulative effect of this approach is to cede market share to competitive products.

ADVANTAGES OF CONCURRENT ENGINEERING

The two major advantages of concurrent engineering are a shortening of the product design cycle and the increased likelihood of a manufacturable product that also meets customer requirements. This is achieved through setting up a dedicated team of representatives from all departments at the very start of the design effort. The team works together from the initial concept stage all the way through initial production and strives to eliminate the over-the-wall mentality and error-feedback loops that characterize the sequential development process.

In its simplest form, concurrent engineering is a team approach to product design, involving marketing, engineering, manufacturing, and other departments at all stages.

Because marketing stays involved all the way into production, it can ensure that the critical features are not compromised along the way. Because manufacturing is involved in the early project stages, it can influence the design for ease of manufacture before it is locked in. Because everyone has a chance to be heard from beginning to end, there is less chance for nasty surprises and finger pointing. Because team members are directly responsible for both the project and the product, accountability is not an elusive issue. Finally, because a dedicated team can provide valid project planning and monitoring input, effective schedule and budget control mechanisms are possible.

OBSTACLES TO CONCURRENT ENGINEERING

Does this sound simple enough in principle? It is, and many companies have used a project team approach with good results. However, these companies have succeeded only by avoiding the hidden obstacles and pitfalls of the team approach. As with other good ideas, there can be a large and dangerous gap between the theory and practice of concurrent engineering. The most common and serious barriers to the successful implementation of concurrent engineering teams are discussed in the following sections.

Weak Team Leader

A weak team leader will doom a project before it begins. A strong team leader must be 100% dedicated, act as a generalist rather than a specialist, and be able to communicate effectively with many different departments and disciplines. He or she must also prevent or overcome corporate bureaucracy and organizational inertia.

Weak Team Members

Weak team members result when departments assign representatives whose main qualification is that they have the time or when team members retain critical duties within their respective departments. Like team leaders, team members must be highly dedicated, able to communicate effectively, and overcome (not become) bureaucracy.

Weak Horizontal Authority

Many traditional manufacturing organizations have strong vertical lines of authority within functional areas but relatively weak horizontal power lines, as illustrated in Exhibit 3. The vertical lines link managers, supervisors, and subordinates in the traditional corporate hierarchy. The horizontal lines represent interdepartmental communication and cooperation. In most manufacturing companies, the vertical lines carry the power of hiring and firing, promotion, and common functional responsibilities, whereas the horizontal lines are relatively weak and informal. However, a project team can operate effectively only when there is strong horizontal influence and communications, which requires nothing less than a culture change in many companies. Sometimes, the culture change begins with management recognition of the connection between the company's vertical and horizontal power imbalance and its waning competitiveness.

Unrealistic Schedule

A compressed time to market is a reasonable expectation and is indeed one of the major benefits of concurrent engineering. However, an unrealistic schedule

Exhibit 3
The Corporate Power Grid

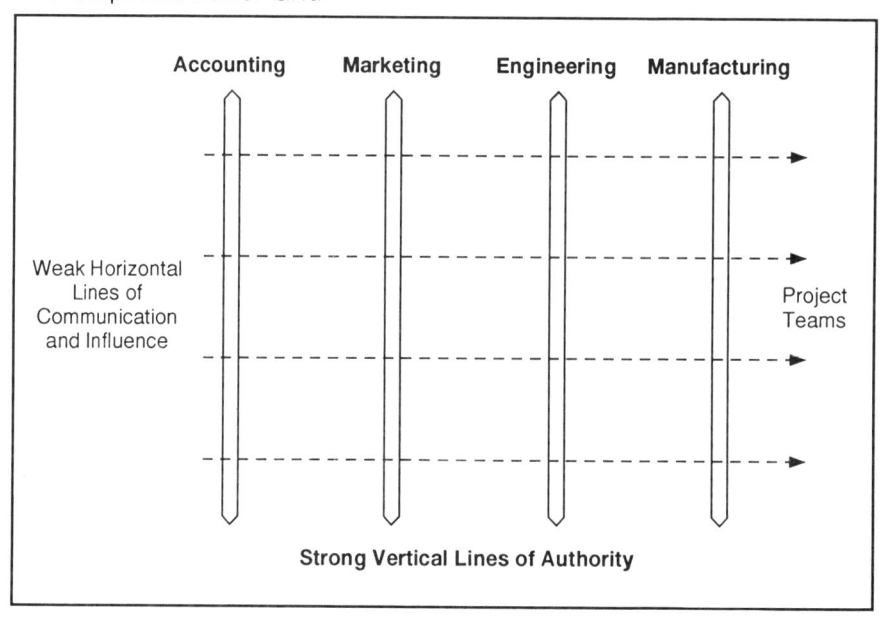

is just as destructive in a concurrent engineering environment as in a sequential engineering environment. Many projects have failed because managers or team members overestimated the time-compression benefit of the team approach, especially when trying it for the first time.

A consequence of the sequential approach is a general lack of accountability for any project schedule or for design errors.

Poor Team Dynamics

If for any reason team members do not relate well or interact smoothly, the poor dynamics will eventually scuttle the project. Common symptoms of poor team dynamics include intrateam power struggles, unhealthy internal competition among team members, a lack of mutual trust and respect, apathy, and confusion over specific objectives.

Poor Team Processes

Even if all team members are enthusiastic, dedicated, talented, and individually productive, the team will still fail without the correct processes. These processes include up-front schedule and resource planning for each development phase, regular team meetings with published minutes, and an intense focus on hitting intermediate project targets and milestone dates.

Lack of Management Support

This is the kiss of death for a concurrent engineering project or indeed any other nontraditional endeavor in a corporation. If management has weak faith in the value and principles of concurrent engineering or condones any territorial behavior, the project will most assuredly fail.

BACK TO REALITY

It is worth emphasizing that the foregoing list describes obstacles to effective implementation of concurrent engineering principles, not weaknesses in the principles themselves. In fact, it might be argued that a properly executed concurrent engineering project would have no problems or drawbacks of any kind. That may be true, but the same could be said of other approaches, including the sequential approach. In an ideal sequential design process, each department would faithfully perform its particular duties in a timely, cost-effective, and error-free manner. Each department would then transfer its work completely and unambiguously to the next department in the chain, thus eliminating the error-feedback cycles. Likewise, in an ideal concurrent engineering process, all team members would contribute their experience and enthusiasm from start to finish, and all departments would fully support the team's efforts. In practice, these ideals are seldom achieved, but the true test of any theory is how practical it is to implement in a real-life, competitive environment.

As vice-president of engineering for a manufacturer of gasoline dispensers for service stations (the company), the author experienced first-hand the trials and triumphs of a large-scale concurrent engineering project. This experience illustrates some of the problems encountered—and their solutions—in implementing a team approach to product development.

PUTTING THEORY INTO PRACTICE

In early 1991, the company officers were debating a major decision to develop an entirely new series of gasoline dispensers using the latest technology and extensive customer input. The company's then-current dispenser model line had been designed several years before, and major competitors had recently introduced models with improved technology and features. In summary, the company was faced with a game of catch-up.

The company had traditionally operated development projects using a fairly rigid sequential approach, with very specific lines of responsibility drawn between various engineering groups (e.g., mechanical, electronic, software, drafting, and model shop) and between engineering and other departments, such as marketing, manufacturing engineering, and quality assurance. There was also a rather elaborate, even convoluted, process for project funding approval, along with the usual amount of shifting priorities, day-to-day emergencies, and product enhancements competing for development resources. In other words, this was a typical manufacturing company environment.

DOING HOMEWORK

In these difficult though not unusual circumstances, the author worked with other managers to begin the imposing task of developing an entirely new product line. Because no project funding had been established yet, an informal group with shifting membership struggled through the early phases of product definition.

Rather than jumping straight into a design effort, the unofficial team recognized a need to do some homework first. This meant getting customer input, performing competitive analysis, and performing regulatory environment analysis.

CUSTOMER INPUT

The first step in any concurrent engineering approach is to get plenty of customer input. In this case, the company sold its dispensers through distributors to independent service station owners and smaller oil companies but also sold directly to major oil companies. These diverse customer segments had differing priorities and requirements for product features, pricing, and delivery times.

Recognizing this, the author's team created concept drawings, constructed wooden models of alternative designs, and took them on the road to use as props in customer interviews. The team members held meetings with each of several customers and customer groups and received valuable feedback that was incorporated into the design. This is the first step in the team approach—making customers part of the team, even if only as outside advisors. A more complete approach is to invite customer representatives to take an active, ongoing role on the actual design team. Although the author's team did not go this far, it benefitted greatly even from the limited approach.

> **The two major advantages of concurrent engineering are a shortening of the product design cycle and the increased likelihood of a manufacturable product that meets customer requirements.**

In parallel with this basic market research effort, the author assigned a small but multidisciplinary team to perform competitive analysis, a second crucial step in a concurrent engineering process.

COMPETITIVE ANALYSIS

In its simplest form, competitive analysis involves getting competitors' product literature, visiting sites with competitive products, and interviewing their customers. The author's team did all that but also purchased competitors' dispensers for closer examination. The author's team tested their performance and dissected them to component parts and analyzed them in various ways, including estimating labor and materials costs. This was followed by construction of a features-comparison matrix in which the team assigned a relative score on each feature for each competitor. Thorough evaluation of all aspects of such a complex product required help from purchasing, finance, marketing, manufacturing, field service, and quality assurance as well as engineering.

Competitive analysis can be accomplished well only by a team with representatives from all company areas. The effort will fail if it is perceived solely as a marketing or engineering function, because some important product aspects will probably be overlooked or underrated.

REGULATORY ENVIRONMENT ANALYSIS

Another vital step in product design homework is to analyze the regulatory and legal environment in which the product must perform. For the company's products, this required consultation with Environmental Protection Agency officials, Underwriters Laboratory (UL) inspectors, and internal and external experts on Federal Communications Commission (FCC) requirements and product liability laws. The product's regulatory environment may be more or less complex, but a team approach must always be taken in this area, because no single department has the expertise in the various technical, financial, and legal disciplines.

This is when following concurrent engineering principles can really pay off, by giving complicated interdisciplinary issues the proper attention early in the design process. The consequence of leaving regulatory matters to design engineers alone is that they often consider them too little or too late in the project.

SELECTING A TEAM LEADER

As the informal homework team made its halting progress, the budget paperwork finally caught up, and the author's team officially launched the design project. This led to the first major decision, selecting a development team leader.

This decision was complicated by the company's time-honored tradition of using a sequential design process. In the prior year, the author had already introduced concurrent engineering practices on smaller projects, with mixed results. The biggest single problem had been getting active participation in design meetings by groups outside of engineering. Although most people in marketing, manufacturing, customer service, finance, quality assurance, and other areas claimed to support the concurrent engineering process, in practice their representatives were often too busy to attend design meetings. In some cases, they attended but did not really participate, only to complain later that their ideas had not been considered. It was also common for attendees to defer even minor decisions because they had to check with the boss first.

Considering these earlier difficulties and because of the bet-the-company nature of this project, the author chose to function personally as the project manager. This decision was not arrived at lightly nor as part of an ego trip. Rather it seemed the only possible way to mount a major concurrent engineering effort in a company with a decades-old tradition of very strong vertical power lines and very weak horizontal ones.

For any concurrent engineering project to succeed, the team leader must have, and selectively use, clout. The weaker the teamwork culture, the higher up the corporate ladder the team leader must be in order to compensate.

SETTING UP THE DESIGN TEAM

For similar reasons, it was necessary to select movers and shakers as team members. Their ability to organize, communicate, and inspire others was deemed at least as important as any technical abilities.

Although the author was in a position to choose strong team members from engineering, getting strong representatives from other areas required some behind-the-scenes negotiating with various department heads. Eventually this effort paid off, and a qualified team was formed.

The team leader must enlist the best influencers and communicators as team members, especially those from departments outside engineering. Their primary roles will be representing and getting support from those departments, more than any direct design-content contribution. Extra effort will always be needed to get the best people, but it is crucial to make that effort.

The next problem was to get a dedicated effort from each team member, given that most of them retained other responsibilities within their functional areas. In an ideal concurrent engineering project, team members are 100% dedicated to the project and are relocated to the same work area. In many real-life situations, including this one, team members cannot be relocated and also have various day-to-day emergencies competing for their time.

> Because marketing stays involved all the way into production, it can ensure that the critical features are not compromised along the way.

SUPPLIERS AND OUTSOURCING

Recognizing the constraints on internal resources, the author worked with the team to identify any holes in the skills and time capacity needed for success. The author then filled those holes by selectively using outside contractors and vendors. For example, the team agreed that the services of an outside industrial design firm could greatly enhance the product appearance. The team visited several candidate firms and conducted a formal selection process, which eventually added this key vendor to the design team.

It is critical for success that the team have skill and time resources adequate to the task at hand. An open and objective self-examination is needed to reveal areas in which the team must be supplemented by outside talent. A complete concurrent engineering team will often consist of contractors, vendors, and customers in addition to employees.

PROJECT DISCIPLINE AND GROUND RULES

The team leader has the unique challenge of propelling this diverse group in a forward direction. This challenge is compounded when team members from different companies, or even from various departments in the same company, have contrasting backgrounds and are not used to working together. In other words, the same diversity that is the strength of the concurrent engineering approach can also be its weakness if not managed properly.

Anticipating the potential for confusion and delays, the author compensated by enforcing a formal project discipline. The centerpiece of this project discipline was a set of ground rules established at the very first project meeting:

- The team would meet once per week, always at the same time on the same day of the week.
- Team members were to reserve these designated meeting times on their calendars for the duration of the project.
- Attendance was mandatory. Team members were to designate a back-up person to represent their department in case of unavoidable absence.
- Team members were expected to actively participate and support team goals while also representing the interests of their functional areas.

- Meetings would follow an agenda (but not a rigid one) focused on schedule, milestones, special problem solving, and major group decisions.
- Minutes would be recorded and distributed within 24 hours after each meeting and would summarize only major decisions and action-item responsibilities.

The general purpose of the meetings was to provide a regular forum for schedule reviews, major group decisions, and problem solving. It was not meant to replace the daily informal communication among team members. In a typically chaotic manufacturing company environment, a design team needs the anchor of a disciplined project methodology. Good design cannot happen without a healthy project environment of stability and control.

DESIGN GOALS

With a good team and an appropriate project discipline in place, the real work could begin. However, a critical step at this point was to achieve consensus on the broad design goals. These goals included cost targets, major feature and functional requirements, operator interface requirements, regulatory and safety requirements, special requirements for manufacturability, standards for quality and durability, and basic model configurations. Other products might have different design criteria, but the key point is the need to establish and document all critical design goals prior to investing heavily in design effort.

This consensus did not come easily, but the author insisted that disagreements on critical design goals be settled before proceeding further. Good concurrent engineering practice demands that sufficient up-front planning and debate take place to form a consensus on all critical design issues. Otherwise, these unresolved issues will return to confound progress at some later point.

PROJECT PLANNING

Another significant investment of up-front effort was needed to establish a project task schedule. For this large project involving many individuals, the author recognized that only an automated scheduling tool would suffice. The team used a PC version of Project Workbench and assigned a project administrator to gather schedule and cost estimates from team members. These estimates were then input to the PC and a project PERT chart was generated. After several iterations, the team established a believable schedule at an appropriate level of detail. This proved to be a difficult and time-consuming task, but it also forced considerable communication and coordination to take place early in the project. This early attention to schedule coordination eventually proved very beneficial in overall time saved. Keeping the PERT chart up to date as things changed proved even more difficult but also paid off on several occasions when the team saw the critical path shift and then adjusted plans and resources accordingly.

A formalized project schedule is beneficial to any large project but really indispensable for concurrent engineering. This is because concurrent engineering requires the telescoping of time frames through parallel rather than sequential execution of project tasks. Without a carefully constructed, documented, and maintained task schedule, the effort will fall apart or, more likely, revert to a sequential process.

> **Because manufacturing is involved in the early project stages, it can influence the design for ease of manufacture before it is locked in.**

PRODUCT PLANNING

As a PERT chart is the road map for a project, so a bill-of-material is the road map for a product. In fact a PERT chart and a bill of material are both structured lists (of project tasks and product components, respectively) that are useful in planning projects and products. The author placed early emphasis on developing a bill of material for planning purposes. This planning bill of material had little low-level detail at first but instead defined multiple dispenser models based on common, modular subassemblies. This provided a high-level guide to help direct and prioritize the design tasks, and the lower-level component details were filled in as the project progressed.

This was the company's first experiment with a planning bill of material on a large-scale development project, and there was considerable resistance to the idea within the project team. However, a few key designers were enthusiastic enough to take on the job of creating, updating, and distributing the planning bill of material to the rest of the team, who then saw its value.

In a concurrent engineering process, a planning bill of material is extremely useful because it focuses attention on the end goal (a finished, manufacturable, serviceable, marketable product) right from the beginning of the project. In contrast, the sequential approach typically leaves bill-of-material construction to a very late stage in the design process, with inevitable errors often requiring major redesign.

RAPID PROTOTYPING

Although it is critical to invest adequate up-front time in homework, team selection, and project and product planning, it is equally critical to move quickly to create a product model or prototype once basic plans are in place. There are three reasons for this:

- A model or prototype is a concrete, if small, evidence of success and sets a positive tone early in the project.
- A model or prototype is a great visual aid for flushing out design errors and misunderstandings among team members.
- A model or prototype is useful in soliciting early feedback from prospective customers.

The author uses the model shop to good advantage in preparation of various models and prototypes, and these generated team excitement beyond what any paper designs could. Good concurrent engineering practice requires rapid development of a design sample, even if only a crude model, to get the juices flowing, set the tone for success, and get critical early feedback from both team members and outsiders.

PRODUCT TESTING

Any complex product design requires extensive testing before being put into production. For the author's dispensers, testing included:

- Performance of electronics and hydraulics over a wide range of temperatures in environmental chambers.
- Compliance of electronics with FCC regulations.
- Compliance of electronics, hydraulics, and mechanical components with UL guidelines.
- Compliance of entire design with safety and quality standards.
- Compliance of software with functional and performance specifications.
- Field testing of prototypes in selected customer locations, representing various use and climate profiles.

This extensive testing cannot be effectively performed by design engineers, programmers, and technicians alone. It requires a collaborative approach involving marketing, manufacturing, quality assurance, field service, external testing and auditing services, and customers themselves. And though effective testing specifically requires a team approach, it is also the project phase in which the value of an overall team approach becomes most apparent. If the team has successfully followed concurrent engineering principles to this point, testing will be a confirmation of the design rather than the beginning of major corrections. The author's testing went smoothly, and though it caught many small problems and several not-so-small ones, all required changes were manageable with no major surprises or project setbacks.

In successful concurrent engineering projects, the heavy time investment in up-front planning and coordination begins to pay off during the testing phase and continues throughout the remainder of the product life cycle.

TRANSFER TO PRODUCTION

The final step of a sequential development project, from engineering's viewpoint, is the physical transfer of all design drawings and bills of materials to manufacturing engineering. This is often the traditional over-the-wall process that actually begins the project from manufacturing's viewpoint. In concurrent engineering projects, however, transfer to production is done in parallel with design.

> **A weak team leader will doom a project before it begins.**

With representatives from manufacturing engineering and purchasing on the design team, the team shared drawings and planning bills of materials at regular intervals throughout the project. In return, the team received valuable feedback on parts availability and cost, ease of fabrication and assembly, and drawing and bill-of-material errors. By the time the design had passed all performance tests, the team already had preproduction quantities of long-lead parts on hand, had developed training videos for critical assembly operations, and had debugged the drawings and bills of materials. Although the first production units were not completely trouble free, no major redesigns were required at any point.

Concurrent engineering principles require close involvement by purchasing and manufacturing people right from the start of the design work. This assures that the design will be cost-effective, based on readily available components, and easy—or at least possible—to fabricate and assemble.

A HAPPY ENDING, BUT . . .

The new dispenser line is a market success and has been extended to many model variations without ever having to change the basic modular core design. Relative to all earlier model lines in the same price class, this new product family:

- Has more features.
- Has more flexibility and extendibility.
- Is easier to manufacture.
- Has fewer parts.
- Costs less to produce.
- Was developed in less time and at a lower project cost.

In summary, the project reached a successful conclusion by all important measures. However, this happy ending was not easily achieved, and there were many difficulties along the way. Many of these difficulties have not been described in any detail, because the focus of this article is on solutions rather than problems. For example, the team fell behind schedule on several occasions, but with the updated PERT chart, the team members could see how far behind they were, why, and what was needed to get back on track.

PEOPLE PROBLEMS

In the early stages of the project, there were also a few people problems. These might have sunk the project before it could build momentum, if firm steps had not been taken immediately. Two examples illustrate some of the dilemmas facing a team leader in a company with strong vertical authority and relatively weak cooperation among departments.

After the first team meeting, some members started arriving very late or not at all. The alternatives of a team leader for dealing with this problem are:

- Start without them (i.e., ignore the problem, allow the precedent, and cripple the effectiveness of the team).
- Wait for them (i.e., punish the rest of the team and abdicate basic project control).
- Complain to their supervisors (i.e., escalate and publicize the problem, at the risk of touching off interdepartmental warfare).
- Quietly fetch and bring them to the meeting (at the risk of embarrassing them in front of the team).

Although all the alternatives were unpalatable, the author chose the last one because it addressed the problem most directly and with the least fuss. It worked very well, too, and resulted in nearly perfect on-time attendance after being applied only a few times.

A second critical problem was the apparent abdication of responsibility by a certain team member who always had to check with his boss (who was usually traveling) before approving any decision. This occurred despite the understanding that team members were to be fully authorized representatives of their respective departments. By digging beneath the surface, the author discovered that the situation was rooted in the desire of the team member's boss to exercise some degree of remote control over the project. Recognizing the potentially devastating effect of delays and indecision on team morale, the author worked directly with the boss to explain the negative consequences of his actions and to obtain a written agreement authorizing the team member to speak for his department. The author also made sure to include the boss in periodic reviews of models and prototypes in order to increase his comfort level and feeling of involvement.

A team leader must address people problems firmly and immediately, putting the interests of the team above any individual member. A team leader must be a coach, cheerleader, judge, counselor, and above all, a facilitator of the team.

LESSONS LEARNED

The lessons from this experience are encapsulated in the following list of advice to team leaders:

- Do homework—in advance—on customer requirements, competitive environment, and regulatory issues.
- Pick movers and shakers as team members.

- Relocate as many team members as possible to the same work area and dedicate them 100% to the project.
- Take an honest, open inventory of team skills and capacity. Supplement the team when necessary with vendors, contractors, and customers.
- Establish a disciplined project methodology at the outset.
- Form an up-front consensus on all critical design goals.
- Establish and maintain a project PERT chart. Use it to run the project.
- Use a planning bill of materials to guide project activity and as a measure of accomplishment.
- Use rapid prototyping as a short-term success symbol and to gain early feedback from team members and customers.
- Perform adequate testing using a multidisciplinary team approach.
- Perform production-transfer tasks in parallel with development.
- Handle all people problems firmly and immediately.
- Put the interests of the team above any individual.
- Always remember that the team leader's role is to serve the team, not the other way around.

In controlling project complexity, concurrent engineering shifts the focus from technical issues to people issues.

CONCLUSIONS

The sequential approach to product development focuses mainly on technical issues while attempting to control project complexity by containing, even limiting, interactions among various disciplines. It applies a divide-and-conquer strategy in a time sequence, with value placed on technical specialties within each functional group.

In contrast, the concurrent engineering approach shifts the focus from technical issues to people issues while attempting to control project complexity through up-front planning and project management tools. It also applies a divide-and-conquer strategy, not in a time sequence, but by breaking the problem into smaller pieces that can be worked on in parallel by functional groups. Although technical specialties remain important, there is a far greater emphasis on communication and people skills.

SECTION 5

TECHNOLOGY TRANSFER

Technology transfer covers various activities, including the internal transfer of technology from the R&D or engineering department to the manufacturing department function within a company based in one country. It also includes a multinational company's transfer of technology from a laboratory or operations in one country to a laboratory or operations in another country. Finally, it includes the transfer of technology from a research consortium supported by many companies to one of its members.

Michael W. Schoonover describes how one organization successfully (if not painlessly) reengineered its technology delivery process.

Don Tijunelis addressed technology transfer during the progress of a project—communication within an R&D organization and communication between a corporate R&D unit and a divisional profit center for the purpose of commercialization.

Keith E. McKee maintains that the design of a product and its manufacturing process must be closely linked. He presents two problems that can occur when the design of a manufacturing process is neglected; he also shows how these problems can be avoided. In addition, McKee examines the problems of transferring manufacturing to the manufacturing plant. He identifies six factors that are critical in making such a transfer effective.

James R. Key addresses a relatively new aspect of technology transfer: transfer of technology from a research consortium to a company. He outlines the several stages involved in such a transfer.

Sakharam Patil describes how organizations can use cross-functional teams to improve product development and to ensure that products are delivered to market on time.

Does the large company approach to technology transfer in simaller companies? Glenn Dugan shows that being small does not mean thinking small.

REENGINEERING THE TECHNOLOGY DELIVERY PROCESS

Although challenging and time-consuming, a well-executed reengineering effort can help an organization streamline its R&D processes and thus reduce time to market for new products and technologies.

Michael W. Schoonover

Lured by the promise of miraculous improvements in productivity with little effort and no pain, many companies joined the reengineering frenzy of the past several years. However, most reengineering efforts fail. In this author's opinion, these failures do not indicate any shortcomings in the concepts of the reengineering process; instead, they demonstrate that reengineering is neither easy nor painless.

Reengineering is not a temporary fix; it is a time-consuming, ongoing process charged with high levels of emotion and conflict. Many so-called reengineering efforts are nothing more than old-fashioned reorganizations and layoffs. Reengineering properly done requires great strength of will and commitment.

Any reengineering process involves the following basic steps:

1. Identifying the critical business issues and the key processes.
2. Documenting and analyzing the current process.
3. Designing a new process.
4. Developing a new measurement system.
5. Implementing changes.
6. Establishing an infrastructure for continuous improvement.

This case study examines the reengineering of UOP's research and development (R&D) organization. Rather than simply downsize or reorganize the R&D organization, the goal of the reengineering effort was to deliver technology three times faster than the current R&D process, without increasing R&D costs. UOP is a successful business; the reengineering effort was not an attempt to revive an ailing company. Instead, the reengineering initiative was intended to help UOP remain successful. The challenge was to convince people of the need for change, before significant problems arose.

MICHAEL W. SCHOONOVER, PhD, is a senior manager in R&D at UOP.

The reengineering effort at UOP required radical changes in thinking and in operating methods. This case study describes the steps taken, what worked well and what did not, and the lessons learned along the way.

RECOGNIZING THE NEED FOR CHANGE

In 1992, UOP senior management decided that the company needed to redesign its R&D process. Changes in the marketplace necessitated a much faster rate of new product delivery. In addition, UOP's strategy for long-term growth required a faster, more responsive R&D function. From a financial perspective, traditional expansion of R&D resources to meet these demands was neither possible nor desirable. Instead, management decided that the company should completely overhaul its technology delivery process.

The *technology delivery process* encompasses all the steps from idea generation to successful installation of the first commercial unit or product. Time is the critical business issue for the technology delivery process. UOP's continued competitiveness depended on R&D's capability to reduce the time required for delivering innovative technology and products, without increasing R&D resources.

GETTING STARTED

The design phase of the reengineering effort started in March 1993. The Rummler-Brache Group served as consultants; they provided a specific methodology for proceeding with the design phase and served as facilitators for this effort. As facilitators, they provided an objective perspective on the organization as it currently existed. This perspective was clearly needed, because many people within the company did not recognize the need for change.

The consultants emphasized the value of setting short deadlines for the completion of tasks as well as the importance of maintaining those deadlines. This kept everyone moving forward despite anxiety over not yet having the perfect system. Just getting started was the key requirement.

The vice president of R&D, who was also the champion of the reengineering effort, defined the critical business issue as reduction of cycle time in order to maintain competitiveness. This critical business issue was translated into a specific objective for the reengineering effort: Reduce the cycle time for technology delivery by a factor of three. As mentioned, the technology delivery cycle spanned the entire range of innovation, from generation of new ideas, through invention and development, to the first successful commercial operation.

Previous benchmarking studies showed that UOP's cycle time for delivering breakthrough technologies (as opposed to incremental product improvements) was seven to 10 years. Consequently, the goal for the reengineering effort was clear, measurable, and easy to understand: Reduce cycle time for commercialization of breakthrough technology to three years.

CREATING THE NEW DESIGN

For effective change, the new design had to come from those people who best understood the strengths and weaknesses of the current system. The reengineering effort also required the support of the current management staff. Without their support, the entire effort could have degenerated into an exercise in protecting the status quo. Several teams were created to ensure buy-in by those people who were in the best position to either support or resist any proposed changes.

> Time to market is the critical business issue for the technology delivery process.

The vice president of R&D and his staff acted as an executive team to oversee the reengineering effort. The executive team set the goal for the reengineering project and championed the process.

The executive team then selected a group of middle managers to act as a steering team. The steering team selected the members of the design team—the people who would develop the new R&D work design. The steering team was responsible for overseeing the work of the design team and contributing knowledge and experience to help the design team formulate new concepts. The facilitators from Rummler-Brache also helped guide the efforts of the design team.

The steering team selected a design team of 18 people, who represented R&D and other functions of UOP that had strong interfaces with R&D. One-third of the team's members represented departments outside the R&D organization; they came from marketing, engineering, and manufacturing. Design team members were chosen for their direct knowledge of the many subprocesses that make up the technology delivery process. Consequently, the design team consisted of technical specialists, scientists, and engineers, rather than managers.

Analyzing the Current Process

The first step in the redesign process was to chart the existing technology delivery process and verify the accuracy of this documentation by comparing it with several previous R&D projects. This was a difficult task. Previous attempts had been unsuccessful, and skeptics believed that mapping UOP's R&D process was impossible.

However, the design team successfully mapped UOP's existing process. Their success was due, in large part, to the consultants' emphasis on business processes. The consultants provided the design team with a preliminary sketch of the current R&D work process, based on interviews they conducted with numerous senior and middle managers. The design team members then added their more detailed knowledge of existing work flows. The product of this exercise was a detailed outline of the current work process, which contained 128 separate steps.

The design team also identified a list of *disconnects*, or inefficiencies, in the current system that slowed the process of technology delivery. These 125 disconnects fell into the following categories:

- Insufficient customer input.
- Lack of long-term strategy.
- Poor project definition.
- Poor prioritization of resources.
- Lack of teamwork.
- Poor communication.
- Duplication of effort.
- Lack of fundamental approaches to technical problems.

Analysis showed that when R&D avoided or eliminated these problems, projects were successful and cycle time was fast. Otherwise, technology delivery was slow or unsuccessful. Most important, the current process did not ensure the prevention of these inefficiencies. They were overcome only by unique, individual efforts or just plain luck.

Designing a New Process

The list of disconnects was an important input to the next step in the reengineering effort: designing the new technology delivery process. The goals for designing the new process were to retain those parts of the existing system that worked well and to eliminate or revise those elements that did not. The newly designed process also needed to remove all the disconnects identified in the analysis of the existing system. The critical design criteria were as follows:

- Customer focus.
- Creativity.
- Teamwork.
- Working in parallel.
- Improved communication.

These design criteria combined lessons learned from the identification of disconnects in the existing process and a set of industry's best practices, provided by the consultants.

The creation of a new design began during late June 1993. This part of the reengineering process required creative, out-of-the-box thinking, which in a team environment requires respect for other people's ideas. The facilitators and the design team leader played pivotal roles in maintaining a positive, cohesive effort.

This stage of the reengineering effort also required effective communication between the steering team and the design team. As the new design took shape, the steering team had to understand the design team's thought process. The two teams held biweekly meetings for reporting progress. During these meetings, the steering team could respond to the design team's requests for input and its members could raise issues that required additional study on the part of the design team.

> **The new process should retain those elements of the existing process that work well and eliminate or overhaul those elements that do not.**

During this phase of the reengineering effort, moments of tension occurred. For example, one concerned manager stated that letting the design team construct a new organization was like letting his 16-year-old son have the keys to a Corvette. A design team member replied, "We are not just driving the car; we're changing the engine!" This summarized the magnitude of the necessary changes. UOP would not achieve its reengineering goals by simply tweaking its existing system.

This phase was successfully completed because of the effective lines of communication established earlier in the reengineering process. The result of this creativity phase was a new design for a technology delivery process applicable to all R&D projects. The new process consisted of five major phases:

- Project selection.
- Project planning.
- Proof of commercial feasibility.
- Customer acceptance of technology.
- First successful installation.

The project selection phase focuses on selecting the right projects before the company commits significant resources. Early and continuous customer focus are important elements of the project selection process. Creativity is specifically recognized as a process step.

The project planning phase involves forming a functionally complete project team at the outset of the project and holding the team accountable for technology delivery throughout the project. The project team has the authority to make decisions and control the resources needed to achieve its specific objectives.

Both project selection and resource planning were considered crucial to the success of the design. Proper decisions in the early stages of the technology delivery process eliminate most of the disconnects that typically arise during subsequent stages.

Proof of commercial feasibility combines the R&D and scale-up functions as parallel efforts rather than sequential tasks. To ensure maximum customer acceptance, the project team delays freezing product specifications as long as possible.

Customer acceptance and first successful installation are the final two steps in the reengineered process. With the new technology delivery process, the project team remains intact until the customer is satisfied with an operational success in the field. This ensures the continued availability of the people who have the knowledge and the ability to provide technical support for the first installation.

The design team also identified 21 supporting components needed for the success of the new process. These supporting components were documented as 21 specific recommendations, which fall into the following categories:

- A new organizational structure.
- Project teams.
- Process measures.
- Customer input mechanisms.
- Manufacturing integration issues.
- Productivity tools.
- Resource allocation.
- Career development for all employees.

The final deliverables from the design team were the new design, the supporting recommendations, and an implementation strategy. The design team's efforts culminated on schedule, in August 1993, with a formal presentation to the executive team and the steering team. This presentation marked the starting point for the next phase: implementation.

IMPLEMENTING THE NEW DESIGN

With the design phase complete, UOP formally disbanded the design team and immediately initiated the implementation effort. The design team's 21 recommendations were grouped into eight categories, and a separate implementation team was chartered for each group of recommendations. These teams averaged between eight and 10 members each.

Experience from the design phase clearly showed that the success of the reengineering process depended on effective communication about reengineering activities, throughout the entire organization. Therefore, one implementation team focused solely on communications.

To ensure that the reengineering effort remained customer focused, the steering team for the implementation phase included representatives of functions throughout the company. Each implementation team had a team sponsor who was a member of the steering team. To maintain continuity from the design phase to the implementation phase, each implementation team included at least one member of the former design team. The implementation teams also included R&D personnel and people from other parts of the organization.

The success of the reengineering effort depended on gaining the acceptance of the entire staff. This necessitated the direct participation of as many people as possible in the change process. For this reason, the work of the design team was intentionally stopped at the conceptual level and a new set of people took responsibility for implementing the new process. Thus, these people also contributed to creating the new organization and developed a sense of ownership for the results of the reengineering effort.

> **The success of the reengineering process depends on effective communication throughout the organization.**

The implementation effort focused first on the recommendations related to organizational changes. Two teams were chartered with responsibility for those changes. The communication team also began its work at this time.

Designing the New Organization

Implementing the reengineered technology delivery process required major changes in the existing organizational structure. The existing, traditional structure involved handoffs between a formal research function and a formal development function. In the new technology delivery process, these activities would be performed in parallel. In the old structure, UOP spread its R&D organization across four separate locations, which resulted in some functional duplication as well as inefficiencies in the R&D work process. These characteristics were inconsistent with the goals for the new design.

The new structure required a skill-focused—rather than a site-focused—structure. In the existing, site-focused structure, the research function and the development function each had pilot plants for testing new product formulations. Each site had its own infrastructure and organization for operating and maintaining these plants. Consequently, when a research prototype was handed over to the development function, that group would use its test equipment to retest and reformulate the catalyst before starting the actual development work.

Consolidation of Functions. The new organization consolidated and simplified the R&D structure from 53 separate departments down to 25 departments grouped into three centers:

- Technology centers.
- Basic and applied sciences skill centers.
- Administrative services.

The employee-to-supervisor ratio increased by a factor of two, from approximately seven employees per supervisor to a ratio of 15:1.

The technology centers house approximately 15% of the total R&D staff. These centers are organized around UOP technologies and provide continuity within each technology area. The technology centers serve as the direct link between R&D and the customer; they also help select, rank, and sponsor the right R&D projects to support their respective business areas.

The skill centers provide tools and resources that cut across technology lines and are shared by all technology areas. These centers are responsible for maintaining and enhancing core skills and tools and providing a flexible resource base. Most of the R&D staff are located in skill centers.

The Administrative Services group provides support for the technology centers and the skill centers. Its responsibilities include training courses, budgeting tools, site management policies, and the health, safety, and environmental group.

Staffing Assignments. The implementation teams refined the design team's recommendations, by determining the number and the types of technology centers and skill centers. In an unprecedented step for UOP, the implementation teams also provided a list of recommended persons to manage these centers. The vice president of R&D made the final selection of managers from this list and then formed a team with those managers, to decide the appropriate assignment for each of the remaining 750 R&D personnel.

For the new managers of the technology centers and the skill centers, this exercise provided their first opportunity to work together as a management team. Faced with decisions that affected not only their respective centers but all of R&D, they turned to the principles of the new technology delivery process for guidance in making these assignments. Specifically, they were able to reach a consensus by putting the needs of UOP first, those of R&D second, and the needs of their respective centers third. Their success in this exercise went a long way toward building team spirit among these managers.

During this effort, the management team discussed various organizational alternatives for staffing the centers. They had to determine how they could best balance business alignment with resource flexibility. An emphasis on business alignment would place most people in a technology center. A focus on resource flexibility would place most people in a skill center. The result favored resource flexibility. In large part, this decision reflected management's confidence in R&D's capability to maintain a strong business focus because such a focus was already a strong tradition within UOP.

Project Teams

Project teams have the primary responsibility for the new technology delivery process. They identify resource requirements to meet their objectives; develop work plans; conduct experiments; and assist in commercialization, manufacturing, and engineering. From its inception, a project team must be cross-functional; this reduces or eliminates handoffs to other functional departments in UOP.

A technology center typically sponsors a project team. Each project team remains intact until it completes the final step in the technology delivery process: the first successful installation. The task of maintaining the technology and providing ongoing support then falls to the continuing services function within each technology center.

Each project team identifies its own resource requirements, develops its own work plans, conducts experiments, and assists in manufacturing and commercialization.

Skill centers sponsor projects aimed at developing new capabilities, new tools, or improvements in work processes. Like the projects sponsored by technology centers, these projects involve cross-functional teams. However, the end users of these developments are internal UOP customers, either in R&D or in other departments.

Project Team Implementation. To quickly roll out the new design, the new management team decided to convert all of the active, existing projects to the project team approach and, when practical, consolidate projects. Although this was a daunting task—R&D had more than 100 ongoing projects—teamwork on the part of all the center managers ensured the success of this effort.

This effort resulted in the development of 80 separate project teams, each with a sponsor and an objective. Each project team member was required to attend high-performance team training, including a session on team chartering. This training course, which was developed in house, emphasizes the new work process, the idea of team empowerment rather than management control, and the qualities of high-performance teams. In particular, the course emphasizes the importance of reaching a consensus on critical team decisions rather than using a majority vote or other methods that might not lead to a complete, honest discussion of the issues.

The first project teams were established during the period from March through June 1994. During July and August 1994, each team clarified its objectives and developed work plans for achieving its goals in the shortest possible time. Slightly less than one-half of the people in the R&D organization were assigned to project teams and were then introduced to the concepts developed by the design team members during the previous year.

Another development of the project team system was the creation of an R&D project-tracking tool, which shows the number of projects in each of the five new technology delivery process phases as well as financial metrics. This tool provides a useful means for monitoring the idea pipeline and gauging the effectiveness of the project selection process. With this tool, R&D will eventually be able to monitor its progress toward the goal of cycle-time reduction.

Project Team Impact. In many ways the project team rollout marked a turning point in the reengineering process. For many members of the R&D organization, the project teams provided their first opportunity to experience firsthand the concepts and the philosophies embodied in the reengineered technology delivery process. The rollout also provided an important opportunity for the members of the new management team to clearly demonstrate their commitment to the reengineered technology delivery process.

Following the rollout and the chartering of the first project teams, morale improved quickly for those employees who were assigned to a project team. However, anxiety levels increased for those employees who did not join a team. Some expressed concerns about their value as employees, wondering whether their job security was threatened because they were not on a project team. Others viewed the objective of reducing cycle time by a factor of three to mean that management now expected them to work three times harder for the same pay.

Following the initial enthusiasm, the morale of the project team members began to falter as the teams ran up against the old ways of working. As the 80 project teams attempted to meet their objectives at faster rates, they became increasingly frustrated with the inability of the old systems to keep pace with their need for data and other services.

The project teams struggled to learn new ways of communicating with the support groups and with one another. In many minds, the new process became synonymous with *too many meetings*. Teams booked conference rooms several weeks in advance, adding to the frustration. Although team members knew how to set objectives and establish time lines, they lacked the training and the tools necessary for effectively managing projects on a daily basis. They also did not know how to resolve conflicts over resources.

New organizations were wrestling with new structures and new supervisory relationships. With half as many supervisors running the organization, people who had been managers were trying to adjust to their new role of technical specialists. With the implementation of the new, skill-based organization, UOP was faced with the challenge of blending the cultures that had developed separately at each location into a single, new culture.

The business units became increasingly impatient for results. Some people called for a return to the good old days, claiming that the reengineering effort had failed. Productivity seemed to be declining rather than increasing. To some, the entire system appeared to be on the verge of collapse.

This period marked the beginning of UOP's *transition phase*. The R&D organization had installed a new system and a new way of thinking about technology delivery. R&D had educated its technical specialists and instilled in them a sense of urgency. Now, the work processes at the next layer needed a major overhaul to support these changes.

THE TRANSITION PHASE

The next major milestones involved promoting the reengineering philosophy to the support staff and implementing tools to help project teams improve the management of projects and the allocation of resources. Management knew that additional metrics were needed to measure improvements in cycle time.

Gaining Staff Support

Slightly more than one-half of the R&D staff typically were not members of project teams. However, the efforts of those people—who were called Team UOP—were essential to supporting the teams' efforts.

> After the initial burst of enthusiasm, morale may decline as project teams encounter resistance and inadequate support from existing systems.

Team UOP consisted primarily of people working in the analytical and testing departments as well as the crafts and shops personnel. Team UOP also included administrative support groups in such areas as training, safety, purchasing, and shipping and receiving.

R&D management eventually realized that they had been completely focused on the technical side of the reengineering effort and had ignored the softer, human elements of this task. These scientists and engineers were less adept at addressing nontechnical issues, and classical reengineering methodologies typically did not provide practical approaches to handle these problems.

Approach. Help came from UOP's Human Resources department. Human Resources specialists met with a small group of R&D staff members who were responsible for developing a Team UOP roll-out plan. From those discussions, the group recommended a half-day course designed specifically for examining the emotional stresses associated with organizational change. The course focused employees on steps they can take to regain control of their surroundings, deal with stress, and help the company move forward.

The R&D group decided to combine this program with an additional session developed in house which would connect the general messages conveyed in this program to the specific problems facing the members of Team UOP. The product of these sessions would be a list of specific actions that Team UOP members could take to start solving some of the problems they faced.

To emphasize the necessity of change and the amount of change required, the R&D group also wanted to supplement these sessions with an overview of the UOP strategic business plan. To ensure that employees believed that management recognized these needs, the strategic plan was to be presented by a senior-level R&D manager.

Results. These roll-out sessions started with supervisors and then worked down to the staff levels. Overall, the results were mixed and depended primarily on timing. For those employees who found themselves in a seemingly hopeless situation, results were positive. For those employees who had already developed their own coping behaviors and attitude changes, the effects were less positive.

From a practical standpoint, the sessions generated internal exchanges within the skill centers for improving work processes. They also led to the creation of several work process improvement teams. These teams attempted to find better ways of providing certain types of services, such as equipment repair and construction of new test rigs.

These teams had high esprit de corps and they produced positive results. An R&D Reward and Recognition Program highlighted the success of these team efforts. Winning teams were picked through an employee-run selection process. The selected teams gave a presentation to R&D managers, describing their results and the keys to success. They were then formally recognized at a banquet at a local resort hotel, followed by an evening of dancing and entertainment.

Improving Project Management

Project management was the next major hurdle. Again, R&D management turned to an external consultant for help. UOP retained a local expert in project management, to analyze the current situation and propose a solution.

Expecting that they would need the help of a software tool to better allocate resources, R&D management was pleased to learn that the project management system selected dovetailed perfectly with the reengineering work that UOP had already completed. Effective project management is a philosophy of operation with some fundamental components; the specific tracking tools are of secondary concern. These fundamental components address two aspects of effective project management: planning systems and scope management systems.

UOP's reengineering project had concentrated on the planning systems— in particular, the organizational issues, the definition of project life-cycles and milestones, the definition of the requirements and the scope of a project, and the establishment of metrics.

On the other hand, UOP's reengineering effort had not focused on the scope management systems, such as recognizing and managing projects in jeopardy of slipping. To date, the reengineering effort had not considered how to take corrective action and make decisions, nor had UOP determined how to devise methods of authorizing work and changing control. Fortunately, some basic, well-established tools were available to address the problems that R&D faced.

Approach. The manager of Training Development worked with the consultant to develop a project management training module that was targeted specifically for project team leaders and team sponsors. The module integrated basic project management tools and ideas with the reengineering concepts and the specific operations of the R&D department. A trial training session was conducted for four project teams. On the basis of their feedback, the training manager modified the module and rolled out training to all project team leaders and sponsors.

Standard monthly reporting formats were defined to easily consolidate all the project team activities into a single measure of technology delivery performance. These reporting formats emphasize the three aspects of effective project management: scope, schedule, and cost. Within this framework, R&D

is now establishing guidelines for deciding when a team's delivery schedule is in jeopardy and for monitoring projects in trouble. The objective is to focus scarce resources on resolving the most critical bottlenecks in the system.

Effective project management is a philosophy of operation with some fundamental elements; the specific tracking tools are of secondary importance.

Implementation. Having estatablished a standard methodology and reporting requirements, the next hurdle for R&D management was obtaining the cooperation of all the project teams. The discipline of comprehensive, formal project management was new to the R&D organization. To obtain reliable metrics on system performance, management needed full compliance with reporting requirements. A group of facilitators now meets regularly with team leaders and team sponsors to help them comply with the reporting guidelines and to remind them in a positive way that they need to report their team's progress.

One current barrier to full compliance is a view by some team leaders that the reporting requirement does not add any value to their team's activities. They see the report as yet another form they must complete solely for management control purposes. Over time, however, more team leaders have begun to understand that the teams themselves benefit from having a sound, working plan. A few team leaders have provided testimonials about how, despite their initial reluctance, the extra effort is paying off. Specifically, effective project management leads to more efficient team meetings, clearer communication between team members, a heightened sense of accountability, and a better view of the overall project goals and their relationship to the goals of the individual team members.

As reported by team members, other benefits of effective project management include improved focus on individual tasks and their relationship to the overall goal and a heightened awareness of the project's critical path. Sound project management has necessitated some difficult decisions regarding commitments made earlier in the programs and it has helped project teams to identify critical resources. It has also allowed teams to concentrate on the best technology options and deliver the right technology faster.

Expected Impact. From management's viewpoint, the standardized reporting system should be a means for clearly and intelligently capturing and quantifying the specific resource bottlenecks. With this information, R&D can embark on high-impact work process improvement projects that speed up the entire R&D technology delivery process. Project management with meaningful metrics is the key to arriving at the last stage of the reengineering process: the establishment of an infrastructure for continuous improvement.

LESSONS LEARNED

The reengineering process has been an organizational learning experience. Reengineering is a means for making fundamental cultural changes in an organization. However, obstacles are present at every turn, and change is not readily accepted. The following sections summarize some lessons learned so far during UOP's reengineering journey.

Identify the Need to Change

Change is the driving force for the success of a reengineering process. A successful reengineering effort also requires a few key individuals who are passionately committed to the change. Commitment comes from a vision of the potential capability of the new system and the faith that the vision can be achieved.

During UOP's reengineering effort, those individuals with a clear vision of a better future stood out from the crowd. They kept the reengineering effort moving forward, despite the inertia of others who wanted to stop and turn back. They continue to act in this role, and as the vision becomes reality others join their ranks. These people come from all levels of the organization, not just management. Every company has these visionaries, who are ready to come forward for the right cause.

UOP envisions R&D becoming a truly knowledge-based system. The old, hierarchical, command-and-control organization has been replaced by an empowered work force. The members of this work force trust one another, act for the common good, and continually strive to eliminate non-value-added work. Rather than build empires, the members of this work force have the flexibility to handle various tasks.

Design and Implement the Process

The reengineering methodology offered by the Rummler-Brache consultants provided the right approach for UOP's culture and needs. Whereas UOP's employees and management tended to overanalyze, the consultants pushed them to act. The consultants also provided the objective, external perspective that was essential for honestly assessing UOP's current work processes.

Trying to implement too much at once is a mistake. The focus of any reengineering effort should be on resolving the one or two most significant problems. The next problem to be addressed will then reveal itself. Furthermore, the solution to the current problem depends on the solutions to previous problems and it affects the solutions of subsequent problems.

Empower Employees

Ensuring that the new structure works requires an evolutionary cultural change: Empowerment with accountability. In other words, the entire organization must make the transition from a dependence on following orders to a sense of personal accountability. This fundamental change in thinking represents a maturation of both managers and staff. This often-painful evolution continues at UOP. During this evolution, an organization must devise new reward and recognition systems to replace those that evolved to support the old structure.

Work in Teams

Teamwork helps keep all employees focused on the company's objectives rather than their own interests. Problem solving with cross-functional teams is a powerful, practical tool to keep all employees focused on the important issues. With an effective cross-functional team, the primary objective remains in plain view.

The benefits of effective project management include improved focus on individual tasks and their relationship to the overall goal for the project.

The formal team approach has been a critical factor in the success of UOP's reengineering effort to date. Staff members continue to learn how to improve their effectiveness as team members, but the debate has shifted from deciding whether teams should be used to determining how to make teams work better.

Communicate

Communication is essential within any organization undergoing significant change. Throughout the organization, communication regarding the redesign and improvement process helps to maintain momentum and document progress.

Another communication issue involves the development of new systems and infrastructures to document the technical activities of the project teams. Faster delivery times necessitate the installation of improved computer-aided technology to share information at increasingly faster rates. UOP's efforts in information technology address the challenge of building communication networks and systems to keep pace with the need for faster technology delivery.

Train Employees

A reengineering effort increases the need for training and skill enhancement. As job definitions and structure change, new skills are required. A successful project team approach requires training team members in project management techniques as well as the softer issues of team dynamics.

The stress created by change necessitates training sessions to help employees understand the reasons for stress and the methods for handling the personal and organizational stress that arises during a reengineering effort. Staff must also be trained in the use of new information tools.

Cross-training of employees helps to create a more flexible work force. A reduction in the number of supervisors and the empowerment of project teams necessitates new training courses on safety in the workplace.

Meeting all of these training needs consumes time and money. Furthermore, without effective management of the training program, an organization might fall into gridlock.

Focus on Customers

Customer focus is essential for concentrating R&D resources on those projects that can contribute to UOP's continued business success. As R&D improves its capability for defining the resources required to deliver specific technologies, the business units continue to improve their understanding of the factors influencing overall R&D output. The reorganization efforts in R&D have contributed to the development of a companywide process for ranking all R&D projects.

Measure Key Parameters

Although managers may find it difficult to establish measures for key parameters of the process, this essential exercise pays dividends by forcing in-depth analysis and discussion of the keys to success. In this way, the development of an effective measurement system yields deeper insight into the means for improvement. Without a measurement system, no one can gauge the effectiveness of the new system.

UOP's R&D organization has developed time and value measures to gauge the effectiveness of the new work process. R&D is also establishing some project-specific measures that will give more immediate feedback to management and the project teams. By establishing performance measures, an organization creates a powerful tool for driving change.

Pursue Continuous Improvement

The success of any reengineering effort requires the pursuit of continuous improvement. However, the organization must avoid discarding existing practices that work well. UOP's reengineering effort identified several practices and cultural traits that were critical to past successes and innovations. These critical elements must be preserved and emphasized as role models. The new design attempts to create an environment that enables the consistent practice of these essential elements.

Continuous improvements build on one another, layer by layer. The improvement process must start at a high level of the work process and work its way down. After establishing new organizations which are based on a general work process concept, the reengineering effort can then overhaul the processes within each department. Eventually, individual job types must be redefined and a plan must be developed for moving from the current distribution of skills to the desired set of skills.

Persevere

An organization in the midst of a reengineering effort must keep sight of the vision, expect improvements, and, above all, keep going. In the two and one-half years since UOP initiated the design phase of its reengineering effort, R&D has implemented the basic changes required for the new system. The organization is now progressing from the transition phase of adjustment and training into the continuous-improvement phase.

As the reengineering process continues, this author expects that the next logical step will involve a dramatic change in experimental methodologies, from experimental design to testing and analysis methods. These adjustments, which will be necessitated by the demand for faster cycle time, will lead to major changes in equipment and job functions. Additional changes to the organizational structure may become necessary. Other required changes may not yet be apparent. Perseverance is essential, because the change process never really ends.

Time-to-market improvements necessitate the installation of improved computer-aided technology to share information at increasingly faster rates.

CONCLUSION

Was UOP's reengineering effort successful? The answer is yes. By early measures, R&D is meeting or exceeding the goal of a threefold improvement in the rate of commercialization of new technology. The rate of new product developments completed in R&D has increased by a factor of 3.3, as measured by project team completion of major deliverable milestones and by the increase in the number of product trial schedules. The number of experimental resources needed to achieve success is also decreasing. R&D is documenting a rapidly increasing number of cases of faster project completion.

Cross-functional teams can reduce the number of experimental resources needed to deliver technology. For example, a new product improvement was successfully completed in just one experiment. With the old system, this same improvement would have consumed at least three to four months of time and dozens of experiments. This striking acceleration in development time was made possible by a cross-functional team that could bring customer needs, UOP manufacturing capabilities, and the technical requirements of the problem together early in the project cycle. The team was able to develop a model, based on both theory and experience, and work quickly toward a solution to the problem. In so doing, the team shattered cycle time records. Although this example may not yet be the norm for the new R&D process, it clearly shows the potential of the new process.

The reengineered changes in R&D benefit the company economically and scientifically. The rate of technology delivery is accelerating, and the new project selection process better utilizes R&D's resources. Cross-functional project teams bring together people with diverse skills and scientific backgrounds to focus on specific problems. The resulting sharing of information accelerates the development of fundamental understanding and therefore technology delivery. Technology delivered faster is of greater value to UOP and to its customers.

THE R&D PROJECT: A FEEDBACK PROCESS

The evolution of new technology through industrial research is a feedback process. This chapter presents a study of technology transfer that sheds light on that process.

Don Tijunelis

The description of technology transfer in the scope of this chapter is the transfer of technology between a corporate R&D unit and a divisional profit center for the purpose of commercialization. More specifically, it is the transfer of technology, whether in the form of communications, reports, prototypes, assemblies, or blueprints, that takes place during the commencement of an R&D project, its progress, and the commercialization of its results. It is the transfer of technology to a project that gets it started and the transfer from a project that results in a commercial product. The transfer includes communications, a set of technical objectives, and physical items or laboratory test results.

Industrial research today is carried out, at least in the larger industries, in elemental units, or projects. A project is defined by specific objectives, scope, duration, resources, and justification so that it can be readily identified and examined. Although organizations vary with respect to forms and procedures, the principal description of a project is the same—the smallest planned, budgeted, staffed, and controlled unit of research.

A study of technology transfer between an R&D project and the commercializing unit sheds light into the process through the simplest cause-and-effect model of industrial research. The technology transfer process on the project level has a complementary effect on the already fairly well-publicized statistical and postresearch studies of total research effort's effectiveness.

NOTHING BUT A FEEDBACK PROCESS

The evolution of new technology through industrial research is a feedback process. The feedback process of an R&D project can be determined by analyzing either the daily activities or the overall flow of information to and from the project. For example, a research engineer collects information in the library as background material for an experiment to develop the corrosion rate of a metal alloy exposed to sea water. In looking through the background information, the engineer ensures that the material is relevant to the specific application requirements for outboard motor carburetors manufactured by a division. The application conditions were given to the research engineer by a divisional engineer, who in turn, received them from a field technical serviceperson over the phone a day before.

DON TIJUNELIS is vice-president of R&D for Viskase Corp. in Chicago.

Having finished the search, the research engineer is apt to start the experiment but, on this particular matter, is not sure whether the measurements in terms of weight loss and change in dimensions will have as much meaning as they would through a visual inspection of the corroded metals. The engineer is planning to use a salt spray test. The experiment is supposed to demonstrate that the alloying recommended by the research engineer retards the rate of corrosion. The final judgment as to whether the rate of corrosion is significant is made by the divisional engineer, who from experience would say whether it is excessive or not; the divisional engineer may wish to touch base with the Gulf Coast serviceperson to confirm this judgment. Thus, the feedback loop is complete, even in this day-to-day example, as soon as the research engineer calls the divisional counterpart to ask whether weight loss or photographs would be more meaningful. In a broader sense, the research engineer controls the background search by the specifications given from the divisional engineer, who in turn, feeds back to the research engineer approval or additional specifications on the basis of the test results.

The general characteristics for the cycle of technology transfer of an industrial research project can be broken down into three phases:

- *Phase I.* This is the transfer of technology to the R&D project at the start.
- *Phase II.* This is the transfer of technology to and from the R&D project during its progress and evaluation.
- *Phase III.* This is the transfer of technology from the R&D project to the commercializing unit for the ultimate application of its results.

The three phases of the technology transfer have a relationship to project progress with respect to time and cost. In a simplified form, this relationship— of the time- or costwise inefficient phases of the project as they pertain to the achievement of the project objectives—is shown in Exhibit 1. The most easily corrected parts of a project, through the elimination of inefficient technology transfer are the onset and the finish of the project.

The interchangeability of time and dollars with respect to the termination and the start of a project is not totally valid. In many cases, the finish of a project takes an asymptotic curve with respect to time toward meeting the final objective yet does not necessarily continue consuming the project budget. Furthermore, inaccurate or ineffective technology transfer rather than just inefficient technology transfer can be the cause of a total misdirection of the project and have a totally catastrophic effect.

There are probably just as many cases of an R&D project failing to get off the ground after repeated meetings and trips to a division by R&D personnel as those getting started quickly but missing a need. Endless discussions sometimes fail to define an objective, stimulate the research staff to undertake some business-motivated challenge, or induce divisional business management

Exhibit 1
The Three Phases of Technology Transfer

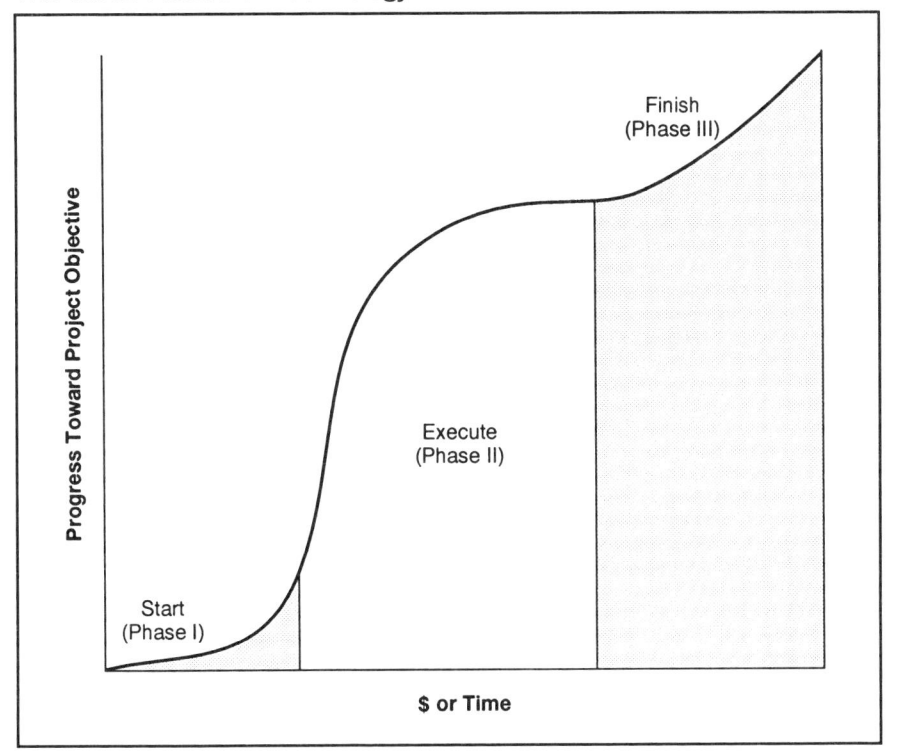

Progress Toward Project Objective (y-axis)

Start (Phase I)
Execute (Phase II)
Finish (Phase III)

$ or Time (x-axis)

to endorse the sponsorship of a long-range theoretical study. Early start advantage over competition is lost. The spark of motivation fails to be exploited. A preproject R&D budget is expended at an intolerable rate. The annual planning process is complicated and reduced in reliability.

On the other hand, at times, the costs of project start-up are negligible, and an agreement on the objective and justification is reached without any debate. Unfortunately, the loss of the needed minimum checks and balances occasionally proves embarrassing, when it becomes apparent that the results meet with the expectations of only a biased few. The analysis of the technology transfer phases therefore addresses itself both to the efficiency and accuracy of technology transfer. An example of a project starting very easily and quickly but not without pitfalls is discussed in the following section.

THE AUTOMOTIVE DRIVE TRAIN SCENARIO

A central research project manager carried out a basic research project for an automotive drive train component manufacturing division. The purpose of this project was to generate cost-saving design procedures, based on effective computer use. It was essentially a mathematical analysis project. The project was considered a great success. It provided the division with cost savings and showed the possibility of incorporating heretofore unexpected performance features in the product.

Central research management was quick to pick up on the opportunity for further exploitation of the basic mathematical approach for other mechanical component manufacturing divisions. If shortcuts in parts design were realized at the automotive division, it should be possible to use similar techniques to design parts for a hydraulic pump manufacturing division. Casual suggestions by central research management for a similar type project for this division were met with casual approval.

At budget time, when project proposals must be matched with available expertise, a project was proposed for the hydraulics division with rather little explanation of what specifically would be done for the division. The objective was stated more or less in broad terms of mathematical assistance for cost reduction in design, with detail replaced by examples of how successful it was for the automotive division. No one could argue with the objective supported by references to the successful work done so far. The project was quickly approved, and yet, cost reduction in design can be interpreted in many ways.

As the time came for the work to take place, there seemed to be an unusual delay in selecting a project engineer and laying out a detailed schedule. Senior R&D management's inquiries into the problem revealed that the project manager of the former related project for the automotive division was not the initiator of this project. The engineer who at the last minute wrote up the proposal at the urging of central research management to match available head count was transferred to a division. The department director could not find a project engineer who could quickly review the objectives with the hydraulic pump manufacturing division's requirements and come up with a valid program. It all seemed to say that it could not or should not be done.

It soon became apparent that the critical aspects of design for the automotive division had far less complicated mathematical relationships than those for the hydraulic pump. Here, not only mathematical but also fluid mechanics expertise was required. The hydraulics project could not be tailored as parallel to the one for the automotive division because of its special features, such as the need for consultants to analyze convective heat transfer through multiphase fluids. Furthermore, it became apparent that the hydraulics division was not interested in the efficiency of designing its parts but rather with design efficiency or, in other words, improvements for better, more efficient performance. The budget proposed as comparable to that of the previous project was by luck quite adequate, but with respect to its objective, the initiated project quickly became a source of embarrassment to those who proposed it and those who hastily approved it. Its start was quick, but it never got off the ground in its original form. It was later deferred, reviewed again, and then clarified in objective and scope.

THE VARIABLE-SPEED DRIVE SCENARIO

Projects that start without adequate technology transfer often are the result of unilateral suggestions by someone of business influence. A major customer of a division can at times easily influence the division if the division is sales oriented. The customer may initiate a project that is significant only to the specific customer. As such, it has a place in a central research mission only as a part of general divisional service.

Often the burden is on the research person to request the sufficient technical and marketing background to make an objective judgment of how to treat such a request for research. For example, a division requested consultation on an unusual, variable-speed drive system for a special industrial drive application of its major customer. Sales to the customer at the time were approximately 7% of the division's total and could be 30% to 40%. The consultation requested, however, was for an experimental setup by the customer to satisfy the engineers on an unrelated matter. The division wished simply to comply with the request to get in favor with the customer through its engineers and was going to make a pitch through the engineers for a greater share of the drive business during the next industry association meeting.

To get some action, a salesperson was smart enough to request research center consultation through the divisional vice-president of operations. That person, in turn, went through the corporate vice-president of research. Thus, by inference, the consultation request came to the section manager as top priority. To please the customer, the divisional sales personnel put the research engineer in direct contact with the customer's technical staff. They, anxious to get the greatest possible technical support to their special project, described its significance totally out of proportion with respect to the interests of their company. In return, of course, they got more consultation than they dreamed of.

The research center engineers came back not only primed with an exaggerated significance of their contribution to the variable-speed drives applications but also with the wrong idea of their own division's long-range interests. They were quick to propose a project for the development of variable-speed drives with features as noted during the customer contact. Their justification was based on the total market potential even beyond this particular customer. The priority was biased by all involved.

The project, not surprisingly, got quick approval at central research but was turned down at the division several months later. Evidently, although there were a few additional special units sold at the industry association's meeting, soon after, with a change in economics, the customer canceled other orders

The critical aspects of design for the automotive division had far less complicated mathematical relationships than those for the hydraulic pump.

projected earlier. This resulted in a net loss of sales, which decreased the division's interest to please them, causing the failure of the proposed project.

It is next to impossible to have a meaningful discussion of the transfer of technology without some reference to the organization within which the technology transfer is assumed to take place. The information presented here is the cumulative effect of experience from several organizational structures, each differing to an extent yet similar enough to be represented by a composite or typical organizational structure.

THE TYPICAL CORPORATE R&D UNIT

A typical corporate R&D unit is directly accountable only to corporate management and yet provides R&D support to multiple autonomous divisions. It is also a centralized R&D operating on a corporate budget, judged for its effectiveness by corporate management on the achievement of objectives for the most part set or at least endorsed by the autonomous divisions.

This type of organization is common. The corporate R&D has no direct cost to the divisions but is supported by their share of general overhead (divisions pay whether they use the service or not). R&D exists independent of individual divisions' particular needs. Effectiveness depends on objectivity at the central R&D to evaluate divisional interests and on voluntary cooperation by divisions to follow through beyond the no-cost R&D objectives and commercialize at their own expense the results of divisions varying in their organizations, both in terms of technical and business competence as well as size. Not the least important parameter of such an organization is the geographic dispersion of the divisions to be served by the corporate R&D. In this climate, the corporate R&D is pressured by corporate management to show its cost-effectiveness with respect to service to various profit centers. On the other hand, because of no budgetary commitment by the divisions, divisional participation in R&D planning is left rather open.

The relationships in technology transfer for central R&D organizations are somewhat different when these are directly paid for out of divisional income. Central research, which is isolated from the divisions and supported by them, is faced with the dilemma of dual reporting channels within it: an administrative channel to keep the research center activity functioning independent of the divisions and a reporting channel linked directly to the divisions for which the work is being done.

Divisional interest in central research activity effectiveness in this case is, of course, far stronger. Here, misguided research results in a direct expense to the division. Research engineers and their project manager in such an organization find themselves torn between suggesting programs and projects that are thoroughly compatible with the mission of central research as a technological pool or long-range technological development center and the shorter-range objectives given to them by the divisions that sign their paychecks. The example of variable-speed drives cited earlier is apropos. Although, philosophically, the divisions agree to the need for long-range research and the maintenance of a technological pool, in their daily activities, they concentrate on that which is more directly related to the immediate sales and profit potential.

The project got quick approval at central research but was turned down at the division several months later.

Thus, whether central research is sponsored by the corporation and is essentially at no charge to the divisions or whether central research is a direct expense to the divisions, their participation in long-range planning is weak. The research center, its directors and project managers, and the project engineer must sell the program and take the initiative of involving divisions in the long-range planning process. In the latter case, the selling process is made only a little more difficult.

TECHNOLOGY TO START A PROJECT

Needless to say, the start of an R&D project in a typical central research organization is a complex matter. The process and the formal require—ments to initiate an R&D project depend on a number of factors, including the subject matter, the specific organization, the budgeting practices of the company, the branch of industry, and even the internal politics within the research organization. The process of project initiation also depends on the initiative and character of the individual involved.

Where there is a will, there is a way. Projects can be started, assuming the objective is valid and there is enough desire to provide adequate justification and, maybe more than that, insight into the circumstances. For example, two chemical engineers had an opportunity to propose a project for material replacement in corrosion-resistant pump components. Both were equally knowledgeable in materials. Both were just as likely to carry out a program of material selection, preliminary testing and evaluation, negotiating with mold makers and raw materials vendors for the molding of prototypes, field testing, and final justification for production based on anticipated cost savings. The intention was to replace a corrosion-resistant metal with a high-temperature, high-priced thermoplastic and thus reduce labor during manufacturing. Both engineers prepared separate project proposals. They were essentially alike, approximately a $25,000 budget for labor and $5,000 for out-of-pocket costs.

One engineer submitted a brief project proposal with few facts about the cost-savings potential and waited for a response from his department head, who in turn took it upon himself to pass it on to senior management at the division for a preliminary opinion on the value of such a project. The objective was stated as that of purely material replacement for cost savings.

The other engineer did not submit his proposal but held it back until, by phone, he contacted several down-the-line engineers at the division manufacturing the pump to feel them out. He sensed very quickly that their biggest concern was that the connotation of plastic is that of a cheap product.

To them, any plastic was just another plastic, and they tended to overlook its properties and price.

He concluded that, the way his proposal was written, without some additional input, it could sound like a reduction of product quality for the purpose of cost saving. This he knew would be unacceptable to the division, which prides itself on product quality and had a bad start in value engineering He prepared a cost analysis showing an overall price advantage through labor savings while not hiding an obvious increase in materials costs. He also did a little library research on other applications of the particular plastic. Together with his proposal containing the cost summary, he included reprints of technical publications expounding the use of the particular plastic material on various aerospace applications.

When he told his colleague that he now would submit his proposal with the aerospace application examples, he found himself faced with a bet that he would lose if the material requirement already proposed was more expensive than the metal being used now. The examples of even more exotic high-cost applications would ruin his chances. There was some sense in this, at least within the context of past research association with government contract work. The reference to aerospace in this particular area of central research had left a bad after-taste because of previous high-cost, unsuccessful projects that had been carried out for aerospace applications. They all ended up with an unreasonable material cost to do a simple job.

The engineer who made the effort of feeling out the divisional climate himself rather than relying on the prejudices at the research center succeeded. He correctly concluded that, most of all, he needed to overcome the psychological implications of a low-cost plastic. A little additional effort went a long way. His approach was not devious; it simply took into account some of the human factors where they counted. The project was approved, and she carried out the program to a successful conclusion. The plastic parts are in manufacture today and the product, though cheaper, is of higher quality.

In spite of the various differences in project initiation procedures, subject matter, or personalities, in all industrial research there must be:

- A definition of the technical problem and a clear research objective.
- Some form of understanding as to its commercial impact, either immediate or potential, depending on whether the subject matter is applied or exploratory.
- Some idea of who is likely to do it, who will evaluate it, and who will make use of it.
- Some idea of its feasibility, degree of difficulty, and time involved for accomplishment.
- A big advantage in knowing how the proposal fits into corporate, divisional, or R&D long-range planning strategies.

- The need for standards. Almost all larger US companies have project title sheets, project approval forms, and R&D job orders. With the completion of these (i.e., full endorsement), the project is considered to be started.

A typical corporate R&D unit is directly accountable only to corporate management and yet provides R&D support to several autonomous divisions.

Depending on the specific corporate organization, approvals may have to come from the central business headquarters as well as the division to which the subject matter applies, or only one and not the other. Nevertheless, some approval is needed from the director or vice-president of central research, who in turn depends on support from, or is accountable to, either the corporate headquarters or the divisional management. Whether the research budget is directly funded by the division's operating income or as part of general corporate overhead, the research director must sell the budget, just as the research department head must sell it to the research director and the project engineers to their supervisor. In each case, to initiate the project, its commercial value must be recognized.

TECHNOLOGY TRANSFER DURING A PROJECT

Don Tijunelis

The most important aspect of technology transfer once a project is under way is to provide proper evaluation of its progress and keep the project on target. This chapter describes how to accomplish that objective.

When speaking of technology transfer in industrial research, people more often refer to the transfer of project results (i.e., commercial use of the product developed). Least often do people think of the communications during the progress of a project as technology transfer. For a centralized R&D of a company with multiple autonomous divisions, communications between research engineers or scientists and the technical unit at the division (the engineers concerned with implementation) are critical. Communications must extend further to business management at the divisions that include sales and marketing, for example. The pattern of technology transfer during the execution of a project in a generalized form is described in Exhibit 1. The most important aspect of technology transfer to and from a project during its progress is to provide proper evaluation of its progress and to keep the project on target.

In a typical industrial research situation, once a project is started, little attention is given to the objective, at least until close to the end of the project, when the transfer of its results must be dealt with. For at least a while, the project manager and the project engineer lose contact with divisional business management with whom they had to negotiate project objectives and the support to justify its initiation.

TECHNOLOGY FOR ITS OWN SAKE

The project staff deals more with the engineering counterparts at the division. As shown in Exhibit 1, communications is generally less frequent in total. At the same time, left alone in the central research environment to proceed with the needed experimental work, the engineers or scientists find themselves highly influenced by their peers, other engineers, and scientists working on other projects. They meet and discuss their work at lunch, on the job, and after hours. They do not have an understanding or insight into the significance of the objectives nor do they have insight into the relevance of the project work with respect to corporate or divisional business interests. The other engineers have even less of a feel for the particular business objective. What they have in common is technology for its own sake. In common among the

---

DON TIJUNELIS is vice-president of R&D at Viskase Corp. in Chicago.

Exhibit 1
Technology Transfer to and from During the Project Execution

Formal Project Transfer

Corporate Research and Development Project

Divisional Management (e.g., Sales, Marketing, Field Service)

Divisional Technical Unit

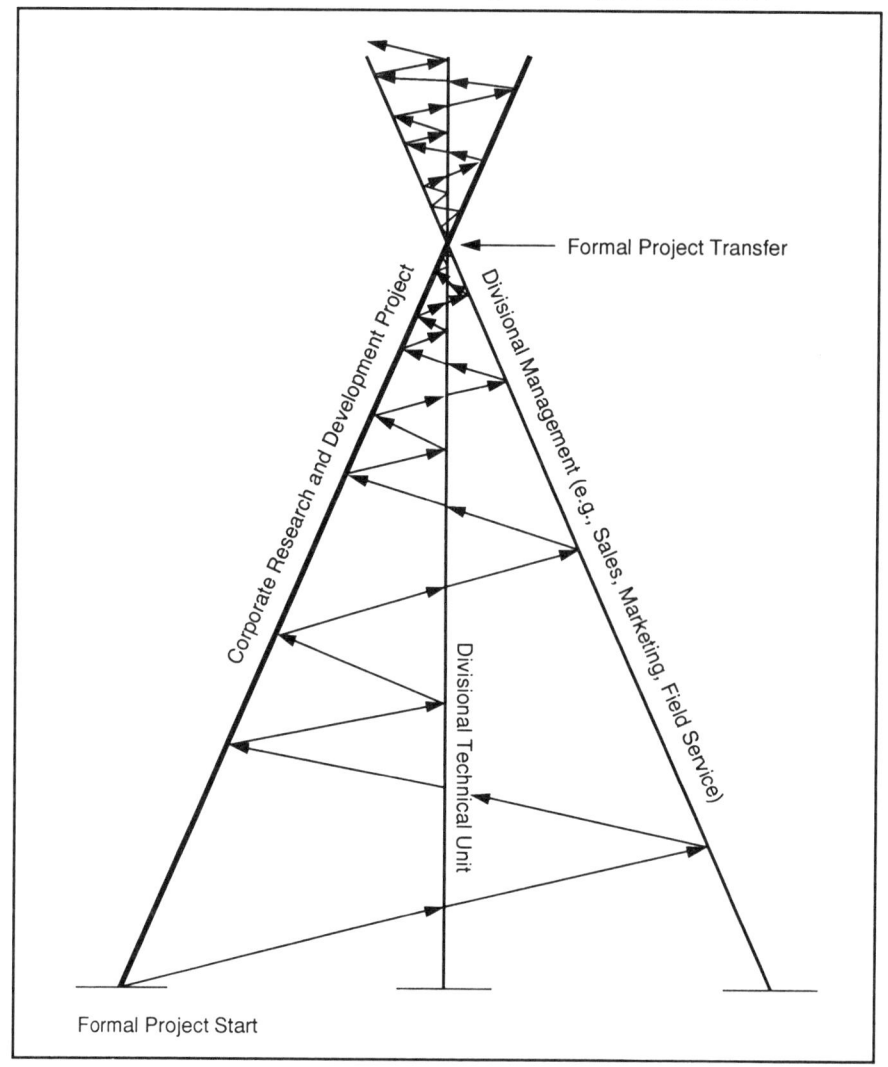

Formal Project Start

peers are techniques as well as basic technical information. Emphasis shifts not to why or what is being done, but how. Who is using the latest technique? Who has the most ingenious method of molecular characterization, phase separation, and surface analysis? With some exaggeration, the discussions are over who can use the most complicated mathematics so as to show his or her skills—who can generate the data with 99.0% confidence limits. Consciously or not, in the day-to-day communications among the scientists and engineers, accomplishment drifts from meeting the objectives to excellence in the execution of the work regardless of the business strategy. In research circles, people can expect to hear it said with almost a sense of achievement that the division never was able to make a buck on it, but it sure was a good piece of research work.

What may be appropriate in an academic environment is not appropriate in industry.

What may be appropriate in an academic environment is not appropriate in industry. Advancing the state of the art for its own sake is not much of an accomplishment in a business climate. Only rarely are the results of industrial research made available to the public if they miss the intended business objective. Industrial research is not benevolent.

Drifting Off Course

As the project progresses, the research staff, if left isolated, drifts toward the purely technical concerns of the objectives, or drifts off the objective with more focus on technology for its own sake. On the other hand, on the business side of the project cycle, the ultimate recipients of the project's results become more anxious. They become more frequently involved in the communications between R&D and the division. The divisional sponsors of the project, who during its initiation agreed to the timetable and the associated costs of experimental data as a necessary evil, develop short memories. Especially if the interim results begin to be encouraging, to them the agreed-on rigor of data gathering to ensure a reasonable confidence level begins to appear redundant. Some of the divisional engineers, even more the business managers, are more concerned with business strategy and cannot appreciate the fine points and critical mass levels of the data needed to ensure the success of the project objectives. They are apt to influence the project engineer by essentially cutting his or her work short. Either directly or through the manager, the divisional engineers will try to convince the manager that accomplishment in terms of business impact is at hand.

Time is money in business. Those charged at the division to exploit research results, if the results are in their interest, will cut short the time. The inertia level at the division is much lower than at central research. Response time is quicker. The division, when interested, will influence the project to accelerate toward commercialization. However, by the same token, if the project should run into trouble, the division would accelerate its cancellation.

IMPROVING TECHNOLOGY TRANSFER DURING THE PROJECT

During the progress of the project, its progress and aim is in a delicate balance between apathetic drift to the archives and emotional propulsion to either success or failure by luck and business speculation. Thus, during the course of the project, technology transfer to and from the commercializing unit must be properly planned to maintain a balance of evaluation of its progress and aim. Technology transfer for both the evaluation of progress and aim can be provided in several ways.

Balanced Communications. The project engineer must be assured balanced communications with both divisional engineers and divisional business management, including marketing and sales. This would help the project engineers keep abreast of the pulse of the business needs, keep those needs fresh in their mind, help them to be more objective, and help them chart the progress of the project.

Cross-Functional Cooperation. As the research project progresses, control of progress must shift to the divisional engineer responsible for implementing its results. This person will be influenced by the business management at the division; therefore, more frequent cross-functional meetings at the division, even informally, are helpful. A trip to the division for the research engineer should be facilitated whenever possible. Three-way meetings, even lunch, involving the research engineer, divisional counterpart and divisional business manager avoid misunderstandings and motivate cooperation.

When Not to Involve Senior Management. Quite often, senior R&D managers travel more and are apt to volunteer to be the technology transfer agents. There is a temptation to agree to this as a form of technology transfer with a mistaken notion that it will save money and time as well as avoid unnecessary additional arrangements. Too frequently, however, replacing the research engineer with a senior staff member demotivates staff, and it could damage essential communications. It stifles the iterative communications between the actual doers aiming at the real technical objective. The responsibility to transfer technology can shift to senior management and no longer reflect the critical details of the technical challenge facing the project. What is discussed in the executive offices may not relate to the laboratory, the factory, or the marketplace.

USING THE TECHNOLOGY REVIEW MEETING TO TRANSFER TECHNOLOGY

One approach to increasing proper technology transfer during the course of a project is to expose the research project staff in a planned manner to the business or divisional attitudes. Another way, or an additional way, is to bring the divisional staff abreast of the technical roadblocks in a more informal manner. Especially in the hustle-bustle climate of the divisions, the preoccupation with expediency is unavoidable. It is difficult, if not next to impossible, to sit business managers down and have them hear out the logic for some decisions based purely on technical evidence. It is easier to gain their attention by first taking them out of their business environment.

Holding a technology review meeting at the central research facility has proved successful in transferring R&D technology to divisional managers during the progress of a project. To be effective, a technology review meeting must:

Advancing the state of the art for its own sake is not much of an accomplishment in a business climate.

- Attract the attention of senior management both at central research and the division.
- Be well timed to have substance for discussion and at intervals adequate for the costs involved.
- Have a formal agenda for the research presentation and divisional follow up.

It is absolutely necessary that such a technology review meeting to transfer technology of ongoing projects receive attention of senior management. If it does not, it will simply propagate mistaken notions about the rate of progress, business impact, changing goals, and resource limitations between the technical staffs. If the meeting were to be attended by divisional business representatives who have no overall divisional budget responsibility, it would only result in an introduction of highly divergent biases. Divergent views may be useful in an absolute sense of the total data available, but in the context of an important infrequent meeting to reestablish direction, they would be ineffective unless mediated by someone of broader responsibility capable of handling decision making for the course of the projects.

It would be ideal if the technology review meeting would also attract the attention of corporate planners, for they would be able to inject an overview that is in tune with the ultimate aims of the corporation. This is often impossible; it is, fortunately, not essential. The presence of the vice-president of corporate research or corporate research director on one side and the divisional vice-president of engineering, general manager, or possibly group vice-president on the other is quite sufficient.

To attract such a select group may be difficult for a project engineer or a manager of a single project, who must seek the assistance of the department manager, research director, or a liaison officer if such exists. For the same reason, the technology review meeting should be set up to cover a group of projects. Related by their technology or discipline (i.e., projects related to the mechanical efficiency improvement of a product, projects related to reliability improvements in a group of products of a single division, projects with possible contribution to cost savings of the division, or projects exploring new product opportunities for a division). It all should be for a single division, but from perhaps more than one department of the central research facility.

The meetings should be at intervals acceptable to the demand for management attention and participation. Semiannual meetings have worked out well in a number of cases. At this frequency, the central research organization and the division can afford the time and associated out-of-pocket costs. A considerable number of involved individuals can participate. The division can bear the travel expense, and central research can bear the disruption in its schedules and lunch costs. A typical meeting reviewing 6 to 12 projects could have an attendance of somewhere between 12 and 24 people. This would include the research and divisional top management, project managers, key project contributors, and interested central research department heads in related fields as well as a corporate patent attorney or market research representative.

Keeping Meetings Short and on Track

When several project status reports are to be presented at one time, some formality is absolutely essential. It is essential to avoid conflict between participants as to the allotted time and order of presentation. However, more than that, it is essential to force the project managers to generate a well-thought-out summary of key elements for the benefit of the general management in attendance.

As mentioned earlier, engineers and scientists are fascinated by technology itself. In a presentation of their work, they often ramble, on speaking to their peers and themselves while losing the rest of the audience. If the discussion were left informal, arguments—sparked by an internal pecking order in the technical community of central research—would ensue on technical points irrelevant to the project objective.

A successful approach to solving this problem is a formal agenda mailed out considerably in advance, with reminders sent a few days before the meeting, giving no more than five to ten minutes per topic. With the agenda is a suggestion to use visual aids, which will be made available at the meeting site. This forces the individual project managers to condense their work with visual aids as an inevitable tool. In preparing these visual aids, the managers must inevitably identify key points and thus make their presentations effective.

In the agenda, discussion of the formally presented project matter is held back until after all have had their say. In many cases, this can be held after lunch for a convenient break-up of attention to ensure that all topics, not only the last topic presented, get a fair chance of discussion and review. Even then, discussions are limited to an hour or less covering all of the topics.

The discussions at such a technology review meeting should highlight only the burning questions that require commentary by upper management. They should not stray into technical details. For that, subsequent communications between the appropriate divisional and research personnel are sufficient, using regular communication channels of telephone, mail, and visits.

What Happens After the Meeting

Decisions made on the burning items coming out of the technology review meeting should be implemented both in the central research organization and down the line at the division. Technical decisions for technology transfer must be enacted. Because the meeting is at the research facility, and both project engineers and management participated, their link is complete. On the other hand, however, the technology is transferred only to the visiting few. The senior managers in attendance may have made some judgments on the most noteworthy points, but can the research staff rely on them to disseminate their decisions for implementation to the lower ranks?

Only rarely are the results of industrial research made available to the public if they miss the intended business objective.

It is very simple to ask each of those who made a formal presentation to summarize and submit his or her subject, including any noteworthy comments and decisions, plus a copy of the visual aids used. This then is quickly compiled, covered by a letter of transmittal from the research director or liaison officer, and sent out to all the participants as minutes of the meeting with a suggestion to disseminate as fit. Usually, on receipt of the minutes at the division, the general manager or engineering vice-president will circulate it throughout his or her staff. Thus, an aspect of the transfer of technology, including key decisions evaluating the progress of a project and its objectives, will have been completed.

Needless to say, in the preceding suggestion for a technique to improve technology transfer during the course of the project, personalities play an important role. The personality of the person chairing the technology review meeting determines the tone of the meeting. Depending on his or her personality, a number of aids can be used to improve the effectiveness of such a meeting, including side discussions over breakfast with the visitors to discuss delicate topics; lunch in the meeting room to change the tone of, and introduce informality into, the subsequent discussions; a tour to view the experiments; or an elaborate advance agenda notice with suggestions for key discussion points.

IMPROVING COMMUNICATION

Much of the communication during the project is a process of transferring increments of the ultimate collection of results meeting the objective. Progress reports and prototypes flow out of the central research for evaluation, critique, and comments in return. Without a sincere critique or objective reporting, a project may go astray or become inefficient.

Experience has been that the form of formal reporting is rarely inadequate or insufficient. However, the distribution of the reports as well as the sincerity of comments in the response to the reports may be questioned. As a rule, it is beneficial to include as broad a distribution of research project reports at the division as possible. In other words, the reports should be sent to as many diverse individuals as is justified by the confidential nature of the report and the cost of making copies. When possible, all attempts should be made to include in the distribution people associated with both the potential business and technical aspects of the subject matter.

An Instructive Tale

Errors in communications during the progress of a project are easy to find. In one case, a young but very well-thought-of scientist, a PhD engineering science major, was successful at developing a water filtration device for drinking-water supply pumps. His idea initially was treated with skepticism with regard to practicality because of his reputation as a scientist. As its performance features became more obvious not only to the research personnel but also to the division management, he gained considerable attention and respect. Coincidentally, an increase in marketplace for greater well-water and an accompanying divisional expansion plan put his development in the forefront. He was enthusiastic and ambitious, hoping for greater and more diversified management responsibilities in the future. In a very proper manner, he realized that the commercial success of the development, a break-away from the pure scientist stigma, could mean faster advancement for him in R&D or new opportunities at the related division.

As the prototypes of his device were prepared and proven in the laboratory, he requested field trials through the division's general management during a visit there. Approvals in principle were granted quickly, but arrangements for specific field testing plans were left vague. He waited for division's engineering staff to pick up the ball. They were slow. So the R&D scientist in his eagerness picked it up himself, met with the division's sales manager, and selected a test site. In addition, he got the division's engineering service group to prepare some auxiliary equipment drawings to adapt his prototype to the specifications of the particular field test conditions. He pulled some strings to get the auxiliary equipment made quickly and sent it out together with the prototype to the test site in California, across the country from his lab.

Most of this preparation was done informally by phone. Everyone seemed content; the divisional sales manager was happy that he was going to get something to talk about and advertise for his market expansion program. He did not care much for the division's engineers and their methodical ways anyway, and this gave him a chance to show how he could handle even the technical matters without them. The division's engineers, to whom the filtration device was an imposition in any event, cared less. In a sense, they just as soon could see it fail. They saw little responsibility being placed on them. To research management, it was a point of pride that everything was going so fast, so smoothly, and that their person could handle the whole matter. Division's managers were concerned with the efficiency of their operation. They, too, were content to hear no complaints—only the assurance that it was getting close to commercialization.

With time, telephone communications between the research scientist and the California sales representative became very frequent and direct. Quite often, the scientist called the customer directly to expedite matters further. The field test was in progress with nothing but glowing reports of success. The filtration device not only did not plug, which was a primary concern of the scientist, but improved the taste of the water as well as its shelf life.

A few months later, a product development review meeting was held between the division and central research covering various projects. The intent of the meeting was to cover the status of projects in progress and the establishment of divisional support for the transfer of technology and manufacturing scale-up next year for those projects expected to transfer into commercial use.

> **The discussions at a technology review meeting should highlight only the burning questions.**

The research scientist in charge of the filter development naturally expected to see his project on the agenda. It was there, but only as a project in progress—no one at the division thought of or cared enough to include it as the subject for manufacturing scale-up. This was a shock to the research director, who was primed by the descriptions of field test success as described by the scientist. She surely assumed the division to acknowledge it likewise and prepare for its manufacture. She wanted a feather in her cap, too; she wanted to tell corporate headquarters that she had another new money maker.

During the discussions, the scientist reported that he now was making arrangements with his California contact for a second field test. He mentioned that the test would start as soon as appropriate piping was delivered to the test site for a particular customer that was using fiberglass pipe exclusively. The divisional engineers were quick to ask whether the first test was at the same site and whether the scientist realized that fiberglass pipe would be prohibited there. The divisional engineers also questioned whether he realized that the customer and the local salesperson may have had an ulterior motive to praise the device for filtration excellence.

The divisional engineers, of course, knew of the trials and most of the results informally. They could have gotten all the details, had they only been asked. However, they were not motivated to get involved and now were only willing to point out problems. Being busy on items of their own invention, they may have been anxious for an excuse to be bypassed.

In the face of the negative feelings created by the possible field trial data bias, all field trial data was seemingly discounted. No one, including the research director, was willing to take note that another set of field trials was in order. Manufacturing scale-up preparations remained off the agenda.

The sales representative soon heard that his particular sales scheme fell through. He realized that the customer with a particular interest for public relations through the purchase of the device for his fiberglass pipes is not apt to be used for future field tests. Others he knew were disinterested in an on-site hearing. Consequently, he too lost interest and was quick to attribute it to the lack of features on the filter.

What good would some letters have done confirming various telephone conversations, especially if they were with adequate distribution lists? Copies to the divisional engineers as well as division and research management could have pointed out the lack of follow-up. By inference, it would have forced the engineers at the division to share responsibility and take an interest. They would have become partners. It could have prevented the sales representative from playing games or the planners of the review meeting from omitting manufacturing consideration for the filtration device. The project would not have faltered, nor would its project manager have lost any time or credit.

POINTERS ON MANAGING TECHNOLOGY TRANSFER

Exhibit 2 shows in an oversimplified way the pattern of technology transfer during the progress of a project to achieve the ultimate objective of profitable commercialization. In many ways, it is an extreme simplification; it does not show the technology transfer loops that occur within an R&D organization, the division's business center, nor the division's working level. A dramatic

Exhibit 2

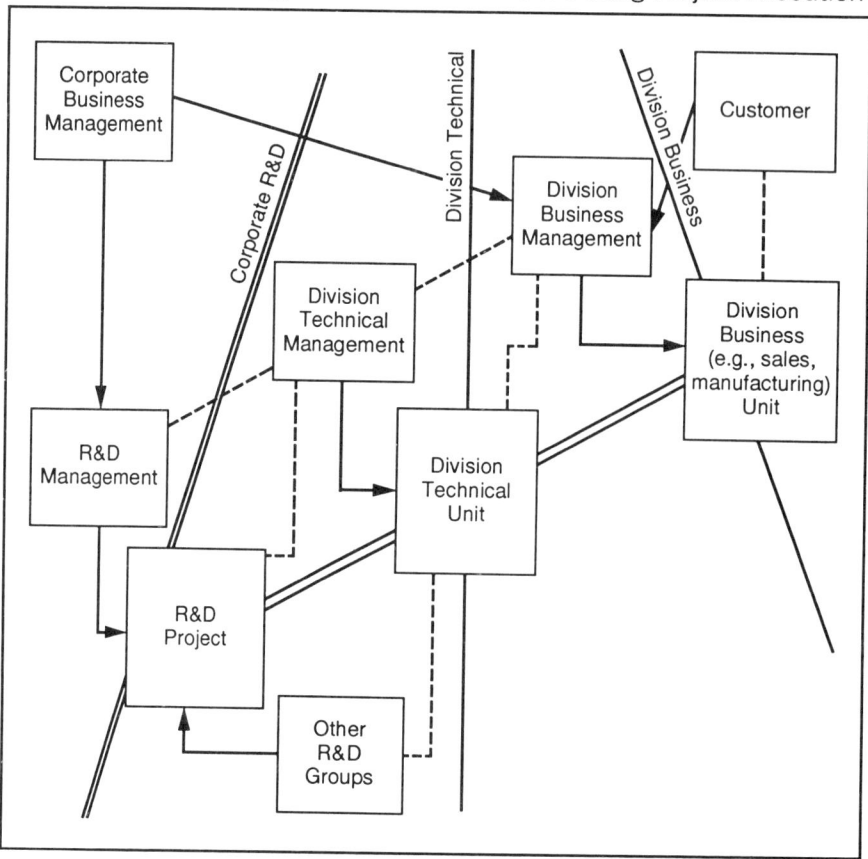

Elements of Technology Transfer to and from During Project Execution

example of the effect of the maze of communications, their subjective nature, and possible pitfalls can be read in an interesting account of an industrial catastrophe at B.F. Goodrich by K. Vandivier.[1]

The personality of the person chairing the technology review meeting sets the tone of the meeting.

With respect to the sincerity of comments and response to the reports from the divisions, it is necessary to read between the lines. Some divisions, based on their work environment and their interest in research support work, are apt to withhold discouragements; others can be unusually discouraging from either an insecure or an extremely nearsighted viewpoint. The interpretation of these comments, therefore, should be left only to the author of the report. Others should have an opportunity to interpret divisional comments.

Several of the many recommendations for consideration are:

- Sending periodic progress reports with the broadest distribution list at the division and at central research.
- Accompanying each report with a letter of transmittal by research management, which would be helpful in bringing the report to the attention of divisional executives and acting as a second screening of the comments as they come back.
- Following up reports with periodic personal communications (e.g., by telephone and short letter) on a purely project management level to review progress with divisional personnel responsible for the business assessment of the ultimate objectives.

There are obviously other options, depending on the peculiarities of each organization and project. A thoughtful review of what is best, based on some related examples to evoke ideas, is the first and the most common step in all cases.

Note

1. K. Vandivier, "Why Should My Conscience Bother Me?" *In the Name of Profit* (New York: Doubleday).

IMPROVING THE ENGINEERING-MANUFACTURING INTERFACE

Thomas B. Turner

Successful organizations are finding ways to improve the transfer of technology from R&D, product design, and engineering to manufacturing. Such practices as concurrent engineering offer great promise for companies trying to break down the walls that have traditionally separated organizational functions.

Organizational units, for various reasons, tend to be segregated, or compartmentalized. Although compartmentalization may help each unit perform more efficiently in some ways, it presents problems with the transfer of technology, especially from R&D, product design, and engineering to manufacturing. This lack of integration between product design engineering and manufacturing is ultimately detrimental to new products, but several forces hold the situation in place, in particular, manufacturing managers' traditional conservatism toward new technology.

Fortunately, many large firms have begun working to improve the engineering-manufacturing relationship. These firms include Lockheed, in its historic Skunk Works operation, and General Electric (GE), with its product management method developed for aircraft engine manufacturing. The trend toward improved relations is also evidenced in the recent interest in concurrent engineering.

This chapter discusses problems involved in transferring technology from R&D and product design and engineering to manufacturing and suggests ways of improving the process of technology transfer. The observations presented in this chapter are based on the author's working and consulting experience in firms that manufacture such discrete products as refrigerators, washing machines, electronic equipment (e.g., large radio transmitters), and heavy industrial equipment (e.g., large fans and blowers).

ORGANIZATIONAL COMPARTMENTALIZATION

Each organization, whether a corporation or a unit in a corporation, has a history of successes and failures in its markets. On the basis of these experiences, the leaders of an organization establish what they consider to be the best way of doing things. To encourage the use of such methods and systems, they usually have established a system of informal and formal rewards. In general, employees learn how to advance in such organizations by following the methods of operation and the reward systems established by senior management.

THOMAS B. TURNER, CMC, retired from the Cresap division of Towers Perrin with more than 45 years of experience in industry and management consulting. At the time of his retirement, he was director of the Department of Defense Manufacturing Technology Information Analysis Center.

The hierarchical nature of organizations often results in compartmentalization of functions. This compartmentalization segregates product development from manufacturing and segregates cost accounting from manufacturing as well as from other operations. Ultimately, this compartmentalization is detrimental to the process of technology transfer, and usually the larger and more complex the manufacturing operations and processes, the greater the gulf between the functional elements of the organization.

A Separate Peace

Compartmentalization of functional units often discourages teamwork between the units because it might be construed as interference. During the 1940s, the author was a methods engineer in a large appliance manufacturing organization. A tooling engineer in the same group was designing a large, complex stamping die. This 18-stage progressive die would be used to manufacture a small part for an electrical switch. The part was a hollow, light-tight box, about as big as an adult's little finger, to be made of very thin steel in one pass through the die. When in production, three such dies would be required, each to be used in a coil-fed stamping press and scrap-cutter.

The annual production requirement was for more than 36 million pieces; the schedule was firm, because the part went in a larger appliance: an electric range. The tooling engineer was well qualified, having been a tool and die apprentice in Europe before World War II, but his design and drafting efforts were not going well and he was obviously frustrated.

The tooling engineer started to complain about the work of the design engineers and their product design. When it was suggested that he go back to the design engineers and suggest design improvements that would make the tool design easier, he exclaimed, "I don't tell the engineers how to design products! And the engineers had better not tell me how to manufacture them!" This is an excellent example of the organizational walls that often develop between departments.

Engineers and technologists learn their on-the-job-skills through practical apprenticeship and seek to demonstrate their ability to handle responsibility within a given function (e.g., product design). Their need to be able to market specific skills if they leave the organization influences career choices and limits their opportunity to live and work on the other side of the wall. The lifetime employment practices of the Japanese might afford an advantage here: If engineers know they will never have to work for another employer, they might be more willing to accept assignments in other functional groups on a project-by-project basis.

In the U.S., product design engineers have traditionally concentrated on developing skills, knowledge, and related experience in:

- Materials, processes, and feasibilities, drawing on experimentation and research for innovation.
- Design feasibilities.
- Product performance, life, and durability.
- Cost analysis and competitive analysis.
- Market demands and needs.

The larger and more complex the manufacturing operations and processes, the greater the gulf between the functional elements of the organization.

Manufacturing personnel generally have different work experience, if not a different technical education, with emphasis on:

- Industrial engineering, in particular the analysis and measurement of opportunities for improvement in shop operations.
- Tool design and manufacturing engineering (e.g., development and specification of such processes as machining or design of proprietary manufacturing equipment).
- Material requirements planning and inventory control.
- Shop operations.

Technology is breaking down some of the barriers between product engineering and manufacturing, especially in organizations that use CAD/CAM systems. If a part is designed on a computer system and the design transmitted to manufacturing in the form of computer data, it is likely that engineering and manufacturing have previously agreed on acceptable and feasible tolerances and available manufacturing methods.

AUTOMATION COMPLEXITY AND PRODUCT LIFE CYCLES

In some industries, if a part or product is well designed and well accepted, it will be used for many years. In a number of shops, parts or assemblies are still in manufacture though the drawings are more than 50 years old. On the other hand, some products have a short life cycle.

In general, different organizations have different traditions and processes that are a result of the products and technologies involved. Often, the level of technological complexity and the desired life cycle of a product determine the level of integration of the product design and manufacturing functions. For example, at one end of the spectrum is the manufacture of incandescent electric light bulbs. Light-bulb manufacturing is extremely sophisticated and highly automated and has been developed and evaluated over many years.

The common household light bulb does not outwardly appear to have changed much during the last 40 years, except when manufacturers dropped the copper base in favor of an aluminum base. The complexity arises in the coils of tungsten filament wire that are automatically picked up and welded into place on the support wires without human participation. Automatic evacuation and sealing are also sophisticated processes developed many years ago. Few product design changes can be made without careful consultation between the product engineering and manufacturing functions because of the highly automated production processes, which use machinery designed and built solely for that purpose.

Another example of product design that is well integrated with manufacturing is found in the manufacture of ranges and refrigerators, in which the body can progress from the coiled steel to a stamped, folded, and welded body without a human hand touching it. In these industries, engineering-manufacturing relationships have been carefully developed over a period of years to achieve the integration that a high level of automation requires. And the corollary, of course, is that the process cannot be automated without the careful integration of product design into the manufacturing process.

At the other end of the spectrum is the job shop where the product engineer says, "I'm going to design this product so that anyone can manufacture it." Having completed the design work in isolation, the product designer then (figuratively) tosses the drawings over the wall to the manufacturing manager, who says, "I can make anything within reason. If I don't have the process in-house, I will find a subcontractor who can do it for me."

Somewhere in between is the typical manufacturing organization, which is product oriented but may be under substantial competitive pressure. Many manufacturing managers feel they have few luxuries, such as adequate time and budgetary allocation for equipment and staff, because most firms are under great pressure to improve margins. Even such giants as IBM are becoming far more cost conscious.

Of Alligators and Conservatism: Manufacturing Management

Product design and R&D are creative processes, as is the manufacturing industry itself. These creative processes provide some of the great satisfaction of working in industry, and manufacturing managers usually appreciate and respect the application of innovative technology. Manufacturing managers, however, seem to be inherently conservative. Many prefer to learn by experience, not by theory. Many seek reliability and repeatability: Be in command; do not make mistakes.

Manufacturing managers typically live by the calendar, having to make the shipment budget of the week, the month, or the year. They have labor problems, supplier problems, reliability problems, and distribution problems. Product design engineers would do well to try to understand the challenges facing manufacturing managers.

The employees in a large truck plant used to speak of "the alligator" in the plant. "The alligator" was the assembly line, which had to be fed so many axles per hour and so many engines per hour. If the alligator did not get fed, there were real problems.

Is there a simple way around the conflict between the innovative spirit of design engineering and the conservatism of manufacturing? One way may be to eliminate separate departments and place the two functions under a common manager. This approach, according to Keith McKee, director of the Manufacturing Productivity Center at the IIT Research Institute of Chicago, has worked for several of his clients. Regarding compartmentalization, he comments:

Many manufacturing managers are faced with inadequate time and insufficient budgetary allocations for equipment and staff.

> It is interesting to watch the interchange between manufacturing people and the engineers. When new product designs are proposed, manufacturing personnel sometimes act as though they don't know they have the right to refuse the new design. Some of them act as though they are "second-class citizens." The engineers have a lot of time to think things through and consider alternatives. The drawings are delivered to manufacturing and they have no luxurious amount of time to think things through. They have no quick way to correctly estimate costs and have to make a full set of route sheets in order to come to grips with the estimated costs. Where is the time to stand back and think about it? A technically competent manager can see such difficulties occurring and intervene to manage the situation more effectively.

MANAGING CHANGE

Although automation could make their lives easier in some ways, most manufacturing managers do not rush to embrace new technology, for a variety of reasons. First, the capital budget may be driven by other matters, such as replacement of depreciated equipment, investment to support new products, or support of manufacturing in other locations. Before accepting new technology, manufacturing managers want to see the technology demonstrated and proved. They usually have little interest in seeking scientific or technological information such as that available from the manufacturing-focused data bases at the Society of Manufacturing Engineers. In addition, overt emphasis on clearing established hurdle rates for return on investment, though not as pervasive as in the past, continues to be a significant impediment to adoption of new technology. Managers are likely to question investments in incremental new technology when returns of 20% to 40% are possible with leveraged buyouts and restructuring.

Despite traditional conservatism about new products and the effect of technology on manufacturing operations, however, most manufacturing managers recognize the potential of continued developments in materials (e.g., plastics, polymers, synthetic fabrics, composites, glass- and metal-matrix materials, ceramics and ceramic-matrix materials, and optoelectronics). In addition, manufacturing managers continue to assess the capabilities and limitations of such methods and technologies as computer-integrated manufacturing, CAD/CAM, just-in-time manufacturing, flexible manufacturing systems, total quality control, Shingo systems, Taguchi systems, and robots.

Although manufacturing managers are faced with rapidly changing products, methods, and systems, many of them think of themselves as conservative and do not always seek change and its attendant risks. On the other hand, R&D departments and product engineers symbolize change to manufacturing managers, and in fact, their function is to create new products and processes with the attendant risks.

TOWARD A SOLUTION: TRANSACTIONS AND NEGOTIATIONS

Several factors affect the nature of transactions and negotiations between the functions in an organization. The first factor to consider is relationships between people—that is, when, how, and between whom negotiations take place. For example, who establishes the timetables, circumstances, and conditions for conducting negotiations between engineering and manufacturing?

If existing facilities and products are involved, negotiations typically take place with existing production and manufacturing engineers; the development process is relatively straightforward. If new facilities and equipment are involved, however, an advanced manufacturing engineering unit might be needed to seek new manufacturing methods. In addition, teaming methods or project management methods might be in order. For example, GE's method, which was pioneered in aircraft engine manufacturing, assigns development personnel and advanced manufacturing engineers to a project team that is responsible for the product throughout its life cycle.

Product or project life may be the key criterion for determining the most effective enginering-manufacturing relationship. If the projected life is short—perhaps a year or two—and requires repeated and frequent work, routine channels should be established and used. On the other hand, if the project is a once-in-10-years or 20-years quantum leap, such methods of organization as teaming and the establishment of an advanced manufacturing engineering unit should be considered.

Whatever the situation, time should always be provided for participation—intensive interchange, debate, and argument. So many times the manufacturing manager is not asked what would work best and is not included in the necessary brainstorming. In addition, the necessary resources should be provided, including allotment for the time and effort required. Key people should be involved, on a full-time basis if necessary, and an account number should be assigned to track costs if necessary.

THE RIGHT WAY AND THE WRONG WAY: TWO CASES IN POINT

A successful example of these principles at work is the Lockheed organization known as the Skunk Works, which developed a long line of advanced aircraft, culminating in the YF-12 and SR-71 aircraft. The YF-12 was a prototype for the SR-71, which was the first production aircraft to fly faster than Mach 3. (Mach 1 is the speed of sound in the atmosphere.)

These all-titanium aircraft were developed and manufactured on an urgent basis by dedicated professionals working in the desert in secrecy. A competitor had already started development of titanium aircraft and agreed to share information with Lockheed. Lockheed started work two years after the competitor and delivered its aircraft two years ahead of the competitor, even though it was dealing with material in which the physical, mechanical, and fabrication standards had not yet been developed and had to be prepared from scratch.

> **Overemphasis on clearing hurdle rates for ROI, though not as pervasive as in the past, continues to be a significant impediment to adoption of new technology.**

Bob Vaughn was chief manufacturing engineer at the Skunk Works from 1960 to 1970. He cited the following factors as being fundamental to the success of the Skunk Works:

- Competent management, with exceptional technical leadership from senior management.
- Efficiency.
- Accountability.
- Integrity.

He further stated:

> The engineers in this highly disciplined organization were considered to be "scientists with screwdrivers." The organization had only 100 people. The number of people involved was limited "almost viciously." The drawing systems were kept simple, and there was great flexibility in incorporating engineering changes.
>
> Bureaucracy was kept to a minimum. The number of reports required also was kept to a minimum. There was strong emphasis on cost control on funds spent to date, funds committed, and cost to completion. The order of the day was "no surprises."
>
> The climate called for rapid decisions and direct decision in a "special situation." There was strong emphasis on personal responsibility for finding the right answer and doing the job with integrity.

Vaughn went on to become corporate director of productivity for Lockheed Missiles and Space Inc. and has since retired. The SR-71 aircraft has just been removed from active service after more than 25 years of outstanding and successful performance. It was flown from Los Angeles to Washington DC, in the record time of one hour and eight minutes on its final trip to enter the Smithsonian Institution. The average speed on this final trip was 2,112.52 miles per hour. This shattered the previous record by two and one-half hours.

In other companies, however, there have been recent cases in which excessive pressure to perform caused management to take shortcuts and perform below their usual standards. For example, GE developed a new refrigerator compressor for use in its top-line refrigerators. It failed in 1988, and since then GE has voluntarily replaced 1.1 million compressors. The pretax charge in 1988 was $450 million. This expensive error was thoroughly discussed in an article in *The Wall Street Journal* that appeared on May 7, 1990. The article points out that the new-product failure cost could have been limited if the new product had been phased into production gradually: "Blunders were committed at practically every level. In designing the compressors, engineers made some bad assumptions and then failed to ask the right questions. Managers, eager to cut costs, forced the engineers to accelerate 'life testing' of the compressor, curtailed field testing, and rushed into production.[1]"

It is difficult to determine from reading the public accounts whether GE could have avoided making these mistakes and accruing these costs through improved cooperation between manufacturing and product development, though certainly its field failure expense could have been reduced by requiring a more gradual introduction of the new product to the marketplace and consequently allowing the engineering and manufacturing departments more time to perform their work.

LEARNING FROM HISTORY: SOURCES OF INFORMATION

When initiating advanced manufacturing projects, managers need to learn from history and avoid reinventing the wheel. The Department of Defense has addressed many of the issues confronting engineering and manufacturing managers and can be a useful source of information about these issues. For example, the Air Force's system planning methods and IDEF (integrated computer-assisted manufacturing definition) language are useful for planning large projects. Another information source is the Department of Defense Manufacturing Technology Informational Analysis Center—MTIAC—operated by the IIT Research Institute in Chicago.

Literature searches can be conducted at MTIAC or through a commercial data base such as DIALOG—originally developed by Lockheed and now operated by DIALOG Information Services of Palo Alto CA—on such topics as the IDEF language, producibility engineering and planning, or concurrent engineering. Another source is the data base at the Society of Manufacturing Engineers in Detroit.

Managers of job shops may find it useful to study such concepts as producibility engineering and planning, which ensures a timely and economic transition from development to production. More sophisticated organizations that already use such concepts as advanced manufacturing engineering may wish to consider concurrent engineering, which has been developed and promoted by the Air Force. Concurrent engineering is a systematic approach to integrated concurrent design of products and their related processes, including manufacturing and support. Concurrent engineering programs call for integrated product and process design teams, designing concurrently rather than sequentially and modeling the design and making trade-off analyses. Dramatic reductions in product development time and cost have been achieved by numerous organizations that have implemented concurrent engineering.

> **Concurrent engineering holds promise for dramatic reductions in the time and cost required to develop new products and their related processes.**

Self-Diagnosis

Managers who do not have a preconceived notion of how to approach the matter of engineering-manufacturing relationships might want to consider a diagnostic approach. This method, which is widely used by management consultants, involves the following steps:

1. Setting objectives.
2. Establishing a plan for conducting the assessment.
3. Identifying possible solutions to the problems in advance.
4. Conducting a fact-finding mission to identify the symptoms of ineffective engineering-manufacturing relations:
   — Poor planning and execution of new-product development.
   — Inaccurate product cost estimates.
   — Poor product performance in the field.
   — Lack of market penetration.
   — Excessive design changes.
   — Lack of mutual professional respect.
   — Insufficient time or personnel devoted to the essential transactions between the functions.
   — Excessively complicated product designs and related processes.
   — Improper organization and staffing.
   — Unprofitable products.
5. Selecting a course of action.
6. Drafting the conclusions and recommendations and establishing a plan and timetable for implementation of the recommendations.

CONSIDER SENIOR MANAGEMENT

An attendee at a professional conference commented to the author that when a company selects a senior officer, the last sincere and honest word that person hears is "Congratulations!"

When asked what he thought of senior executives, he responded, "Probably 25% of the chief executives in this country are competent, experienced, capable, educated persons who are well qualified to hold their positions. The balance of the CEOs or general managers are in their positions because of some abnormal drive that forces them to wish to dominate people."

Although this may be overstating the case, these comments point to the importance of understanding senior management's attitude toward improvement efforts. Significant change can occur only with the full understanding and support of the organization's senior executives.

Note

1. "Chilling Tale: GE Refrigerator Woes Illustrate the Hazards in Changing a Product," *The Wall Street Journal* (May 7, 1990), p 1.

TRANSFERRING NEW TECHNOLOGY INTO THE MANUFACTURING PLANT

Keith E. McKee

Manufacturing, typically considered in the past an isolated black-box function of a company's production, is becoming increasingly influential in the product development process. To reflect its changing role effectively, manufacturing must become more integrally involved in the early stages of product development. At the same time, companies must carefully examine their attitudes toward changing technology and their use of it.

This chapter addresses both the manufacturing department's role in introducing new products and materials into production, and the process of introducing new manufacturing technology into the plant. For the purposes of this chapter, technology refers to any scientific or engineering knowledge. Technology can be considered new in either of two situations—when it is being applied for the first time anywhere or when a particular company is first applying it (after other companies have already successfully applied it).

Technology need not be sophisticated or expensive to be of value. The technology involved in changing a fabricated metal case to plastic is simple. Yet this change in technology can make the difference between a product that sells and one that does not. The technology in a roller conveyor is relatively simple and inexpensive compared with the technology in a gantry robot, but in some cases it is the more effective approach.

A MANUFACTURING OVERVIEW

Manufacturing is a process by which material, labor, energy, and equipment are brought together to produce a product that has greater value than the sum of the inputs. This can be shown as a system, as indicated in Exhibit 1. The inputs are material, labor, energy, and capital (or equipment). The output is a product, but there is always some undesirable output—waste and scrap—that should not be forgotten. In addition, such external influences as government actions (e.g., those taken by the Occupational Safety and Health Administration or the Environmental Protection Agency), or natural occurrences (e.g., storms or floods), and the actions of competitors all have effects.

KEITH E. McKEE is director of the Manufacturing Productivity Center at the Illinois Institute of Technology Research Institute, Chicago.

Exhibit 1
A Manufacturing System

[Diagram: Inputs — Material, Labor, Energy, Equipment (or capital) — flow into the Manufacturing Plant, which is also influenced by External Influences from above. Outputs from the plant are Product (to the right), and Waste and Scrap (downward).]

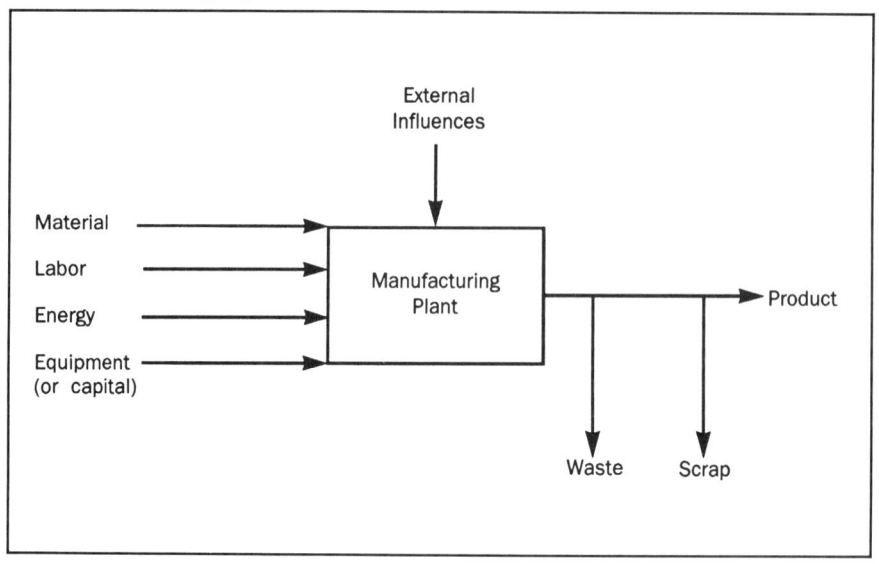

Exhibit 1 presents a broad view of the manufacturing system, based on definitions used by economists and the federal government. The manufacturing plant shown in this exhibit could be a steel mill, a machine shop, an oil refinery, an automotive plant, or a textile mill. Those inside manufacturing rarely have such a broad perspective. To them, the production of automobiles, airplanes, refrigerators, and electronics is manufacturing; manufacturing concerns the production of mechanical and electrical parts and assemblies.

This narrow perspective excludes many industries unnecessarily. Therefore, despite the fact that many of the examples in this chapter come from industries within the narrow definition of manufacturing, the chapter itself addresses manufacturing in the broadest sense.

Exhibit 2 presents a view of manufacturing that emphasizes input, output, and feedback. Quality control occurs after manufacturing and provides feedback; feedback also comes from customers. This exhibit illustrates that manufacturing is a closed system that responds to external feedback related to quality and to the quantities that are required.

In both Exhibits 1 and 2, the manufacturing operation is treated as a black box. In reality, this box contains a wide variety of activities, and within the plant there are usually many interrelated activities.

Over the past decades, many more activities have been introduced into manufacturing. Traditionally, manufacturing was based on one process (e.g., machining). A plant produced a limited set of products with a specific type of equipment (e.g., machine tools).

TECHNOLOGY TRANSFER

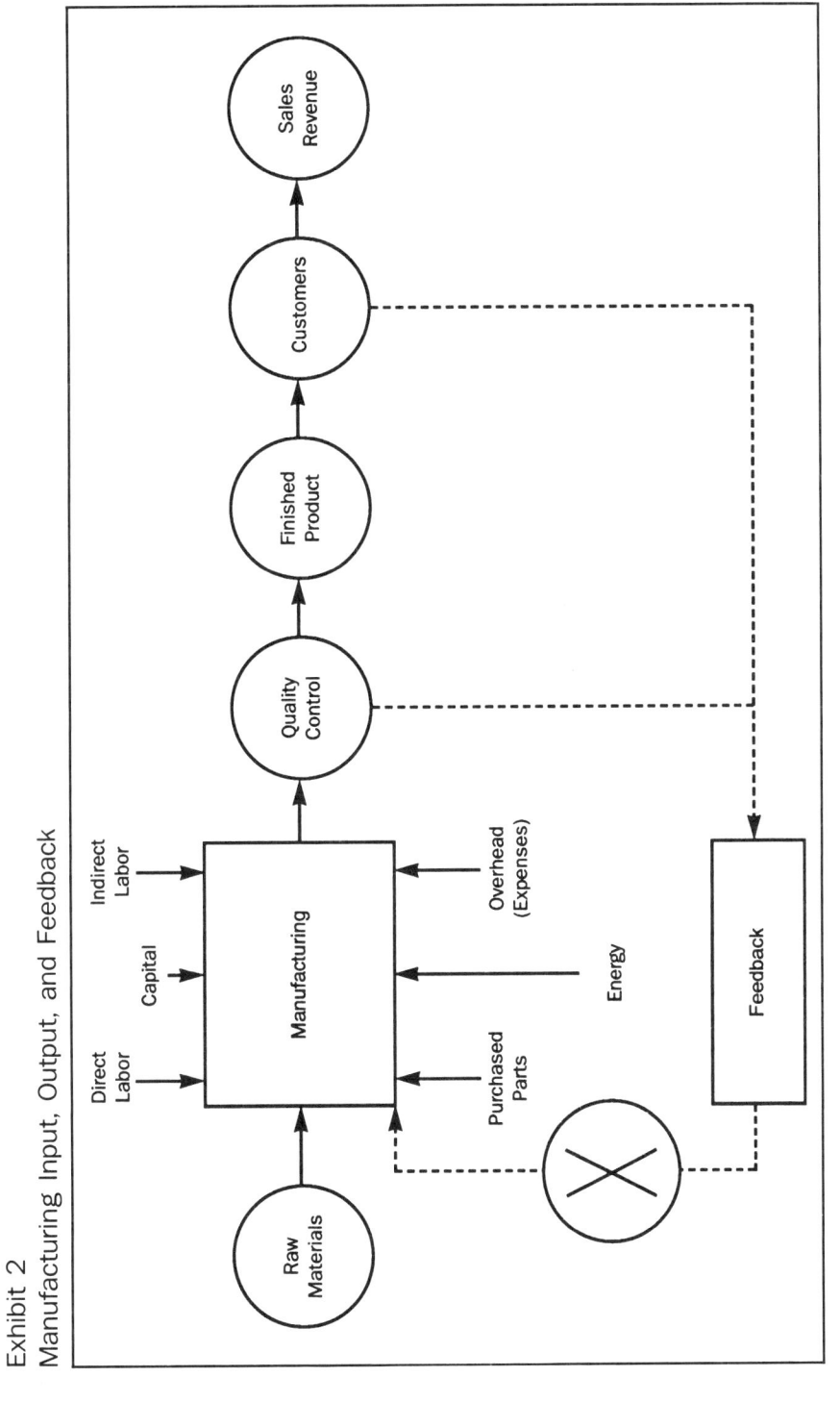

Exhibit 2
Manufacturing Input, Output, and Feedback

TRANSFERRING TECHNOLOGY TO MANUFACTURING

A More Comprehensive View of Manufacturing

Today, however, the typical modern plant consists of many processes. Exhibit 3 shows the processes involved in metal cutting or forming. Besides the principal processes, three other process levels are involved—preprocesses, secondary processes, and finishing.

Preprocesses are operations performed in-house or by vendors to prepare the material for the primary processes. Although these processes could be considered as going all the way back to the mines, the preprocesses in this example are all related to the product being made (i.e., forging, casting, powdering metals, and cleaning).

Secondary processes are performed on the products while the principal processes are going on. Examples are heat treating, welding, and soldering. Secondary processes may be applied by vendors at remote sites or may occur at a remote site within the plant. Whether applied externally or internally within the plant, secondary processes account for an ever-increasing portion of the production cycle.

Finishing refers to those operations performed after the machining. Finishing can include painting or plating, final inspection, assembly, and packaging. All of the operations that are or could be done at the same site as the machining can be included under finishing.

Principal processes are the machining operations themselves. Many such advanced manufacturing systems as numerical control machining centers and cells, and automated tool changers are designed for improving principal

Exhibit 3
A More Comprehensive View of Manufacturing

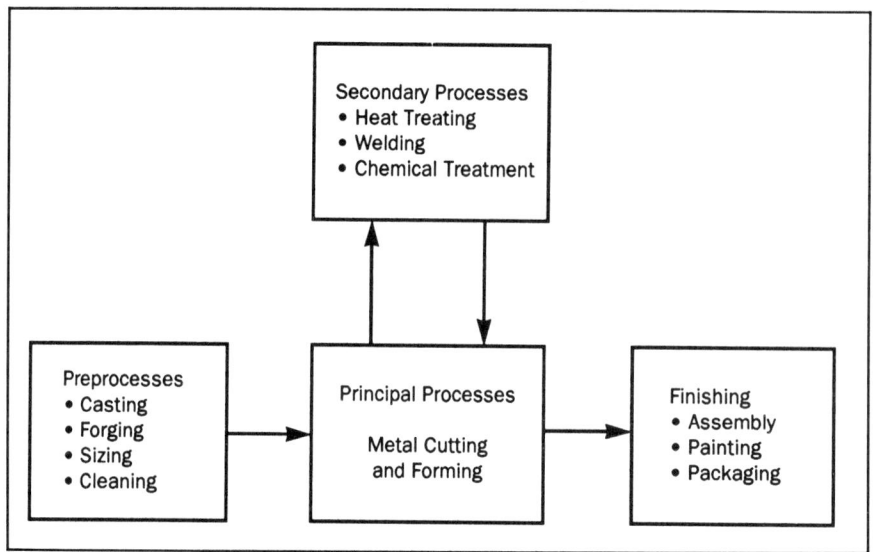

processes. Similar developments on tool materials, high speed machining, water jet and laser machining, machinery and tool sensors, and improved structures for machine tools are also directed at improving these processes.

To be competitive, manufacturing companies must be able to use a range of previously unavailable technologies.

Within and connecting each of these processes are such manufacturing systems as material-handling devices (e.g., automatic guided vehicles and robots) and data control systems (e.g., manufacturing resource planning). Improvements are continually being made to these supporting systems, and more improvements are anticipated. Nevertheless, the optimum approach is to minimize the amount of material handling and data control needed—for example, by integrating preprocesses, secondary processes, and finishing operations with the principal processes.

In Exhibit 3 and the preceding discussions, machining operations are considered primary processes. However, one plant's primary processes can be the next's preliminary processes. For example, in an operation in which assembly is the primary process, machining could be a preprocess; similarly, in a welding shop, machining is likely to be a secondary process. Whatever the primary process, there are distinct advantages in integrating as many operations as possible.

Manufacturing can no longer be equated with machine tools. In fact, even in the industries in which machine tools perform the principal processes, machine tools represent but a small part of the capital invested and even a smaller portion of the technology involved.

To be competitive, manufacturing companies must be adaptable in their ability to use a range of technologies that were unheard of until recently. CAD, robots, AGB, and automated inspection have all emerged in the past few years; manufacturing materials are also changing. Computers are important in manufacturing, but so are such changes in process technologies as new tools, net shape manufacturing, energy beams, chemical machining, and waterjet material removal.

The Factors Changing Manufacturing

Manufacturing is now in rapid flux. Until recently, the main mission of a manufacturing department was to maintain stable operations. New plants were built simply as modern versions of previous plants. This manufacturing model, however, is no longer appropriate because the following factors are forcing changes in manufacturing:

- *Quality.* Customers, particularly other companies that are customers, must have parts that are 100% usable in order for them to implement automated assembly and just-in-time inventory and production.
- *Flexibility.* To be competitive, a supplier must accept product design changes to produce varied quantities and meet flexible delivery times.

- *Cycle time.* The time from the release of engineering requirements to the delivery of the product must be reduced greatly.
- *Inventory.* The previously mentioned factors have significantly affected the handling of inventories. Raw material, work-in-process and finished goods inventory must all be reduced.
- *Costs.* Cost now must be contained and reduced for all elements of manufacturing.
- *Competition.* Every company, however small and specialized, has or will have competition. Safe niches are no longer available.
- *Internationalization.* Companies must continuously evaluate the potential for international cooperation and the threat of international competition.

All of these factors require companies to change their approaches to manufacturing. In the past, the policy of company managers in stable companies was to limit the amount of new technology used in manufacturing and to require extensive justification before introducing more. However, technology has now become a critical factor in making a company competitive, and the analytical techniques traditionally used to justify technology are no longer appropriate.

THE EFFECT OF R&D AND DESIGN MANUFACTURING

All phases in technology development—research, development, and design (RD&D)—influence manufacturing. Over the past few years there has been an emphasis on improving the manufacturability of product designs. Improvements here can significantly increase the producibility of a product, yet design for manufacturability (DFM) involves only the final phase of the RD&D cycle. Decisions made in R&D can limit the options available during the later design phase. Nonetheless, advantages gained through DFM can be significant.

For example, the Illinois Institute of Technology Research Institute (IITRI) recently conducted a DFM project focusing on a disposable medical product. This product was made from seven parts that were manually assembled; when the product was first designed, the company was producing it in small quantities. The objectives of the IITRI project were to help the company redesign the product and then to develop automatic equipment for assembling the product.

The first phase of this product involved improving the design for manufacturing. The actual design included only three components, which had originally been designed so that they could be assembled easily through automation. When the project was half over, IITRI proudly reported the results and recommended the development of the automated equipment. The customer was pleased with the results but did not immediately authorize the next phase.

A few weeks later the customer announced he had decided to forgo the automated equipment. With the DFM results, he really did not require automation, because manual assembly was now so much easier. Thus, by improving the design for manufacturing, the customer had achieved cost targets that originally had been established for automated assembly.

DFM refers primarily to the assembly of components, but it can also include reviews of manufacturing procedures for making parts and assemblies. For example, a DFM project can involve using a molded plastic part instead of a metal assembly or using a single chip to replace an electronic assembly. With regard to continuous processes, DFM specifies the design processes in such a way that they are insensitive to parametric variations.

Typically, it is impossible to select parameters to encompass all anticipated variations throughout production; thus, designs should include process controls that work despite unplanned variations.

For example, a company whose manufacturing operations included the extrusion of foamed plastic insulation was encountering frequent defects in its product when continuous processes were operating at the specified design conditions. Using the Taguchi approach, IITRI reviewed the design parameters and redefined them so that almost all defects in the product were eliminated. To do this, IITRI selected the design parameters that would allow the system to tolerate the widest possible range of parametric variations. The goal of a DFM effort for continual processes is to select design parameters that maximize the robustness of a production system.

It is not typically possible to select design parameters that produce a system so robust that all anticipated variations can be tolerated. Consequently, the design should also include process controls that work despite unplanned variations. In addition, as variations occur in one parameter, other parameters should be able to be varied to maintain the integrity of the product.

As an illustration, a pharmaceutical company operated the same process at three plants, where yields ranged from 35% to 70%. Because all three operations were working with the same design parameters, the differences in yield presumably resulted from differences in the knowledge and skill of the operators. As a result of an IITRI study involving all three plants, a procedure was developed for optimizing process control throughout the entire company. When the company implemented this new procedure at the plants, the yields at all three plants exceeded 80%.

Interaction of Product and Process Design

Products and processes must be designed to maximize all aspects of the manufacturing operation, including costs, quality, cycle time, and flexibility. In particular, the design of products and process can significantly affect process control and the manufacturing process.

Exhibit 4 shows how decisions about different manufacturing factors have markedly different amounts of influence over the effectiveness of a manufacturing process. Even though materials constitute one-half of the total product

Exhibit 4
Comparison of Four Factors' Influence on Product Cost and Effectiveness of the Manufacturing Process

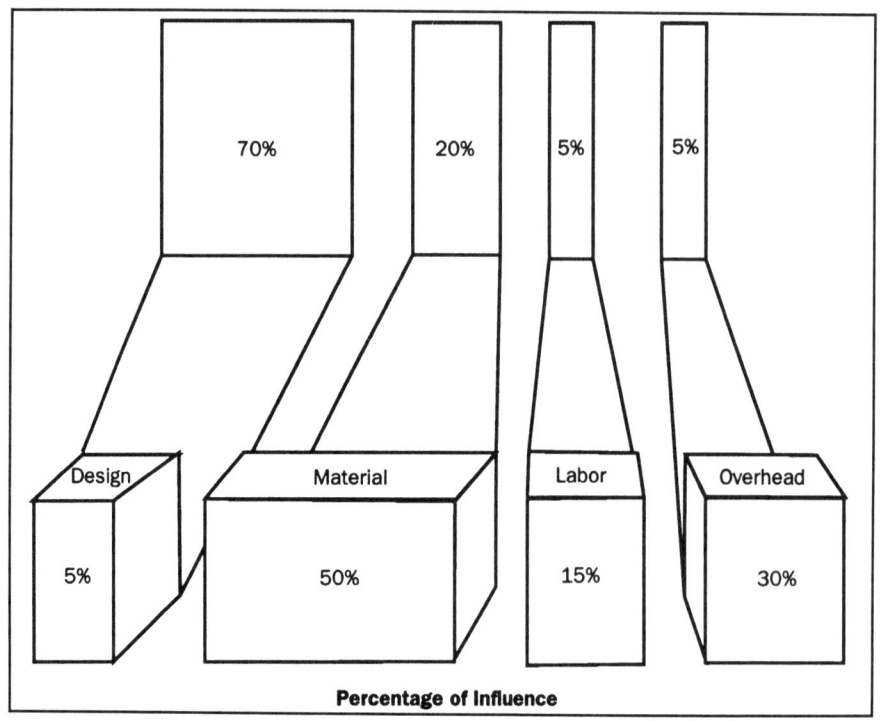

	Design	Material	Labor	Overhead
Percentage of Influence (top)	70%	20%	5%	5%
Percentage of Influence (bottom)	5%	50%	15%	30%

cost, decisions about materials have only 20% of the total influence on the effectiveness of a manufacturing process. On the other hand, cost of design is only 5% of product cost, but decisions about design have 70% of the influence on the effectiveness of a manufacturing process. Thus, focusing attention on manufacturing during the design process can result in cost savings.

Just how can a product's design affect a manufacturing process and how can a manufacturing process be improved through redesign? Two case studies illustrate some answers.

Two Case Studies

The first case concerns an electromechanical product that was designed by a R&D-oriented staff. The company anticipated that it would produce 15,000 of these products over a 5-year period. The electronics components in the product were designed for manual assembly with wave soldering; this design was similar to the one used in developing the company's other products, and the company produced less than 1,000 of these products annually.

The company manufactured the electronic components of the existing products internally, even though it could have had them assembled by companies that specialized in this process. The company also made the circuit boards of these products in-house to maintain control and facilitate repair of boards that needed rework after testing burn-in (more than 80% of them did).

> Too often, personnel choose new materials without considering their decision's potential impact, only to find too late that the material is unavailable.

Surprisingly, the company produced the circuit boards of its existing products with almost no process controls, and it did nothing to improve controls when it produced the new product. Interestingly, however, the company did conduct formal design reviews for the new product. The manufacturing department accepted the proposed production approach, presumably because the department was already familiar with it. At least in this instance, therefore, a design and manufacturing review did nothing to improve process controls. The company did not think seriously about process controls for the new or existing products.

The company designed mechanical components of the new product to handle web materials that could vary in width. The designers of these components had no relevant experience in this area but did have an eminently successful, top-of-the-line design to use as a model. With it, they designed a sophisticated mechanism for relatively simple web handling. To compound the problem, they specified tolerances that were as tight as they could justify. One of the company engineers had found a vendor who said that he could satisfy these tolerances for each of the purchased parts.

Unquestionably, the mechanical system would work, but it was clearly overdesigned in terms of the design approach and the tolerances allowed. The mechanical system cost approximately double what it should. Equally serious, problems in producing this mechanical system were ensured because of the tight tolerances; even the vendors who had claimed that they could produce the parts had problems. Consequently, the vendors' parts were continually out of tolerance, which affected in-house production and ensured problems in assembly. Thus, poor product design can derail manufacturing.

A second case concerns a company that had been producing a thermal sensor for five years with a very poor yield (only 40% of the yield was satisfactory). The sensor was built in accordance with specifications and manufacturing process sheets. After the sensors were cured, they were tested; the sensors that did not pass were scrapped. The company sold 400 sensors per week. To have this many good sensors ready, the company produced 1,000 sensors each week in batches of 100.

When IITRI evaluated sensor tests from the previous two years, it found that the yields had ranged from approximately 20% to 70%. After further examination, IITRI found that these yields were an improvement over the yields of the initial production runs, when the company had considered yields of 20% to 25% to be satisfactory.

This product was a financial disaster for the company. The inconsistent yields generated numerous customer problems. The manufacturing department had worked from the beginning to improve the process controls but had not been successful. Five years after initial production, two manufacturing engineers were assigned full time to review the materials and the manufacturing procedures being used.

The manufacturing department had claimed from the beginning that the design of the product was the root of the problem, but its claim had not been taken seriously. RD&D argued that because the product could be successfully produced at least part of the time, the problem was clearly in the manufacturing process and quality control.

The design consisted of a flat and essentially rigid metal cap that sealed against a machined ring of the same diameter. This metal-to-metal seal was the basis for the design. Presumably, had both metal surfaces been free of nicks, dents, or dirt, this seal would have been adequate. Tight dimensions had been specified for these parts, and inspection was done with considerable care. When the design was reevaluated, two options considered were developing a clean room for the assembly and establishing even tighter tolerances for the mating parts.

When IITRI reevaluated the design, questions arose about design requirements and the design itself. Several alternatives were considered and evaluated. The option selected was to add a sealing material to the cap that would force the metal ring into the seal. The tolerances on the metal ring were reduced, and the edge of the ring was rounded to avoid cutting the seal. The company introduced this design modification within two months, after which the yield rose to 100%.

This case illustrates the intimate interaction between product design and production. Researchers and designers largely establish the parameters in which manufacturing must operate. Manufacturing processes, however well executed, cannot overcome such a deficiency in design as the one just described. The fact that a product can be made with a given manufacturing process does not prove that the process is optimal.

Designers, developers, and researchers all can introduce production problems. Indeed, technologists who in any way influence product decisions affect production. Much of the literature on product and process design interaction suggests that only the designers are involved; this is far from true, because designers are frequently limited by decisions made earlier in the R&D cycle.

Introduction of New Materials

Technologists who develop new materials are usually highly aware of problems that can occur. They are familiar with the many new materials that are developed, characterized, and publicized—but that never become commercially available because the market for them is not sufficiently large. These new materials would become available if there were a sufficiently large market for them, but companies considering using these new materials will do so only if they become commercially available.

Establishing the future availability of new materials can be difficult. Because materials suppliers try to develop markets for their new materials, they tend to be optimistic about future availability and costs. The

Often, the R&D department released a prototype to engineering, expecting to have the item available for sale within six months.

more research oriented the technologists, the more likely they are to accept these claims; such technologists would naturally like to be the first to use a new material. Seat-of-the-pants designers, on the other hand, are interested in using only materials that have been commercially available for some time and have been successfully used in other products.

The optimal approach, which lies somewhere between these two extremes, is a difficult path to follow. Many products have never appeared or have disappeared because the materials on which they were based did not become available or were withdrawn because there was not a sufficient market for them. On the other hand, using only those materials that are commercially available and have previously been used greatly limits the potential performance or quality that a designer can design into a new product.

Consider the case of the National Aerospace Plane (NASP), which had been proposed as an alternative to the space shuttle. Because NASP is launched from a runway rather than a rocket launch pad, it could have made space travel more practical. Production of NASP, however, depended on the availability of a high-performance material under development. Because this development progressed much less rapidly than had been expected, the NASP concept will not now be pursued. Several major aerospace firms have lost considerable money because of this failure, but in the world of R&D this loss can be considered part of the game.

A similar failure at an individual firm can result in financial disaster. Consequently, companies should generally avoid the risk of using new materials. If a new material is used, material development must be included as a critical path on the PERT chart. Too often, RD&D personnel choose to use new materials without considering the potential impact of this decision on production. Then, when the new product is to be manufactured, the material is either unavailable or available only at a premium price.

In a discussion of the introduction of new materials into manufacturing, the term *materials* can refer to anything that is purchased. For a plant that assembles cars, materials include engines, seats, brakes, or windshields. One company's products, therefore, are the next company's materials. Occasionally a company can be misled with regard to the performance and availability of new materials. Sensors might not work as promised, or a computer program might not run properly. Although this can often happen, the most common problem with materials is that they can be phased out. Castings, motors, electronics, and fabrics that at one time were easily available, frequently turn out to be no longer produced.

Organization of R&D and Manufacturing

Many companies are still organized in a traditional organizational structure, which is illustrated in Exhibit 5. In this structure, the various groups are relatively independent of each other, and interactions between the groups are quite formal.

R&D and marketing interact, but they do so almost without regard for manufacturing. The rationale behind this organizational structure assumes that R&D operates more effectively if it is disconnected from day-to-day company operations. In many instances, the R&D department is located far from manufacturing, on the presumption that an R&D department is more effective if it is located in a remote facility.

Occasionally R&D solves one of manufacturing's technical problems, but as a favor, not an assignment. When R&D releases a new product or process, it passes it off to engineering or manufacturing. In most instances, R&D develops the concept for a new product, but engineering develops the product and provides manufacturing any necessary information about it.

When the engineering group in one such company was given a new product to design for manufacturing, it would often find that the R&D department had already built and tested a prototype of the product and that the R&D and marketing departments had already decided what the new product should be like. Often the R&D department released the prototype to engineering, asking it to make the new product available for sale within six months. Sometimes engineering could not reproduce the materials or components used in the prototype because R&D had used materials and components that were not commercially available. The engineering group would then do the best that it could to design an equivalent product using available materials and components. Sometimes the equivalent product did not work because there were

Exhibit 5
The Traditional Organizational Structure

```
                        Corporation
            ┌──────┬───────┼────────┬─────────┐
        Marketing  R&D  Engineering Manufacturing Administration
```

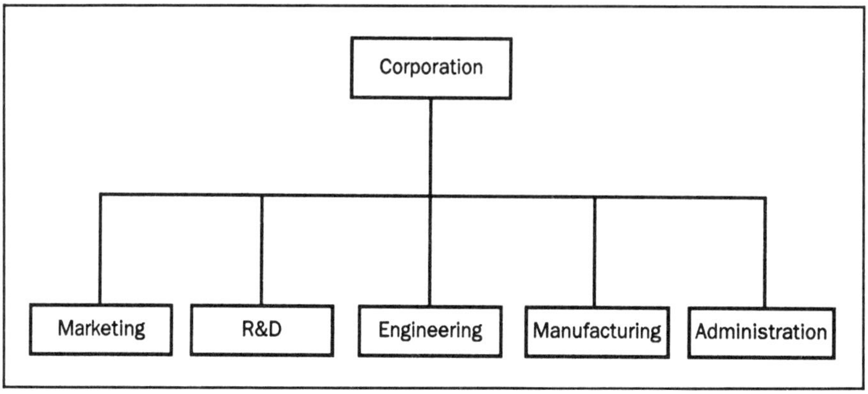

too many variations between the R&D prototype and the new product. Then the R&D, engineering, and manufacturing departments would have to work together to design a new product that could be produced. Occasionally, this would take as many as two years to accomplish.

Many alternatives to the traditional organizational structure facilitate interactions among company groups to a greater extent. For example, engineering and manufacturing may report to one person, or the engineering or the manufacturing department may assign someone to work in R&D. Within R&D, a special group can handle manufacturing technology or design for manufacturability.

MANUFACTURING TECHNOLOGY

Manufacturing technology refers to all technologies relating to manufacturing, not just the application of computers to manufacturing operations. Although different industries use different manufacturing technologies, they share many technologies relating to materials handling, inventory, quality control, and plant layout. In fact, more manufacturing technologies are common to all industries than specific to just one.

A study by the US Bureau of the Census examined how companies used any of 17 different manufacturing technologies. (Exhibit 6 lists these

Exhibit 6
Manufacturing Technologies

Computer-aided design (CAD) or Computer-aided engineering (CAE).
CAD output used to control manufacturing machines (CAD/CAM).
Digital representation of CAD output used in procurement activities.
Flexible manufacturing cells or systems.
Numerically-controlled or computer-numerically-controlled (NC/CNC) machines.
Materials working lasers.
Pick and place robots.
Other robots.
Automatic storage and retrieval systems (AS/RS).
Automatic guided vehicles.
Automated test performed on incoming or in-process materials.
Automated test performed on final product.
Local area network (LAN) for technical data.
Local area network (LAN) for factory use.
Intercompany computer network linking plant to subcontractors, suppliers, or customers.
Programmable controller.
Computers used for control on the factory floor.

technologies.) The industries polled were manufacturers of fabricated metal products, industrial machinery and equipment, electronic and electric equipment, transportation equipment, and instruments and related products. Of the companies from these industries, 76% use one or more of these 17 manufacturing technologies; 23% use five or more. Most manufacturing technologies are appropriate for a wide range of industrial companies.

Organization of Manufacturing Technology

An enormous variety of organizational structures are used to help companies improve the introduction of manufacturing technology. For example, in some companies, manufacturing technology groups that introduce new technology into manufacturing are located in R&D, engineering, or manufacturing departments. In other companies, manufacturing engineering and design engineering are combined. Several companies have search and development groups that search the world for product and process technologies that could benefit the company.

No organizational structure best facilitates the introduction of manufacturing technology. There is even one company in which a manufacturing technology group is successfully operating as part of the marketing department, on the assumption that new manufacturing processes are associated with new products.

Although organizational structure is not critical to the introduction of manufacturing technology, the following other factors are:

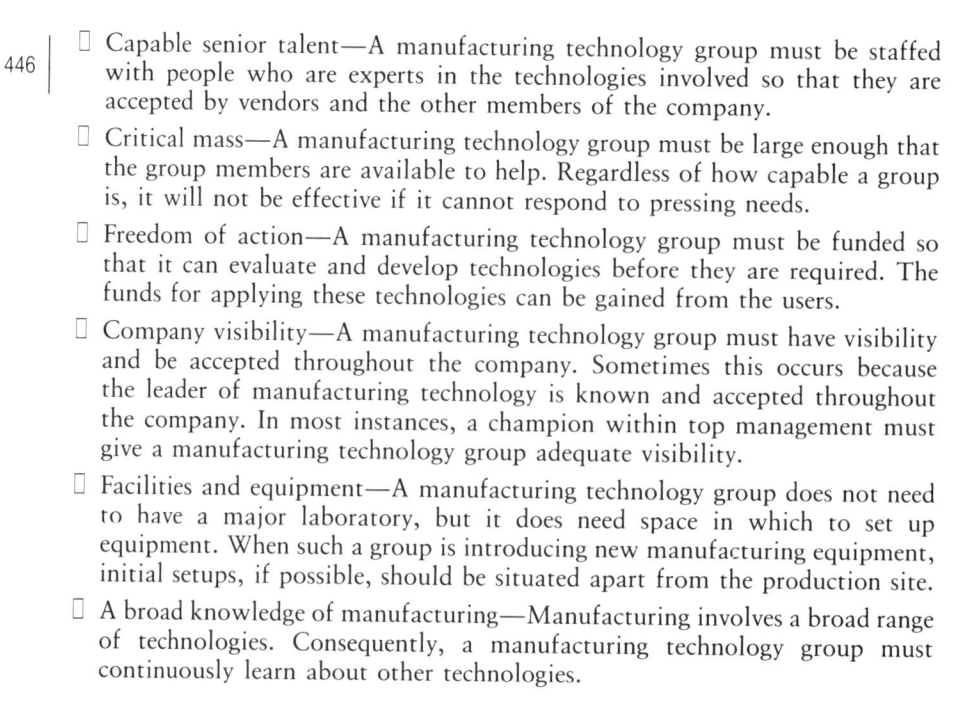

- Capable senior talent—A manufacturing technology group must be staffed with people who are experts in the technologies involved so that they are accepted by vendors and the other members of the company.
- Critical mass—A manufacturing technology group must be large enough that the group members are available to help. Regardless of how capable a group is, it will not be effective if it cannot respond to pressing needs.
- Freedom of action—A manufacturing technology group must be funded so that it can evaluate and develop technologies before they are required. The funds for applying these technologies can be gained from the users.
- Company visibility—A manufacturing technology group must have visibility and be accepted throughout the company. Sometimes this occurs because the leader of manufacturing technology is known and accepted throughout the company. In most instances, a champion within top management must give a manufacturing technology group adequate visibility.
- Facilities and equipment—A manufacturing technology group does not need to have a major laboratory, but it does need space in which to set up equipment. When such a group is introducing new manufacturing equipment, initial setups, if possible, should be situated apart from the production site.
- A broad knowledge of manufacturing—Manufacturing involves a broad range of technologies. Consequently, a manufacturing technology group must continuously learn about other technologies.

One notably effective manufacturing technology group operated as part of the R&D department of a $1 billion company. The group had a staff of 25 and a leader with more than 25 years of experience in manufacturing, engineering, and R&D. The staff was recruited from manufacturing, R&D, and organizations outside of the company (notably vendors of automation equipment). This group started with a three-pronged approach. First, they established a rapport with the operating units to identify their needs. Second, they visited companies, universities, and other organizations capable of revealing the state of the art in various technologies. Third, they set up facilities for developing a computer-aided design and manufacturing system (CAD/CAM). Within two years, this group had designed and developed an effective, state-of-the-art CAD/CAM system and delivered it to the operating units. After the CAD/CAM system was installed, the operating units could manufacture products within 10 minutes of entering data into the system.

Another effective manufacturing search and development group has existed for more than 10 years in a European-based company. This group developed a directory of all the manufacturing technologies that its company needed, and three staff members regularly tour the world looking for new technologies that can help their company.

Much of the success of Japanese companies has been attributed to their searching worldwide for new technologies. Although a few US companies have manufacturing search and development groups, this approach has not been accepted by most US companies.

Even in well-organized companies, opportunities for using manufacturing technology are often ill-defined.

AUDITS OF MANUFACTURING PLANTS

One of the major challenges facing those who are interested in introducing new manufacturing technology is to identify areas within the plant in which manufacturing technology could be used effectively. Each of the parties in this process has its own perspective. The manufacturing plant staff usually is interested in modern versions of what it already has (e.g., equipment that is faster or has more bells and whistles). Technologists are usually interested in advanced technology, particularly in their area of specialization. Consultants and equipment vendors almost invariably promote the approaches with which they are most familiar. Financial people emphasize return on investment. They usually pay little attention to the long-term impact of technological improvements and focus instead on direct cost reductions.

Even in well-organized companies, opportunities for using manufacturing technology are often ill-defined. Management must consider the entire operations of a plant and other issues (e.g., human resources, quality, and strategic plans) in determining whether a manufacturing technology will be beneficial.

At IITRI, we have found that an effective approach for evaluating opportunities for introducing manufacturing technology is to audit the manufacturing plant (see inset). Since 1976, the Manufacturing Productivity Center (MPC) at IITRI has conducted many audits each year. Depending on the company and the focus of a given audit, some facets of the manufacturing operation might get only casual attention. However, assessment of quality strategies, products and processes, materials management, management accounting, and information management is vital for evaluating the status of technology in a plant and for identifying opportunities for introducing new manufacturing technology.

AUDIT CRITERIA

Each plant audit that MPC conducts is different. The type of audit depends on the nature of the plant and the experiences of the company. In each audit, several areas must be considered methodically.

Quality Strategies

These strategies include the following:

- The company's effective use of:
  - Labor.
  - Capital.
  - Energy.
  - Materials.
  - Appropriate technology to increase quality.
- Management for quality and feedback of information:
  - Responsibility for quality.
  - Quality standards.
- Design verification:
  - Function effectiveness.
  - Reliability.
  - Manufacturability.
  - Testability.
- Manufacturing for quality:
  - Process capability.
  - Statistical process control.
  - Traceability.
  - Diagnostic guidance.
  - Performance prediction.

- Quality Contribution:
  — Inspection and audits.
  — Process control.
  — Design of experiments.
  — Measurements.
- Vendor capability or procurement quality.
- Distribution:
  — Quality perception.
  — Warranty.
  — Repair or replacement.
  — Field service.

Products and Processes

A capable audit should include a comprehensive look at the factors affecting products and processes, including:

- Technology's effect on profits and the workforce.
- Barricades to using technology.
- The extent of manufacturing automation.
- The status of preventive maintenance.
- Machine loading and unloading.
- Machine tools changes.
- How parts are held for machining.
- Machine control processes.
- Shop data collection methods.
- Parts loading and unloading.
- Assembly operations.
- Painting or finishing.
- Feedback controls.
- Parts inspection.
- Monitoring layout of departments.
- Work in process.
- Interdepartmental movement of parts.
- Input and output controls.
- Bill of materials generation.
- Use of CAD/CAM, CIM, FMS, robotics, and group technology.

Materials Management

The audit of materials management must include a thorough assessment of:

- Inventory cost and turnover.
- Systems for managing inventory.
- Order processing.
- Transportation.
- Warehousing.
- Sequence of operations.
- Handling by direct labor.

(cont'd)

AUDIT CRITERIA (cont'd)

- Travel distances for materials, equipment, and personnel.
- Opportunities for group technology, product, and process layout.
- Uses of storage.
- Expansion capability.
- Schedule of arrival of out-bound and in-bound carriers.
- The physical environment.

Management Accounting

Areas to be assessed in an audit of management accounting include:

- Management accounting information.
  — Does the information exist?
  — Is it well thought out?
  — Is it used on the shop floor?
  — How well is it used for pricing and for make-or-buy decisions?
- Validity of existing information (e.g., if profitability information is available, how good is it?).
- General questions:
  — Is a standard cost system used?
  — Is it adequate?
  — How are standard costs developed?
  — Does the cost build-up process make sense?
  — What goes into it?
  — At what level are actual costs collected (e.g., factory or cost center)?
  — How is cost information reported?
- Analysis of performance data.
- Budget and consequences of budget performance.

Information Management

Several aspects of the information management system should be scrutinized:

- The adequacy of existing information systems.
- The necessity of reports.
- Speed in obtaining needed information.
- The strategic use of information.
- The cost effectiveness of information systems.

Labor Characteristics and Training

The company's labor environment should be assessed to productively characterize the relationships among workers and management. Important aspects of this environment include:

- The labor-management climate: cooperative or adversarial.
- The degree of idea sharing between hourly workforce and supervisors.
- Employees' perception of management and employment security.

- Employees' feeling of general job satisfaction and morale.
- Employees' understanding of need for quality and productivity improvement.
- The workforce's readiness for work redesign.
- Definition of skill and knowledge needs of company-allocated budget and training cost.
- Adequacy of worker training for current and future work.
- Adequacy of supervisors' training.
- New employee orientation.
- Career development in management, technical, staff, and hourly personnel.

Reward and Incentive Systems

Company methods for encouraging their workers are of importance in the audit. Subjects that should be discussed are:

- Financial and nonfinancial rewards.
- Employee involvement and recognition systems.
- Future plans for further improvement.
- Motivational value of the existing reward and incentive system.
- Opportunity for promotion.

Hazard Control

The workers' exposure to workplace hazards should also be considered in the assessment. Subjects to be addressed include:

- Accident cost analysis.
- Ergonomic considerations of materials handling and repetitive motion.
- Tool, equipment, and workplace design.
- Environmental stressors.

Product Liability

Another area that should be addressed in the audit is product liability. The integrity of the product should be assessed by means of the following criteria:

- Company policy for product safety.
- Safety in product development, in all phases: conceptual, design, manufacturing, transportation, marketing, use, and disposal.
- The degree to which employees are prepared to testify in litigation cases.
- Protection against discovery.

Productivity measurement

Finally, the company's methods of measuring its own productivity should be analyzed in depth, including:

- Factors considered in measuring productivity.

(cont'd)

AUDIT CRITERIA *(cont'd)*

- Measurement data to produce information for:
  — Labor tracking: time and attendance.
  — Machine operation monitoring.
  — Material tracking.
  — Plant maintenance.
  — Quality monitoring.
  — Facilities monitoring.
  — Job assignment.
  — Stores control.
  — Energy control.

LAYING THE GROUNDWORK FOR TECHNOLOGY TRANSFER

James R. Key

> Implementing tools and technologies that originate in a research consortium lab is a demanding process. Because both physical and psychological barriers can impede successful technology transfer, all levels of management should be dedicated to overcoming them.

Decreasing an organization's product development cycle through increasingly effective use of technology is a major goal of management, whether the product is generated in the company's research, development, or manufacturing department. An important element in this effort is controlling the amount of time an organization takes to get a new product to market. Being late to market causes a company to miss the initial, high-margin phase of a product and to fall behind faster moving competitors who are raising manufacturing yields and lowering factory costs. By improving the rate and efficacy of technology transfer from external sources, particularly from research consortia, management can enhance company productivity significantly.

Careful selection of new technology and tools can substantially help a company reduce its product development cycles. A company may find the new technology it needs for product development either in commercially available tools or through its own research efforts. To select tools effectively, however, a company must know exactly what is available; staying abreast of every technological development in the world of R&D is a difficult task. As the fruits of the technological age continue to become less expensive and more generally available, companies must work ever harder to stay ahead—or to break even.

The breadth and depth of the research needed for competing in a global marketplace have increased well beyond the resources of any individual company; accordingly, many companies have pooled resources to support research consortia. Cooperative research can achieve a number of objectives beyond the reach of individual companies' efforts. By rapidly attaining critical mass, a consortium avoids needless duplication of efforts in isolated companies. Consortia allow companies to focus more internal resources on applications, precluding replication of generic research results that can be obtained elsewhere. In addition, the amount of technology and knowledge needed for competition are beyond the resources of any individual company's capabilities. These crucial advantages of cooperative research efforts, though recognized by many companies,

---

JAMES R. KEY has extensive experience in integrated circuit technology, high-end computers, and communications. He is with the Engineering Research Center of the University of Maryland, College Park, and is an industry consultant in technology transfer and research consortia.

are understood by few. Consequently, the transfer of technology from external sources is often not as effective as it could and should be.

PRELIMINARY CONSIDERATIONS

Although technology is transferred at the level of technologists, its successful transfer depends on the actions and attitudes of each level of management in each company function. In addition, if a technology is to be successfully acquired from an external source, the project must have sufficient attention and commitment from all levels within the corporation.

Technology transfer can be improved only by a company's conscious and integrated commitment. As in a health club, merely owning a membership does not lead to physical fitness', hard work is necessary to achieve results. For technology transfer to succeed, it must be woven into the fabric of the company's long-range goals and strategic plans by top management. Middle management needs to consciously provide technology support; at this level, a lack of support yields a weak link in the product development chain. Line management must allocate the requisite internal staff and funds for proper operation of the received technology, as well as have sufficient staff and funds to commission and acquire external technology development.

A major responsibility of corporate leadership is to manage risk. In contemporary, lean organizations, many of which no longer have research departments or exploratory technology development groups, many managers fear that if a high-risk development effort runs into trouble, their companies will not have enough scientists and engineers to rescue the effort. These managers therefore try to avoid taking development risks.

At the same time, technology transfer is still largely an art rather than a scientific discipline. There is no definitive yardstick to gauge the individual steps composing a successful transfer of technology. The flow from concept to production-ready technology is rarely an unbroken, single path. Although research consortia measure dollar leverage, quality and quantity of research output, number of contacts with members, and numbers of reports requested, among other items, there is very little concrete data on the actual transfer of technology. Companies are reluctant to release such information, which might decrease their competitiveness, and engineering departments do not usually document the sources of their ideas and technologies. The technical staff is not rewarded for revealing sources; however, there is potential liability if, when a company claims credit for a development, a university or competitor contends a violation of its intellectual property rights.

BIRTH OF A CONSORTIUM

US corporations first began to look seriously at cooperative research efforts during the early 1980s, when several trends and events forced a paradigm shift in the way integrated circuit and computer manufacturers developed new technologies. A number of events conspired to threaten individual and private research. First, federal research funds from the National Science Foundation (NSF) decreased and Department of Defense (DOD) support shifted from silicon-based technology to focus on more exotic semiconductor materials. The breakup of the Bell Telephone monopoly then effectively restricted outside companies' access to Bell Labs technology. Another factor, the increasing disparity in pay between starting salaries in industry and those for new university professors, made an academic career less and less attractive to job-market entrants, creating a faculty shortage just as computer science and electrical engineering departments experienced massive enrollment increases. At the same time, tightening economic conditions caused companies to significantly reduce support for corporate research programs.

The success or failure of the transfer process is critically dependent on the actions and attitudes of the company's entire management chain.

The net result of these trends and events was a dearth of university and private research in the US; concurrently, Pacific Rim and European countries sought to gain market share by investing massively in their national programs to nurture specific industries. US electronics executives, alarmed that their reduced access to research would mean a reduced supply of new technology just as global competitors accelerated their own technology development efforts, began to look for alternative sources of technology. The amount of expertise and technology that companies needed to remain competitive was escalating astronomically. For example, the cost for a new semiconductor fabrication line rose from $50 million in 1982 to $500 million in 1995 and was projected to rise to $1 billion in 1999. It became clear that no US enterprise, not even such giants as IBM or AT&T, had the intellectual and financial capital to meet all technology needs solely through internal development—particularly when the competition included foreign governments' national programs.

Faced with these threats to their survival, several US electronic companies in late 1982 initiated cooperative research consortia to support research and technology development, sharing rising risks and costs. Having operated over several years now and currently producing steady streams of research and technology, these consortia have proved a successful choice for conducting high-risk, long-range research. A benefit that participants had not anticipated was their new-found forum to set long-range goals for research and to map these goals into their companies' strategic plans.

However, each of the member companies now found itself facing the unexpected difficulties of transferring technology from the research consortium to its own internal development organization. And technology transfer is definitely not a simple matter of acquiring a new piece of equipment, ready to operate; it is a complicated do-it-yourself job, with few or no instructions included.

TECHNOLOGY TRANSFER GROUNDWORK

STAGES OF TECHNOLOGY TRANSFER

Many managers do not understand the processes of technology transfer and technology hardening. To some, technology transfer should be like just-in-time delivery of commodity parts for a high-volume manufacturing operation—simply a matter of transporting the technology (i.e., parts) from the originating lab (i.e., supplier warehouse) to the receiving lab (i.e., assembly line). Unfortunately, technology from a research consortium is rarely in a form that is directly useful to a product developer or manufacturing line. Transport and delivery is only the beginning of the transfer process. In managing technology transfer, a company must undergo eight stages to acquire useful technology and tools:

- Acknowledgment of the recipient's awareness, interest, and tracking of technology.
- Transportation and delivery of the originator's technology to the recipient's lab.
- Confirmation of the recipient's expertise in and understanding of the originator's technology.
- The recipient's evaluation of the technology's suitability for eventual application by product development or manufacturing.
- The recipient's modification and hardening of the technology to meet anticipated needs of product development and manufacturing.
- Demonstration and sale of the recipient's version of the technology to a client.
- Adoption of technology by an end user, who furnishes specific requirements.
- The technology support group's maintenance and modification of the recipient's product in response to the end user's ongoing needs (the recipient usually does not perform this function).

At all stages of technology transfer, the recipient company can encounter company obstacles that menace the success of its project; managers must beware of these barriers during all phases of the transfer.

BARRIERS ON THE HOME FRONT

The first barrier to technology transfer is the would-be recipient's lack of awareness of a technology's existence—still a major problem in many research consortia. Often only a small percentage of the technical staff of sponsoring companies is familiar with the consortium, its goals, and the status of its various research projects. The consortium manager here must assume some responsibility, disseminating information and ensuring that each employee has full access to the research output. Unless this task is part of the manager's performance plan, it will not receive the requisite amount of attention. Company managers should budget time and travel funds sufficient to allow their technology scouts to track external technology developments. The resources for this task, often buried in other projects, tend to get cut early when overall budgets are cut.

A necessary ingredient for the success of a technology transfer is the recipient's faith in the involved technology. A transfer will fail if the recipient does not believe the effort will work or be commercially workable. This uncertainty can also arise if the recipient fears that proprietary company information may be revealed or that a liability may be incurred during the transfer process. If management is not sufficiently confident about the technology transfer, the effort will not attain its potential.

Technology from a research consortium rarely comes to the development lab in a form immediately useful to a product developer or manufacturing line.

To give the project the essential impetus, individuals must be designated champions of the technology on both sides of the transaction. Champions are needed both to push technology out of the originating organization and to pull new technology into the receiving organization. In addition, the risk and reward structure of the receiving organization must provide sufficiently attractive incentives to encourage its champion to proceed.

Transportation and delivery of technology can be a straightforward process, particularly when the recipient already has the equipment necessary to the technology's immediate operation. If the recipient owns equipment identical to that of the technology's developer, the technology needs no modification. However, the recipient frequently lacks or does not want to purchase the same equipment, and modifications must be made in the recipient's process, tools, or software to accommodate the technology. A common example of this scenario for software occurs when different hardware platforms or operating system versions are involved. This can be a substantial obstacle, and managers need to determine the strategic significance of the project to judge the most efficient allocation of resources.

Before a new technology can be used by a product development group, it must be brought up to industrial strength to meet the group's requirements. The recipient may perform this function and then pass the technology on to an internal tool and technology development organization; external vendors can also accomplish this phase of the transfer.

If the product development group accepts the technology, a technology support group must maintain it in a stable state. Production environments change, and the support group must develop the technology to meet the requirements of production. The development of industrial-strength tools and technologies can be developed by third-party companies, strategic partners, or vendors. This arrangement is insufficient for product development groups, however, which also need the tools and technologies to be supported while they apply them. External support groups tend to be less responsive than internal support groups, and managers must take this lack of response into account when planning for completion of the technology transfer process.

Successful technology and tools go through a life cycle, beginning as fragile, untested concepts, moving onto prototype stages in which they sometimes work, and developing into robust, mature tools whose characteristics are well known for a given class of applications.

PITFALLS IN THE TRANSFER PROCESS

When considering the acquisition of technology from external sources, the manager must be aware of the difficulties involved in the transfer process. As more companies begin to seek such acquisitions from external consortia, the science of technology transfer will likely become much simpler. Currently, however, a number of potential problems still exist in matters of timing, corporate structure and communication, and legal issues, which the manager must take into account before beginning any transfer process.

A Question of Time

New technology, unlike just-in-time deliveries, can have long periods of gestation, taking years to develop to the point of commercial utility after its acquisition by a company. In addition to the time spent advancing the technology to a useful stage, other time factors affect this period of maturation. The attitudes and stability of the company itself can affect technology transfer drastically. For example, companies in the electronics industry frequently reorganize. Managers are moved or leave for other companies; their annual pay structure is based on quarterly and annual results. How do these divergent time scales affect the perspective and actions of managers with respect to technology transfer? What incentives do companies put in place to reward technology transfer that has taken place over several years?

Successful transfer is also very dependent on other types of timing. If it arrives too soon, new technology will likely languish on laboratory shelves; if it arrives too late, it may be unusable. Product development groups that neglect planning do not provide enough lead time to anticipate their own needs; they tend to react quickly, looking for off-the-shelf solutions to their problems of the moment. These organizations look at research and product trends, then jump directly to planning for the next product—considering as scarcely more than an afterthought the tools and technologies that they will need. By that time, crucial decisions have already been made, limiting the range of choices available to the organization and sometimes making the product too costly to develop and manufacture.

Lines of Communication

Communication is an important aspect of a company's operation that cannot be neglected during technology transfer. Attitudes such as not-invented-here often preclude successful transfer: The recipients may be too reluctant to change or invest the necessary effort to adopt the new technology, or the external version may conflict with an internal effort. The external work may not clearly map onto the company strategic plan, either because of incompatibility or lack of a workable strategic plan.

From the initial stages of planning, the manager must ensure that the company employees involved in the transfer are aware of the ramifications of acquiring the technology. Goals and expectations should be communicated clearly, leaving no questions in employees' minds about the propriety of the technology and about possible risks and benefits.

In many research consortia, only a few of the companies' technical staff members are familiar with the consortium's goals and the status of their various projects.

A workable technology transfer strategy must differentiate among invention, basic innovation, and improvement innovation and address the distinctions between them. The strategy must overcome the major barriers of uncertainty in risk, future cost, future performance, and future availability of a proposed technology. Development and manufacturing organizations must correctly anticipate the evolution of a technology that might exist today only in a lab or in a vendor's marketing brochure. Product groups must deal with uncertainty, fear of risks and failure, and the investment needed to get a payback in the scheduled time period.

A Legal Matter

Legal problems can pose a complex challenge to the manager. Antitrust laws, which were originally drafted to protect US companies from unfair competition, actually handicap many research efforts by preventing companies from transferring the very technology that would help them maintain their competitive edge. Recent developments in antitrust legislation have somewhat relaxed the laws' grasp on research consortia.

Before the 1984 National Cooperative Research Act was adopted, it was virtually impossible for companies to jointly fund research in the US. In Europe and Japan, where such restrictions did not apply, many precompetitive cooperative research programs sprang up; some of the best known of these are the Cooperative VLSI Laboratory and the Fifth Generation Computer Project in Japan and ESPRIT and JESSE in Europe.

When Control Data Corp. (CDC) sought to follow the lead of the Japanese and European models, attempting in 1982 to form such consortia as the Microelectronics and Computer Technology Corporation (MCC) and the Semiconductor Research Corporation (SRC), many prospective partners cited antitrust fears as reasons against even exploring joining the consortia. Although the Justice Department ruled in 1984 that joint research does not violate the law, US cooperative ventures were still lagging behind those in other industrialized nations.

In 1986, CDC's William Norris and other corporate leaders determined that US firms had a low rate of adoption of research results and technology from universities and consortia. Realizing that the high costs of development and manufacturing were blocking sponsoring firms from aggressively using the results they had funded, Norris and others proposed that antitrust laws be revised to allow companies to share the costs of development and manufacturing. In April 1990, the House Subcommittee on Commercial Law voted to provide antitrust relief for joint production ventures. This legislation, a decided victory for joint ventures, is only slowly starting to visibly affect the joint industry.

Other developments in the legal arena are encouraging cooperative research efforts. More than 700 federal laboratories develop technology for the public good in the fields of defense, agriculture, forestry, and electronics. In the past, legal and administrative barriers have been a bottleneck to the diffusion of this technology into the commercial sector. Through the 1986 Federal Technology Transfer Act, which amended the 1980 Stevenson-Wydler Technology Innovation Act, Congress encouraged access to this large technology resource. The National Competitiveness Technology Transfer Act of 1989 further encouraged close alliances between government labs and industry for the purpose of transferring commercially valuable technology. Subsequently, the National Cooperative Research and Production Act exempts certain types of research and manufacturing ventures from specific antitrust challenges in order to encourage the sharing of scarce resources.

Because of the importance of computers to daily life, an increasing percentage of research and product development budgets goes into software development and maintenance. This trend favors transfer from consortia because software is an excellent medium for transferring technology. It is also relatively difficult to protect intellectual property rights in cases involving software.

HOW DOES A COMPANY KNOW IT IS SUCCESSFUL?

One of the most significant issues facing companies today is the lack of quantitative metrics for evaluating the output of consortia and other technology transfer efforts. Most metrics measure activities or anecdotal evidence rather than bottom-line figures. Nevertheless, companies without access to advanced technology resources will lose market share to market leaders. The technical community is struggling to discover the elusive measure that tells a company whether its investment has paid off. The accounting profession could try to create dollar metrics to measure the value of research and development (and the trade-off of developing technology internally or acquiring it through consortia) and the value of licensing the best available technology of those willing to sell (a strategy that fails when a critical technology is not for sale but the competition already has it).

PREDICTING THE FUTURE

Rarely does a consortium produce results in a form directly applicable by product development groups. The following cases illustrate situations in which company management had difficulties predicting the impact of consortium results on their company's bottom line. Evaluating the value of a technology with undeveloped applications is a difficult task, often through most phases of a project.

> To be useful, new technology must be understood, adapted, hardened, and supported by internal groups.

Case 1. The research department of an electronics company received software tools for thermal and vibration analysis from the research consortium it had joined recently. This research department approached its company's product development division, offering to apply these tools to the design of a major new product. Responding that such tools had not been needed in the past, the product team declined the offer.

Eighteen months later, the product was late. Caught in an economic downturn, the company badly needed to complete the development of this new product. Despite the fact that budgets were being cut companywide, manufacturing was still begun. After volume production began, field tests revealed a severe vibration problem that a crash engineering effort could not solve. Finally, special shipping arrangements had to be made and special field installations became necessary.

The company learned from this that a research group, at the cost of a few thousand dollars per month, could have been dedicated to solving a problem that eventually grew to cost several million dollars per month in manufacturing. Eighteen months after the research department's offer of new tools had been turned down, the company was experiencing severe cash flow problems. The attendant expense and sales lag far exceeded the total amount that the company had invested in the consortium and research department for those eighteen months.

In its analysis of the problem, management identified the following factors as having prevented the consortium results from being used in a timely manner:

- Insufficient communication between research and product development teams.
- Reluctance of product developers to commit to untried tools from the research department.
- Lack of understanding of the time frames involved for each group.
- The research department's lack of aggressiveness in selling its problem-solving skills to product development.

Case 2. A consortium developed a prototype for an advanced software tool at considerable expense. Having participated in the consortium, the research department of an electronics company decided that this tool was sufficiently robust to be useful. The research group then tried to convince internal product development divisions to try the tool.

After several months, one of the product teams invited the research team to see a new commercial software product from a new company. This software had equivalent performance claims to the tool that the research group was supporting. After much internal discussion, the research department dropped its promotion of the consortium tool, recommending instead that the commercial tool be purchased. Their rationale was that the commercial tool would receive ongoing vendor support, freeing the research team to work on other, less-developed areas of their research.

Critics of the consortium project claimed that the commercial tool would have been developed without the company's investment in the consortium. Consortium supporters claimed that the consortium's work and subsequent publicity actually spurred the formation of the new company and its product. Furthermore, these supporters maintained, the familiarity that the research group gained with the consortium's tool provided an in-depth knowledge of the subject, allowing the researchers to make a quick and informed decision on the purchase of the commercial product. Researchers pointed out other occasions in which the company had purchased untried commercial software that had not performed as advertised, leading to schedule slippages and budget overruns.

When analyzing the situation, management concluded that the knowledge gained had justified the company's investment in the consortium. The goal was to obtain software tools with the right performance characteristics. It was unnecessary that the consortium provide the actual tools that were to be used in the company's production environment.

Case 3. Consortium members had struggled individually for years, with limited success, to develop software that would prototype electronic designs operating in severe physical environments and that would support the rapid comparison of expected performance with regulated guidelines. Once consortium members realized their common need for this software, an effort was started. Progress was rapid, and defense contractors placed a suite of tools into service. Currently, the tools are now in growing use by designers in the commercial aviation and automotive industries. They have reduced products' time-to-market, and equipment reliability is now designed in, rather than tested in.

In analyzing the success of the program, members agreed that its task was very complex and included many disciplines. No single member could have made the substantial investment in personnel and equipment necessary to develop the software tools it developed, but the consortium could and did. Comparable commercial tools have not appeared on the market, therefore members continue to support the consortium strongly as it develops new software tools.

TECHNOLOGY TRANSFER

SUMMARY

Each consortium member must make two investments—one in belonging to and operating the consortium and the other in tracking, evaluating, transferring, and adapting the technology the consortium develops to the company's internal development and manufacturing process. A consortium can develop excellent tools that its member companies dismiss because they do not appreciate the tools' usefulness.

Ignoring or undervaluing the knowledge that dedicated research consortia offer is a dangerous mistake.

Although various mechanisms and processes can provide equitable access to research results for companies, many participants do not invest enough internally to absorb the output from research consortia; nor do industrial members deal adequately with the output of consortia at all levels of management. Management can improve the productivity of technology transfer through several actions:

- Conducting a thorough evaluation of the role of tools and technologies in the company's strategic plan, and integrating the requirements for external tools and technologies into the plan, rather than handling them on an ad hoc basis.
- Reexamining the company's process for bringing in technology, particularly those that need modifications to meet any unique company specifications.
- Reexamining the significance of technology and tool development in each division. The roles of research, product development, and manufacturing are well understood; the roles of technology and tool support are not, and this factor can have a severe impact on budgets and schedules.
- Consciously identifying champions for bringing in technology. Company pruning and restructuring has seriously depleted the reservoir of individuals who carry out this function.
- Recognizing the considerable amount of time needed to bring a new technology up to speed, and differentiating between technology development cycles and product development cycles.
- Recognizing that significant resources are required for the receiving organization both to receive and to harden technology. To be useful, new technology and tools must be understood, adapted, hardened, and supported by internal groups, even if third-party developers and vendors must be used.

Ignoring or undervaluing the knowledge that dedicated research consortia offer is a dangerous mistake that can hamper company operations. In both of the preceding case studies, the research consortia's findings could have been advantageous to their sponsoring companies. If consortia results are either applied or dismissed too quickly, however, a company can easily lose the advantage it would gain from participating in a consortium. In both cases, the important precept is long-range thinking. Severe myopia—the lack of proper focus on research proceedings and findings—can often result in a companywide headache.

USING CROSS-FUNCTIONAL TEAMS FOR TECHNOLOGY TRANSFER AND NEW PRODUCT DEVELOPMENT

Product development and the critical delivery of products to the market on time are the key ingredients to an organization's success. These goals can come about through a change in the organization's mindset and the best use of cross-functional teams.

Sakharam Patil

The long-term competitiveness of any business, especially a manufacturing business, ultimately depends on its ability to get products to the market on time. The saying that "time is money" should now be changed to "time is more valuable than money." According to one study, a company loses an average of 33% of after-tax profit when it ships products six months late, compared with average losses of 3.5% when it overspends by 50% on product development.

Cross-functional teams—often called multidisciplinary teams—are the secret of developing successful products and getting products to market on time. The development of a company's product-development capabilities contributes to an organization's market position, financial performance, and the continuous renewal or improvement of the business.

Technological changes, government regulations, and other demands have created a tremendous opportunity for growth in new products. In the author's industry—the food industry—consumer demands for healthy and nutritious foods, convenience, and variety have created a need for a dynamic program to meet these ever-changing needs. There is tremendous segmentation and fragmentation based on age, ethnicity, and health. Manufacturers of consumer products in this industry are under increasing pressure to meet the new demands of consumers. Companies must meet this challenge competitively with continuous growth and improvement. It once took the author's company two to three years to develop a product, but now this can generally be done in two to 12 months. New product development is an important strategic step toward overall business success.

Cross-functional teams that work together to develop products can create a learning organization that continuously develops the expertise, know-how, and leadership of the company's employees. With these teams' integrated

---

DR. SAKHARAM PATIL, PhD, is vice-president of marketing and commercial development at American Maize in Hammond, IN.

expertise, their functional areas create a unique system that is highly productive and competitive.

Cross-functional teams typically consist of representatives from marketing, sales, manufacturing, research, technical services, engineering, finance, and other areas that contribute to the development process. Any organization, private or public, can start a cross-functional product development program to improve its productivity, long-term competitiveness, and growth.

Such programs also instill a belief in all employees that they are leaders who can make things happen, especially the growth of new products and services.

Cross-functional teams can speed up product development, solve complex problems, maintain a clear customer focus, improve organization, create potential, and continuously improve the know-how and capabilities of an organization. This chapter describes how a company can recognize that it needs to change and how it can initiate and develop a successful product-development program.

JUSTIFYING THE NEED FOR CHANGE

A company's first and foremost challenge is recognizing the need for significant change that must occur from the top on down. A change must be made in the mindset of many people and ultimately the company's culture—that is, a change in the modus operandi of status quo. Top management must assign clear responsibilities and stay deeply involved in the process. Management should clearly establish the strategy for product development, which should be based on growth, technological and service competencies, organizational skills, and the linkages within and outside the company. In addition, management must focus on managing the boundaries or interface between the key functional areas of business. The objectives should be spelled out based on the company's capabilities, and management must continuously build these capabilities and support of the learning (know-how) process.

Overcoming Resistance

Michael Robert has discussed seven deadly "sins" of companies that are not on the cutting edge of new product innovation (see Exhibit 1). The company's ability to be innovative—to create, and introduce a continuous flow of new products—depends on how well the organization manages and eliminates these common sins. The sins that Robert cites are excellent examples of excuses that management might use to resist change in this dynamic world market. If a company hides behind these excuses, stagnation and the eventual demise of the business could ensue.

PROGRAM FOUNDATION AND FRAMEWORK

Cross-functional teams are groups of people from a variety of functions who join to accomplish a clear purpose. Successful teams have two to 15 members,

Exhibit 1
Seven Deadly Sins of Corporate Stagnation

- We must protect our cash flow at all cost, or else we will perish.
- Our industry is mature. No more growth or innovation is possible.
- We are in a commodity business (in which innovation is not necessary, because the product sells itself).
- Only entrepreneurs in small companies can innovate. Large companies stifle risk-taking and new product creation.
- Innovators are born. It's a personality trait, and we just don't have any of these people around.
- New product creation is too risky.
- We don't have the resources necessary to innovate.

depending on the project's breadth and complexity. These teams consist of a smaller number of people with complimentary skills, who are committed to a common purpose, performance goals, and an approach for which they hold themselves mutually accountable. The complementary skills are technical or functional expertise, problem-solving and decision-making, and interpersonal skills. Teams hold themselves collectively accountable.

Management's Role

The company's management must lay out two elements that are essential for the success of the program: the program's foundation and framework. The management processes and the program's organizational structure need to be interrelated. The program's foundation includes the program's structure, organization, authority level, and project-initiating process. This foundation can also be referred to as the ground rules or guiding principles of a very important facet of a business. The formalized system creates a framework for the program. This is essential for the program's success.

The program must define how the project will be accomplished and the project breakdown (i.e., which person will do what). In addition, the program must define its measurement system (i.e., the milestones that are used to monitor the program). This system should define team operations and the specific operating procedures that apply to an assigned team project. Management must spell out the specific objectives for an assigned project and the teams. The program's foundation and framework should be documented in a formal organizationwide program manual.

Top management must be involved in and actively participate in the program. Those who want to initiate such a program are advised not to do it if they do not have the involvement of top management, a clear foundation, and the framework.

PROGRAM ORGANIZATION AND STRUCTURE

The program's organization is designed and built by the top management team, which is called the product-development steering team (PDST). This team typically consists of managers from marketing, sales, manufacturing, product development, finance, engineering, and quality control. The formal management system should be designed and outlined in the program manual.

The Steering Team's Role

The PDST is headed by the division president or a vice president of marketing, and it is this team that assigns a program manager. The program manager oversees, monitors, manages, and sponsors the whole system or program. In the author's company, the program manager is the director of commercial development because the majority of developmental activities are initiated and conducted by commercial development.

Role of the Program Manager. The program manager is a sponsor or champion of the cause. This person gets involved when needed (e.g., when conflicts arise) and supports the teams and team leaders in their efforts. As a member of the PDST, the program manager is committed to the team's success through the following actions:

- Set team goals and expectations
- Communicates between the PDST and project teams
- Negotiates the assignments of team members by functional department managers if necessary
- Links business issues to team objectives
- Assesses team performance
- Assists in the appropriation of resources to teams
- Acts as a champion of the cause and champions team recommendations
- Participates in team meetings as needed
- Makes sure the teams are meeting the targeted timetable

Project Assignment. The PDST must meet once a month to review the individual project milestones and assign new projects presented by the marketing and product-development managers to project teams, which are discussed later. At this time, the PDST assigns priority to the new projects and reviews and reassigns the priorities of ongoing projects. Each project is also assigned a target date, which is based on the need of a specific customer market or market segment and the probability of the project's success, creating the sales and the growth of new products.

The Project Team

Members of the project team are selected once a project is approved and assigned. Selection for the project team is based on the team members' skills, experience, and functional involvement. The project team typically consists of product development, marketing, sales, process development, and manufacturing. Other project team members, who typically come from basic research, finance, engineering, and other areas, are assigned based on their expertise and the project's complexity.

> It once took the author's company two to three years to develop a product, but now this can be done in two to 12 months.

The Project Team Leader. The leader of the project team is assigned based on his or her expertise, interpersonal skills, and functional area. The team leader has the overall responsibility for making sure the project is completed on time. He or she is also responsible for outlining the project plan, assigning responsibilities, and making sure the plan milestones are accomplished. Milestones are tangible deliverables (i.e., results) and are the performance criteria or the targets that must be accomplished. Establishing these milestones can help the team to feel that its members have contributed to tangible results.

A team leader should have the following skills:

- The ability to admit ignorance
- Knowledge of when to intervene
- The capacity to truly share power
- Concern for what he or she takes on, not for what he or she gives up
- The ability to get used to learning on the job by doing and making decisions with people

Following the assignment of the project and the project team, the team leader should convene a project team meeting to develop a project plan and identify tasks and responsibilities. The team leader will, in discussion and cooperation with the department manager, allocate resources to complete the project. The other responsibilities of a team leader are conflict resolution and communication with the program manager and other functional department heads. The team leader publishes milestones or a Gantt chart that clearly delineates tasks, responsibilities, and the project team's timetable.

The Team Meeting Agenda. At the first project team meeting, the following steps should be completed:

- Establish the objective.
- Do the planning.
- Develop a work breakdown.

- Lay out a time schedule.
- Estimate costs.
- Clearly outline work assignments.
- Define the documentation of the project's progress.
- Agree on milestones and deliverable results.
- Agree on specifications and performance criteria.

Reports and Documents. The project team's reports and documents must be written and addressed to the program manager. Copies must be sent to management, the PDST, and team members. Examples of these reports and documents are:

- Initial meeting minutes
- Completed need analysis
- Milestones—charts and progress reports
- Specifications and performance criteria
- Final report
- Product story to educate sales, commercial development, and marketing on the new product's benefits, which these departments will use to build a customer base for new products

All of the documents should be maintained in a central project file located in the product development laboratory. The project team leader is responsible for ensuring that all documents pertaining to the project are copied and filed in the central projects file.

Team Presentations. The project teams periodically make a presentation to the PDST. This presentation helps in the reviewing and monitoring process, and it also ensures that the project team and management agree that the project is on the right track. The presentation also recognizes the project team's effort and allows a clear understanding of any difficulties and successes. It is absolutely essential that the PDST and the project team openly and honestly discuss the project and maintain a clear vision of the mission of the business.

THE PRODUCT-DEVELOPMENT PROCESS

The essential elements of the product-development process are recognizing customer needs and establishing a procedure for bringing the products that meet these needs to market.

Identification of Customer Needs

A product to be developed is based on a customer need. The product can be developed successfully and faster if this need can be clearly identified and understood in time. In this fast-changing dynamic world, the window of opportunity quickly opens and closes. The company must have a system to develop and commercialize the products rapidly. It has to spend significant resources to do market research, talk to customers, and evaluate whether the scarce resources of development are not wasted. Marketing, sales, and commercial development should work together in identifying and defining the need.

> **Product-development programs instill a belief in all employees that they are leaders who can make things happen.**

Product-Development Procedure

The author's company has developed a 12-step product-development procedure that is unique to the food industry. This procedure guides the project team through the entire process, keeping the development process on track. The process is not sequential; several tasks are completed simultaneously.

It is at this point that the synergy of experience and expertise sharing is most valuable. The speed of development depends on elimination of duplication and working together as a team. The project team should develop a status sheet that outlines the status of projects at a glance. It is absolutely essential that the team leader and the members of the project team communicate freely and openly and develop trust and respect for each member's primary job and that member's contribution to the project.

The 11-step product-development procedure is presented in Exhibit 2.

HOW SUCCESS IS MEASURED

The following are the three key measures of success:

- *Time to market.* When the base technologies are widely available, getting to market quickly is essential. Doing this can improve volume, efficiency, and the all-important profitability.
- *Range of markets.* If the product can cover a wide range of applications, it leverages the core technology across wide markets and produces low-cost benefits.
- *Number of products.* The market segmentation creates opportunities to fulfill niches and increase sales volume.

CONFLICT RESOLUTION

Conflict can arise because of a team member's difficulty in balancing a primary responsibility to his or her functional department versus a given team project.

Exhibit 2
Hammond's 11 Steps to Product Development

1. Product idea based on need
2. Preliminary technical and marketing assessment
3. Concept development
4. Testing, applications, possible formulation
5. Prototype development in pilot plant
6. Trial with customer or initial test at the company's location
7. Scale up in the plant if a product from the plant is required
8. Delivery/shipping
9. Follow-up and fine-tuning based on customer feedback and sale
10. Final project report (product "story")
11. Continuous service and promotion to other potential customers

Differences of opinion can also arise as a result of personal (emotional) issues. Team leaders have to skillfully handle these differences to keep the project on target. The best strategies for handling these differences are:

- Define the problem precisely
- Maintain a concentration on the overall mission
- Separate logic from emotion, if possible
- Deal with emotion (e.g., by asking questions, letting the person vent his or her emotions)
- Handle intervention by skillful interpersonal skills and documentation of changes in decision, direction, and the like
- Focus on situation, issues, or behavior, and not on the person
- Maintain constructive relationships
- Take the initiative to make things better
- Lead by example

EXAMPLES OF CROSS-FUNCTIONAL TEAM SUCCESSES AND FAILURES

In one case in the food industry of a successful team, a team comprising people from product development, technical services, sales, marketing, manufacturing, and quality control had a clear objective to win the business of a large customer within six months. A product with very specific viscosity and stability characteristics needed to be developed for the customer's process, which was unique to the food industry and involved specific temperature and heat-processing equipment. The finished product also had to have unique gelling property to meet the consumer demand.

This team was also given a long-term objective to make a change to the company's manufacturing process so that it could produce a consistent product. This improvement in the process could also be useful in the manufacture of other ingredients requiring similar chemical modification of the starch food ingredient.

The company's ability to be innovative depends on how well it eliminates excuses that are used to resist change.

The team met its stated objective within the expected time frame. The clarity of the team's objective and the timetable helped it to accomplish its goal. The team was required to plan and publish milestones with specific responsibilities for each member. The members worked concurrently and met regularly to stay on schedule. Another positive factor was the management involvement in reviewing the project. The monthly project review by the management steering committee helped the team by supporting and providing appropriate resources as needed. In this case, the process improvement was a major capital project that was approved quickly because of the support of the steering committee. Another very important factor was continuous customer feedback on the prototypes that the team created.

A second example of a successful cross-functional team is an international customer that needed a product design with a very specific function in a specific period of time. Again, specific objectives and a clear understanding of the customer's need were very important. The product had to be manufactured by utilizing different chemicals and processes. In this case, each team member's clear understanding of their role made a positive contribution to the team's process.

Team Failure Through Poor Communication and Vague Goals

Disastrous results occurred in two cases in which cross-functional teams did not work well. In one case, the team worked on the project longer than the allotted time and management did not get involved to drop the project. The team had a clear objective, but because of a lack of team "harmony" and especially a lack of accountability, the team did not accomplish the goal of a reasonable production rate and thereby incurred higher product costs.

The manufacturing and development people did not clearly communicate to produce the product within a reasonable cost structure. A lack of clear accountability by department managers and the team dynamics of working together probably caused too many diversions and slow progress. The team also attempted to reengineer the product to existing plant capabilities and ended up studying it way too long. The company, however, allowed the team to continue.

Another example that involved a joint project with a partner, another food ingredient supplier, almost ended up in failure. However, two key managers got intimately involved, clarified the objective, and steered the project to some success. The original goal—to jointly develop and commercialize the product(s) in the next three to six months—was too vague. The managers provided direction in a timely manner consistent with this project's marketing aspect. Role clarification was also an issue, but it was resolved. There was a lack of understanding of who was doing what, and passive participation breeds resentment among team members who take their role seriously. Accountability, or lack of it, was again a common theme. In general, where goals, roles, and accountability lack clarity, teams are destined to fail.

CONCLUSION

This chapter has tried to outline some of the most important steps in cross-functional team programs. This by no means is the final answer to product development. The joint working of people with various skills, expertise from diverse functions, and a clear understanding of the mission are essential for success. Speed, flexibility, and capabilities have to be continuously developed to improve the batting average of new-product development. A formalized program developed and supported by top management is also essential for success. The program should be continuously refined and improved. The management throughout the company must follow through with initiative to build cross-functional skills and remove obstacles in order to guide decisions and actions on commercialization projects. Without management's active participation and direction, an organization will find it next to impossible to kill or postpone projects or to resist the short-term pressures that drive the organization to spend most of its time and resources fighting fires. Creating an aggregate project plan gives direction and clarity to the overall development effort and helps lay the foundation for outstanding performance. The vision, mission, commitment, clear rules, and the execution make the whole process a tremendous morale builder across the whole organization and ensure long-term continuous success.

TECHNOLOGY TRANSFER FOR SMALL COMPANIES

Glenn Dugan

> Smaller companies can be ideal settings for using many of today's productive management techniques. This chapter focuses on how smaller companies, such as Sears Manufacturing Co., can take advantage of technology transfer.

During the last decade in the US business world, a great deal of focus and print has been dedicated to the transfer of technology from the design and development areas to manufacturing. And to be sure, the ways of achieving this are changing tremendously, encompassing all areas of competitiveness, such as time to market, cost-effectiveness, and quality control. Most stories and lessons are about the largest and most recognizable companies. However, what about smaller and mid-sized manufacturers? No one would disagree that these concepts have an important place in the business philosophies of smaller companies, but does the large company approach necessarily apply?

Those who have experience at both large and small manufacturers have no doubt felt some striking differences in the ways of doing things. This is perhaps no more apparent than in the area of technology transfer, when engineering wants to release new designs and ideas to production. Many large manufacturers focus heavily on the timeliness of technology transfer, organizational structuring, and production readiness. Although smaller manufacturers are concerned about these things as well, they will often have more basic struggles with the coordination of the entire effort, assuring that the proper information is communicated, that designs meet everyone's needs, and that all possible considerations are accounted for. There are a number of drawbacks most small companies have that can contribute to these struggles. Typical problems revolve around a lack of depth in laborpower. Having too few people to face the tasks at hand often leads to a high-paced environment, an inability to attend to all the details, and the ability to handle only the most urgent matters. Coordinating and planning efforts suffer as a result. Proper employee training sometimes gets pushed aside. Some areas of responsibility may go undefined in certain situations. An ineffective system of checks and balances, procedures, and discipline can be more the rule than the exception.

But for all the advantages and disadvantages of life at small and large manufacturers, this much can be said: the smaller companies can actually be ideal settings for using many of today's productive management techniques. This chapter focuses on what smaller companies can take advantage of in the area of technology transfer. There is no intention here to belabor the principles of such management ideas; there is already a large collection of printed material covering this. Rather, the goal is to show that a small company,

---

GLENN DUGAN is chief engineer at Sears Manufacturing Co. in Davenport IA.

because of its nature, has excellent opportunities to make use of many technology transfer strategies, ones that its large-scale counterpart can find awkward or difficult to implement.

One way or another, all companies go through difficulties in the process of design transfer from engineering to production, and much effort has been spent during recent years on working through the various problems in this area. As companies go through these situations, one theme comes through: mind-set and attitude are the main factors feeding the main principle—up-front communication. This includes researchers and designers communicating concepts back and forth; designers and production people communicating about what is workable and what is not; designers communicating with toolmakers and suppliers about what is makeable and what is not; material control, scheduling, and production people communicating about what should be available and when; and designers, production, and quality people communicating about product acceptance criteria. Armed with this theme, a small manufacturer should be aware that it need not try to emulate those larger companies that are spoken so much of. Things need not be so structured. The plan can be much more informal.

BEING SMALL DOES NOT MEAN THINKING SMALL

Many of today's management strategies regarding technology transfer can be brought to the surface in the comparison between larger and smaller firms. Product development teams, concurrent engineering, employee empowerment, brainstorming, consensus reaching, and departmental interfaces are all involved. These are all powerful and useful tools, and the list is much longer. All manufacturers regardless of size owe it to themselves to become familiar with these concepts and use them. Now think for a moment about how a small company in general differs in other ways from a large one:

- The organization is less complicated.
- The structure is less formal.
- The staff is smaller.
- People often have multiple areas of responsibility or perform more than one function.
- The people know each other more readily.
- People are more aware of how other areas of the company operate.
- Information spreads more quickly.
- Communication is more prevalent.

Referring back to management strategies, it is easy to see that a small company is their perfect breeding ground, where they can be given the room and freedom to grow. The innovative and situational management style needed to carry out many of these concepts is so much easier to suggest, approve, and adapt at a smaller company (provided managers with the necessary mind set are available, of course). Management is simply more accessible, and it takes fewer approvals to get things done. Champions are easier to find and recruit because the staff is smaller and more familiar to management. Because talent is more concentrated and identifiable, teams are easier to select. The smaller and more flexible corporate structure makes it easier to reorganize laborpower for particular projects or teams. The smaller company generally has a more open, creative, and cooperative setting. Each team member can make more of an overall impact with his or her contribution. A smaller operation usually lends itself to better being able to focus on the scope, progress, and results of projects.

> In smaller companies, management is simply more accessible, and it takes fewer approvals to get things done.

Small companies have other advantages as well. Generally, the product development groups are located in the same facility as production; often they are in close proximity. This offers great opportunities for cross-exposure and communication, unlike the problems many large corporations struggle with because of interfacing groups being in different places. Because the hierarchy is less formal, fewer territorial problems arise at a smaller company regarding team members reporting to team leaders or distinguishing traditional team member responsibilities from new ones. The reporting structure at many large companies can be restrictive and political enough to make the special team concept ineffective or strained. The simpler environment at smaller companies is also conducive to short-notice meetings, tests, changes, or new directions. This flexibility is perhaps the biggest advantage in technology transfer projects that small companies have over their larger counterparts. Smallness also affords faster information sharing, progress reporting, and decision making at the top.

Of course, large companies have their advantages too—depth of labor power, expertise, and resources, to name a few. But the general philosophies behind these technology transfer improvement principles are based on less structure, more openness or flexibility, and closer relationships between people. The larger company can find ways to accommodate this trend, but it often requires great and painful changes to the way things are traditionally done. In most cases, however, the smaller company is already set up this way.

ALL IN THE MIND

These factors and more can be discussed and compared to the point at which managers at small companies will see that they already have many of the conditions needed to breed good technology transfer practices. However, in the focus on doing the job right the first time, which is mostly about communicating and working together, managers must go back and consider that final factor—mind set and attitude. Here, no particular size of company or any specific organization for that matter holds any advantage. Mental outlook must be an individual matter before it can become a group matter.

No matter where people work, they must find managers, leaders, and team members who have the ability to be open-minded. This means the acceptance and consideration of other people's ideas. It also means being willing to try new approaches, going with the consensus, or seeing the need to learn something new. And it can mean admitting that they do not have all the answers; others may have more, and that is always a good thing. The people wanted for these roles wish to help each other and share what they know. Knowledge is power only when it is given to everyone who needs it. The successful team concept depends on people with these abilities. It depends on leaders who have these qualities and who can get them to rub off on those around them. Assuming that these types of people are enlisted, the following discussions will further illustrate that small companies again have a hand up in the ease of operating under these principles.

When it comes to mind set, what is required more than anything perhaps is a kind of humbleness. This applies to managers in that they need to be more of a helper and coordinator and less of a director or ruler. Again, a good deal of literature has been generated on this topic. It also applies to team members, technical types especially. Unless engineers or designers have spent time in the factory putting the designs together, they really do not have the extent of knowledge about the product that production workers do. They may think that they do, but they do not. They must understand this and look to production workers for information, insight, and recommendations. Again, a humbleness is involved. It would be great if design managers could send their people to spend time working in the factory to gain a perspective that is greatly needed in design. This is not always possible. But the designers can spend time watching, asking questions, and soliciting input from production. They can learn which features can be designed to accommodate welding labor by asking what kind of difficulties they have in loading these parts into their weld fixture. They can learn how dimensions and shapes can be used to better assure conformance to critical properties by watching a sheet metal part being formed through the various stations. The need for tedious operations or special tools can be eliminated if designers watch assemblers and find out how clearance around fasteners can be improved. If designers ask which specs are the most difficult to hold, any number of improvements can result. All kinds of knowledge can be gained, and that will make designers better at what they do.

Teams can help to build working relationships between design and production people, but why wait to be on a team to do this? Engineers and designers need to get into the shop and get to know the people working on the products they create. Whatever the project goals are—reducing costs, designing for manufacturability or assembleability, improving quality—the relationship that engineers and designers build with production people will help both parties. Smaller companies are ideal settings to do this; finding a balance between what needs to be formal and what can be informal is easier.

At first, the technical types may resist approaching production in this fashion, and it may take a while to break through once they try. But they should persist. Leaders must encourage listening and asking. There is nothing people appreciate and welcome more than having their viewpoint sought after and deemed valuable, when the requests are sincere. Over time, what may start out as an exercise in frustration will develop into a rewarding partnership, more so than anyone could have realized. The goal in this relationship building is to come up with products that are better and easier to make. By focusing on communication and cooperation now, lower costs, timeliness, and better quality will in the long run result naturally. Both the technical and production mind sets are drawn toward achieving the general goal (whether the company sponsors a formal project or not), and folks learn that they need each other's help to do it.

Champions are easier to find and recruit because the staff is smaller and more familiar to management.

CASE STUDY

Technology transfer cannot be better assisted than by such partnerships between design and production. The team approach draws all the parties together early in the stages of development and gives them the authority to do what needs to be done. And ongoing informal exchanges throughout the company build that spirit of communication and cooperation. That is what it is all about.

At Sears Manufacturing of Davenport IA, a mid-sized designer and producer of seating products for the agriculture, construction, and on and off-highway markets, progress in this area has made some big differences. Design and production people talk all the time. After spending several years developing cross-disciplinary relationships, communication has become commonplace and expected. If a designer has a question or concern about a particular process, that person will go out to the shop and talk to someone who is familiar with it. Even if a designer wants to make a design change or improvement, it is passed by people familiar with the production aspect of that product. Likewise, if production workers have an idea or problem regarding a product, they will make their way in to talk to a designer that can go over it with them. It is quite informal and impromptu. A mutual understanding about each other's trade has developed, and this not only benefits the projects that are currently under way but reaps dividends on difficult projects waiting on the horizon. People feel encouraged to express their concerns and solve problems. The resulting improvements will be too numerous to count. This is something small companies can learn to excel in.

At Sears Manufacturing, technology transfer has progressively become smoother and more enjoyable because many of these project management principles have been introduced. Typically, once production begins on a new product or design, certain changes, fixes, or improvements are required. In the days before good teamwork and communication, projects required a good deal of rework because of the lack of communication, lack of foresight and understanding, and frankly an ignorance of some aspects involved in successfully executing an effort of that type. Designers often would not be aware of what dimensional tolerances the various processes were capable of. This led to nonconformances, tooling revisions, design revisions, or all three. The knowledge was not shared because no one asked. Occasionally subassemblies were designed and structured awkwardly for manufacturing, not taking into account a particular staging of operations in the factory. Again, no one talked, no one watched. Often parts were designed that could not go together properly with conventional assembly methods. The process was misunderstood because there was no dialogue. Things like these caused little missteps to become big problems.

One way of measuring progress in technology transfer at Sears Manufacturing is monitoring the number of engineering releases required to issue designs, both leading up to and after production start up. Releases issued in this time frame are particularly noted because they primarily pertain to changes or corrections. And experience shows that most changes or corrections are called for because of unnecessary mistakes, incorrect information, or lack of knowledge. By progressively involving and empowering the people throughout the company in recent years, both on project teams and in traditional functions, the number of these releases has decreased dramatically. In Sears's case, the design and development of a new large-scope seating project could entail a time frame of one to three years and should culminate in the issuance of 5 to 10 engineering release packages of the various aspects of the design.

At a time before much communication and teamwork were used, a major project underwent the following number of change- or correction-related design releases:

- Twenty-six releases within two months of production start-up.
- Eighty-four releases since production start-up (four-year period).

After a limited amount of teamwork and communication had been practiced, another major project underwent the following:

- Seven releases within two months of production start-up.
- Forty-five releases since production start-up (three-year period).

After experiencing a year of generous teamwork and communication, another major project underwent the following:

- Ten releases within two months of production start-up.
- Six releases since production start-up (one-and-a-half-year period).

Now, after two years of widespread teamwork and communication, yet another major project underwent this:

- Four releases within two months of production start-up.
- One release since production start-up (six-month period).

Because talent is more concentrated and identifiable in smaller companies, teams are easier to select.

A change in mind set is paying off. Of course, no mention has been made of time savings, financial savings, quality improvements, or product improvements, but the main goals of efficient technology transfer, which is everyone understanding and agreeing to what is being done between engineering and production, are being achieved. Sears Manufacturing is a small enough company that many of the previously mentioned management principles were easily adaptable. In general, people throughout the company now enjoy working together, whether they are involved in special projects or not. And they are commonly interested in making each other's job easier, which of course greatly benefits the company as a whole. It is a process, once accepted and experienced, that cannot be undone.

For all the technological hurdles that need to be faced and jumped in the commercialization of new designs and technology, the lack of good communication and cooperation is a poor reason to miss goals. Although smaller manufacturers may or may not have the edge in developing new technologies, they certainly have the advantage in being able to get people together in ways that can overcome almost any other obstacles they may have.

SECTION 6

NURTURING INNOVATION

An important issue underlying effective technology management is the nurturing of innovation. Most US companies, however, do a poor job of encouraging innovation. Why? L.Y. Stroumtsos and R.W. Cohen compare traditional organizational practices with a new organizational process model aimed at successful technological innovation. They also propose a structured approach to innovation, derived mainly from recent case experiences at Exxon.

Some companies look to scientific networks for new product ideas. Gerald L. Majewski describes how one has used a scientific network.

Henry LaMuth discusses how to acquire and use technology competitively. His chapter discusses the sources from which companies can acquire new technology or technological ideas and describes how companies can determine their own strengths and weaknesses in developing and using technology.

Russell Horres, who as worked at both small and large companies, contrasts the speed and agility of a small, start-up company with the slowness and established procedures of a large company.

Horst Geschka, Mopnique Verhagen, and Barbara Winckler-Rub present a case study that illustrates the successful application of the scenario technique. Scenarios are pictures of the future built under certain plausible assumptions. The scenario technique is an effective planning tool when the time horizon is long range and uncertainties are high.

ESTABLISHING A SCIENTIFIC NETWORK

> Companies of all sizes are straining to keep a competitive edge in a rapidly changing business world. The firm in this chapter has succeeded by forming a scientific network, which extends the reach of its R&D organization by giving it an invaluable set of eyes and ears.

Gerald L. Majewski

With technology in a constant state of change and global competition becoming increasingly intense, many companies—including those with large, well-funded R&D organizations—are supplementing their internal product development programs with external resources. For example, they are licensing technology, acquiring product lines or complete businesses, and collaborating in joint research programs with other companies. These approaches are often the most appropriate ways to augment internal product development from both a business and technology management standpoint.

The information science explosion is also changing the business landscape. The volume of information and number of publications and meetings in which this information originates are growing exponentially. Consequently, companies are now subscribing to a vast variety of services that track publications, patents, research grants, and meeting proceedings. Their information is available to the public at large, however, including competitors.

Ideally, a company should have access throughout the world to product ideas that are in their infancy. It should monitor these ideas as they are developed and translate them into new products when their value is verified. No organization, of course, has the resources to come anywhere near to this ideal, but any organization can extend its reach dramatically with a well-planned scientific network.

ONE COMPANY'S REASSESSMENT OF ITS NEEDS

This chapter describes how one company organized such a network. The company develops and manufactures products that are used in hospitals and independent reference laboratories to test bodily fluids in vitro to diagnose disease and monitor therapy.

GERALD L. MAJEWSKI is vice-president, Research and Development for INCSTAR Corporation and has fifteen years of technical management experience in the in vitro diagnostics industry.

The R&D Structure That Was in Place

The company supported an internal R&D effort with a budget comparable with that of organizations of similar size in the same industry. When it was expedient and financially justified, the company licensed technology from universities and other organizations. It entered into joint product development programs with pharmaceutical and other diagnostic companies when feasible, and its track record in this area was successful. The company also acquired complete product lines with their associated technologies and assimilated them into the overall organization.

Use of Consultants

The company regularly used consultants, primarily for specific projects, and sometimes developed their ideas into products. These consultants were also used to evaluate new products and sometimes to provide or develop raw materials.

This chapter describes a three-tiered scientific network in which advisors, consultants, and contacts are all used. The following sections define these three roles. Depending on the situation, a person may serve in all three. Indeed, there is considerable advantage to having an advisor who also acts as a consultant.

Consultants and contacts can be added to the scientific network without advisors. However, advisors are usually the richest source of leads for the best qualified consultants and contacts in a particular field.

Advisors. An advisor is well established in the scientific community, preferably has an international reputation, and has successfully advanced his or her field for many years. This individual must be very familiar with trends in the field and, most importantly, must be able to predict the changes that emerging technologies and associated products will bring. The advisor's role is primarily strategic, suggesting where the company should invest its resources to succeed in three to five years.

Consultants. A consultant provides expertise on a project or area of knowledge as well as on developing an idea into a commercial reality, sourcing and evaluating raw materials, or performing pilot tests or clinical studies on new products. These activities are primarily a means of achieving a company's strategy in a particular field. A consultant should also understand what customers need not only today, but in the next one to two years.

Contacts. A contact can provide information or material or can direct the company to such sources.

The Need to Expand the Network

The company's senior management decided that this combination of internal and external resources was not enough to make the company an industry leader. The network of consultants already in place was productive, but it was not as wide, diverse, or interconnected as it could be.

> Any organization, regardless of size or resources, can extend its reach dramatically with a well-planned scientific network.

This author, as vice-president of research and development, was chartered to establish and expand the functional scientific network. The efforts of external and internal resources were to be coordinated closely to optimize the new-product pipeline. The key objective of the network would be to identify and foster new ideas up through the feasibility stage. These ideas would then be transferred to the company to complete the final stages of product development.

THE INITIAL STEPS THAT WERE TAKEN

The company initially needed to determine the type of network that would be appropriate to its business. In deciding the best way to make this determination, several alternatives were considered.

Option One: Assessing the Company as a Whole

The organizing of a scientific advisory board was one of the first alternatives that the company considered.

A scientific advisory board is a group of thought leaders who meet regularly but infrequently to review the company's current and future technical direction. The board may also perform due diligence for proposed major technical strategic decisions, such as embarking on a new technological program or making an acquisition.

This board would clearly have been very beneficial for the company in the long term. However, the following factors caused the company to place this board's formation on a low priority.

Distinct Businesses. The company contained five distinct businesses or fields that were primarily characterized by the specific medical discipline they served and by the use of particular technologies for each product line. Each business was at a different level of maturity in terms of experience and expertise within the corporation. The five businesses also had different strategic implications for the company's future. The company had targeted some of the businesses for continued growth, while a few were positioned to provide steady cash flow.

In addition, although all of the company's products were intended for use by hospital or reference laboratories, the products within each business or field were largely separate from the others because the labs were, in many cases, segregated by medical discipline and associated products, and the physicians at these facilities did not necessarily communicate with physicians in different medical disciplines.

Board Manageability Concerns. The size and manageability of the board was a potential problem because the company's board could accommodate no more than ten members. That meant there might be only one or two people at most with expertise in one of the company's five particular fields. The diversity of input to be obtained would therefore be limited, because members who were not experts in a particular field would most likely defer to the more knowledgeable members. Setting corporate strategy with such an approach could be very risky and ill-advised.

In addition, the board would have had to concern itself not only with technology, but also with the global economic, regulatory, and political issues having an impact on the company's technological future. These areas are more generic in nature and have significance for the company's overall direction rather than for a specific business. As a result, the best candidates for board membership might actually have had limited expertise in any one specific field, but a very broad background in a general area, such as technology or health care.

Option Two: Assessing the Company's Individual Businesses

The company decided it would not be fruitful to begin discussing overall corporate strategy without first assessing each individual business. A thorough analysis of the technological trends, challenges, and opportunities in each field was needed.

Selection of Panel Members. As a start, two of the company's more strategic businesses were selected for expansion of the scientific network. For each of these two businesses, a scientific advisory panel comprising four to six advisors with expertise in a particular medical discipline were formed in order to provide more detailed and focused views on current and future developments. The combined expertise of the two panels would be diverse enough to ensure that all important areas of each field were represented.

ESTABLISHING THE ADVISORY BOARD

The selection of appropriate people to serve as advisors is crucial to the success of the scientific network because the advisors serve as the underlying foundation upon which the system expands. A number of avenues can be pursued to identify potential advisors. Many technology-based industries have associated scientific organizations whose membership comes from industry, academia, and government. The key officers of these groups generally have outstanding credentials and are well-respected members of their scientific communities. They can shape the future direction of their particular fields.

Another source of candidates can be individuals who serve on the editorial boards of scientific journals. These people review research findings that are about to be published, giving them access to the newest developments and future trends in a particular field.

A firm's current advisors or consultants can also be used to identify potential candidates. For this company, this approach proved to be very suitable for the following reasons:

The key objective of the network would be to identify and foster new ideas up through the feasibility stage.

- The consultants who were asked for recommendations were being used as consultants in many different projects, but they did not have a vested interest in any particular product or idea. This allowed them to be objective.
- These consultants were also established in their respective scientific communities and were very familiar with other experts in various areas of that particular field.
- The consultants, through experience gained in outside meetings, groups, etc., realized that an essential factor for the success of any panel was the ability of the group to work together.

When a potential advisor is identified, it is recommended that key company representatives visit the candidate's institution, and that the candidate visit the company. These trips are important for the following reasons:

- The trip to the advisor's institution allows the company to assess the overall environment, including the facilities available, key people working within the advisor's group, and the additional support that might be available elsewhere in the institution. The relationship that is developed and maintained with the advisor will usually involve one key contact at the company and, at most, a handful of other people. The nurturing of that personal relationship is probably the most critical factor for a mutually productive outcome.
- Through the advisor's visit to the company, the advisor can evaluate the company and assess the support behind the relationship. The visit also allows senior management to emphasize the company's commitment to the use of outside expertise in advancing the company's future. It is also helpful in generating the advisor's interest and getting his or her commitment to the advisory role.

In assembling panel members, it is extremely important to obtain as wide a range of views as possible. However, each advisor must respect the views of other panel members and not dominate the group. Experts who have been successful in their respective fields for many years will often have large egos. As experts, they can be expected to have strong views on certain issues. This is acceptable as long as the advisors realize that the company is relying on their input to set strategy and tactics. Objectivity is crucial.

Experts also tend to be very intelligent and curious to learn new things. If they believe their participation in an advisory panel will provide a learning experience for them rather than a recitation of their own thoughts and views on various subjects, the experts will be more inclined to become and remain involved.

Retaining Panel Members

Once these panels are formed, a company should try to instill enough professional and personal interest in the advisor to encourage his or her participation on a long-term basis. Membership on the panels will sometimes change due to shifts in the company's direction or the advisor's interests and other commitments. Nevertheless, a company should foster a certain level of continuity within each panel so that it can adequately assess the accuracy of the advisors' forecasts over a period of years.

An advisory panel should meet at least two to three times per year. The company in this case study decided to conduct two annual formal meetings with all panel members and hold one annual informal session for individual meetings with advisors or subgroups of the panel.

Should a Consulting Arrangement Be Used?

When a company wants to set up an arrangement with an advisor, it has the choice of establishing either a separate arrangement or a consulting arrangement. For a multitude of reasons, a company will often find that added value is provided by setting up a separate consulting arrangement.

First, although an advisor is expected to share his or her own knowledge and relevant views, this communication does not automatically extend to other people within the advisor's organization. Other people in the advisor's group, for example, might have a narrower focus in the field but more detailed and specific information about emerging developments. A consulting arrangement with the advisor's group would fund specific projects that require effort by one or more of the advisor's associates. This arrangement could theoretically provide the company with access to all of the talent within the advisor's immediate area of responsibility or even within other parts of the organization of which the advisor is a part, such as other departments or divisions. An advisor, in short, is linked as an individual to his or her organization. A consulting arrangement links both the advisor and the advisor's organization to the company that has set up the arrangement.

Second, as the advisor becomes more familiar with the company's technology, products, and personnel by completing specific projects, he or she should then be able to provide more informed recommendations on overall strategy by taking the company's interests and capabilities more into account.

Third, a consulting arrangement demonstrates that the company that has established the arrangement has an active ongoing interest with the advisor's research and organization. This demonstrates an investment in the advisor's current and future interests, not just the utilization of the advisor's expertise and experience for the company's purposes.

Fourth, having as many eyes and ears available to identify potential opportunities is a key element of the scientific network. If most of the people within the advisor's organization have been informed of the company's interests, they should be easily able to inform the company of developments that should be followed up, based on meetings they have attended or literature reviewed. This is particularly valuable when a company has a limited budget for attending meetings. Even if funding is adequate, a company member's attendance at a meeting may not be justified when only a small percentage of the program is relevant. However, some consulting funds could be earmarked to support travel to such meetings by the advisor's associates. This can be very cost-effective and time-saving for the company, and it can generate excellent goodwill with the advisor's organization.

Finally, the consulting arrangement offers the company a way to validate an advisor's ideas before expending considerable effort and funds. A person who is in an advisory role may provide the company with many potentially useful ideas, but there may not be sufficient resources to adequately investigate the value of each idea. If the advisor also has a consulting arrangement, the company may request the advisor to perform additional research to prove that an idea has sufficient value before the company invests resources for its further internal or external development.

The selection of appropriate people to serve as advisors is crucial to the success of the scientific network because the advisors serve as the underlying foundation upon which the system expands.

ADVISORY TOPICS

The following are examples of questions and issues addressed to scientific advisors and possible responses. Although the specific issues are related to the health care industry, similar generic questions can be developed for any technical industry. Advisors can be queried on a wide range of topics running from specific products to emerging technologies and methodologies to broad policy.

- *How important is the product's adaptability to automation?* The extent of diagnostic product automation runs a wide gamut. Certain medical disciplines use products that require extensive manual involvement from the user. Other medical disciplines may have associated products that are essentially totally automated. These products would require minimal effort from the user.

    The ability to provide a product that can be automated may be as important, if not more important, than delivering a product that has a low cost and high quality. Some technologies are more amenable to automation than others. This factor would definitely affect the company's product development strategy and its use or development of new technologies.

☐ *What impact will new technologies have on the business as a whole or a specific product or products in particular?* The introduction of new technology into a market can radically alter a company's prospects, potentially enhancing or destroying the company's competitive position, depending on the source of the technology.

Current technology, as well as products, can become obsolete more quickly than the company had forecast. Transitions to new technology will inevitably occur, but a company must be able to predict when this transition will most likely happen and take advantage of the change, rather than becoming a victim of it. The availability of newer technologies may dramatically change the targeted market. The market could expand considerably, depending on affordability and applicability. Contraction in the market could instead occur if more complex and expensive products that benefit a smaller user base are introduced.

The company may need a significant effort to adequately educate customers about the new technology and associated products. These efforts would allow customers to become familiar with the benefits of the product and gain a thorough understanding of the product's use. If a company is already developing a new technology and the advisors feel that other technologies could be introduced by competitors either earlier or at the same time, the overall company R&D strategy might require significant adjustment.

☐ *What effect will new therapies have on diagnostic testing?* As new ways to treat, cure, and prevent disease become available, the testing conducted to diagnose these conditions may expand greatly or cease to exist entirely. Diagnostic testing might increase if the testing helps monitor the patient's response to therapy. The physician would request the test during the course of recovery. On the other hand, the therapy might eliminate the need for any testing whatsoever. This would occur, for example, in the case of a vaccine, which would prevent an infectious disease.

The issue of new therapies is separate from that of new technologies. New technologies provide better ways of achieving similar results to the results obtained with currently available products. The benefits of new technologies might include, for example, speed, cost, quality or automation. New therapies, on the other hand, enable one to meet the customer's underlying need from a totally different perspective than is presently possible. An example would be a method to eliminate the underlying cause of a disease rather than treating symptoms.

Developments in the area of new therapies should be monitored even more closely than those of new technologies. Because such advances are more likely to satisfy the customer's requirement more directly than possible with current methods of treatment, they have the potential to revolutionize a field rather than merely add to its evolution.

☐ *How will managed health care affect the type and quantity of diagnostic testing performed?* Managed health care is the type of global policy issue that can become a significant determinant of the acceptability of various technologies and associated products in the marketplace. In this company's case, it determines who actually requests testing, which tests are ordered, and how often that testing is performed. To contain costs, the trend is toward patient management by generalists rather than specialists. The most inexpensive, fastest, and simplest product may have clear advantages over a product that has superior performance characteristics but is not as cost-effective or user-friendly. These issues have political and economic aspects. Therefore, changes in a field that these issues affect will occur more slowly than in a field affected by purely technical developments. However, once adopted, the changes can be mandated as the standard.

> **In assembling panel members, it is extremely important to obtain as wide a range of views as possible.**

IMPLEMENTATION

Establishing a scientific network is a serious undertaking. It requires a considerable initial effort to develop and build relationships with people who are quite successful and have many commitments. In the case of the company in this case study, this author has spent almost six months in this endeavor on a full-time basis and expects that a similar effort for another six to twelve months will be required before the network and projects are crystallized and coordinated with internal resources and programs.

Cost Issues

The cost issues involved in establishing and maintaining a scientific network are a primary concern of the implementation process.

Cost and Value. Planning the budget for external scientific network activities is quite important, because that budget will dictate how extensively the network can be developed. When the network is functioning smoothly, potential opportunities requiring funds to initiate relationships and projects will arise. The budget should be flexible enough to accommodate these situations.

The cost of scientific network activities can be greater than activities done internally. In the case of advisors, the company is paying for expertise and experience that do not exist internally and may be extremely expensive, if not impossible, to obtain otherwise. Consultants should be required to complete the requested work more effectively and efficiently than if it were performed internally. When the company uses the multiple resources of an advisor or consultant, the funds expended can be considerably greater than if internal employees are used. In some respects, however, the combined talent and expertise of the outside organization can be considered a company asset for as long as the relationship exists. In addition, a company can arrange to have right of first refusal of new ideas and/or exclusive rights to these ideas for commercialization. The investment in this arrangement can be quite small when compared with the valuable potential that might exist.

Advisory Fees. Advisory and consulting fees would be expected to be commensurate with time and effort expended. Additional factors to be considered are the advisor or consultant's level of expertise and reputation, as well as any experience he or she has had with other companies. Obviously, the more recognized and successful the individual, the higher the expected fee. This is particularly true if the company wants an exclusive arrangement during the term of the relationship that precludes similar agreements with other companies. Individuals who have had prior relationships with other companies will most likely expect to be compensated at similar levels for comparable effort.

Anticipating Customer Reaction

People chosen as advisors are obviously more on the cutting edge of developments in their field than their average colleagues. These advisors have a better appreciation of possible future requirements and developments. However, being on the leading edge is not always what the average customer of the company's needs or expects. This fact must be kept in mind, because customers are reluctant to change when they are comfortable with current products and procedures. It takes considerable effort to convince them that making such a change will be beneficial and worth their time and effort. Therefore, to achieve the highest probability of acceptance, a company needs to develop its instincts concerning the timing of new product delivery to the customer. The obvious goal is to introduce the right product at an appropriate time.

NETWORK CARE AND MAINTENANCE

No matter how much planning and effort are put into the scientific network's establishment, it will not remain productive without continued attention. In this respect it is like a garden, which requires watering, fertilizing, pruning, and weeding to continually provide the best effect. In a network, the relationships developed need to be kept alive through regular ongoing communication among the parties. During periods when the need for outside advice or support is minimal, discussions with advisors and consultants about the company's progress and direction are important. These discussions serve to maintain interest and demonstrate the company's commitment to a long-term relationship.

The scientific network is a living creation. Over time, some people will outlive their usefulness and should be diplomatically released from the network. As the company moves forward, its needs and interests may differ from some of the people who have previously served it. If the relationship has been open and honest, it should be apparent to both parties that continuing the arrangement is no longer beneficial. For optimum results, the network requires continual pruning and shaping to match the company's requirements. Otherwise, the company becomes burdened with relationships that are not rewarding financially or intellectually, and the network's functioning can be hindered.

> A separate consulting arrangement could provide access to all of the talent within the advisor's immediate area of responsibility.

Results Versus Expectations

The company in this case study has assembled two scientific advisory panels, and the vast majority of the advisors that are a part of the company's scientific network have also agreed to separate consulting arrangements. The arrangements range from specific product evaluations to funding a full-time person in the consultant's lab for product feasibility or research. These people have already directly produced a considerable number of excellent ideas for new or improved products. For instance, through a development agreement initiated with one advisor, ten new products could result from the feasibility efforts of the advisor's lab. In another case, a novel product idea has surfaced from the lab of a person associated with one of the organizations of an advisor. As this process continues, additional relationships will be formed with people connected to or recommended by advisors, thus further extending the network.

Early discussions with scientific advisory panel members have already had an impact on determining the strategies necessary to introduce new technology to customers. In one case, the company has reached a greater understanding of how to position a newer technology with customers so that their resistance to changing a long-established methodology will be overcome.

Although this process has been in progress for only six months, its value to the organization has been considerable and the company's strategic technical direction has become much clearer. Both employees and management feel that this approach is extremely beneficial and that it should have been pursued much earlier.

Benefits coming from the process were forecast to materialize approximately one year after its initiation. The attainment of tangible results in half that time is predictive of a much greater overall return than originally expected.

A NEW APPROACH TO SUCCESSFUL TECHNOLOGICAL INNOVATION

Drawing from recent experiences, Exxon has developed a new model for bringing ideas to the marketplace. This chapter explains that model.

L.Y. Stroumtsos and R.W. Cohen

It is widely acknowledged that US companies do a poor job of commercializing technologies and using disciplined processes for turning ideas into commercial successes. Reflected in the lagging competitiveness of many US industries, the problem is rooted in factors both internal and external to the corporation. This chapter addresses mainly the internal factors, comparing traditional organizational practices with a new organizational process model aimed at successful technological innovation. It also proposes a structured approach to innovation, derived mainly from recent experiences in Exxon. The new model applies to problem-solving, opportunity seeking, major innovation, and the generation of long-term strategic business options.

This chapter first explores the shortcomings of two models for commercializing technologies. Then, it explores an approach Exxon has developed.

Exhibit 1 depicts the hand-off model of product development. Moving an idea to commercialization was envisioned as a linear, unidirectional process. It was idea-driven, the idea typically coming from research. The fledgling innovation would next move to engineering, where it would undergo analysis and design. Development groups would scale up the embodiment to commercial dimensions. Manufacturing would take the commercial version and use it to make a new product or process. For new products or product improvements, marketing would then determine appropriate prices, placement, and promotion.

Bringing in the best and the brightest people was regarded as the key to getting good ideas from Research. Astute management, alert to the possibilities inherent in an idea, was another important factor. However, ideas that emerged from this model were often risky and speculative. Furthermore, companies'

LORRAINE Y. STROUMTSOS is project leader, innovation and quality, corporate research, Exxon Research and Engineering Company in Clinton NJ.

ROGER W. COHEN, PhD, is currently senior planning advisor for Exxon Research and Engineering Company in Florham Park NJ. Before this assignment, he was director of the physical sciences laboratory and manager of technology liaison, corporate research, Exxon Research and Engineering Company in Clinton NJ.

Exhibit 1
The Hand-Off Model of Product Development

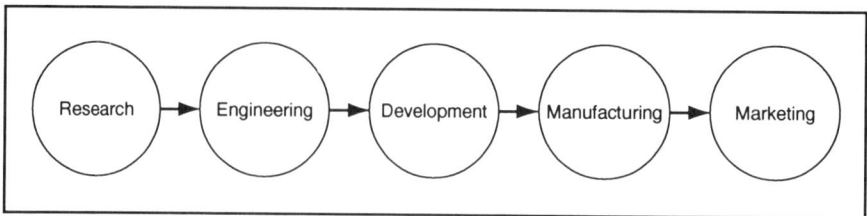

Research → Engineering → Development → Manufacturing → Marketing

most profitable ideas often come from marketing and customers: the hand-off model provided no coherent way to use ideas from these sources.

Each idea from research entered a process that called for the next link in the chain to accept the idea from the previous department, know what to do with it, and hand off to the next link. However, the functions in the sequential process operated with considerable insularity and little, if any, feedback. Department structure separated the functions required to advance the technology, and each tended to focus only on its job. Departments down the line felt little responsibility for research's risky, speculative ideas. Such ideas, good as they might have been, never made it to commercialization or took too long to do so, missing windows of business opportunity.

THE BUSINESS-DRIVEN MODEL

Widespread frustrations with the performance of the hand-off model led in the mid-1970s to the development of a counter model in which business had a much greater level of control over the research, development, and engineering functions. The transition to a business-driven model took place at greatly different times in different industries and corporations. The transition was often accompanied by substantial decentralization of research, with individual units becoming organizationally tied to product line organizations.

If the hand-off model operated in an idea-driven, linear, feed-forward manner; the business-driven model (see Exhibit 2) moved an idea to commercialization in a more collective fashion. The business-driven model also tended to focus on business needs (e.g., manufacturing wanted to reduce costs, or marketing wanted another advertisable feature). Business called the shots; engineering and development reacted, translating the business need to technology requirements and feeding the results to research.

Ideas emerging from the business tended to be extensions of what had already been done; the technology translation tended to result in incremental gains in technical knowledge. Because research was isolated from any feedback loop, few paradigm-breaking innovations occurred. The desire for more certain outcomes yielded valuable, but incremental, advances and encouraged risk aversion. The roots of this incremental orientation lay in the fundamental

Exhibit 2
The Business-Driven Model of Product Development

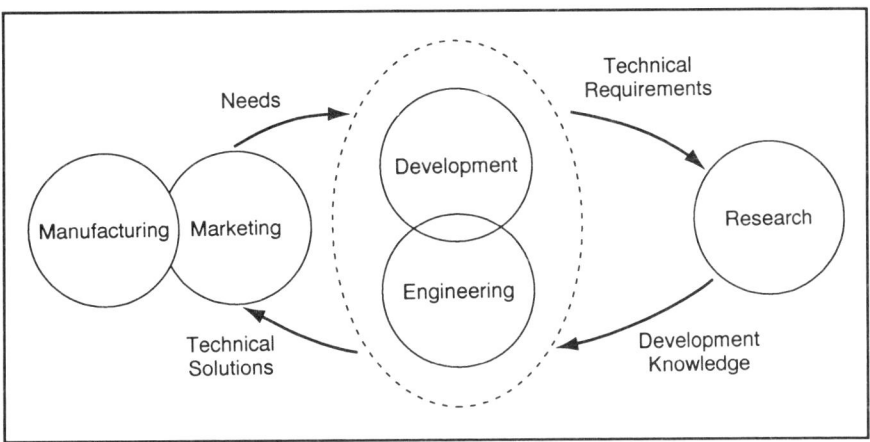

situations faced by business operating functions. Marketing, faced with intense competition, looked for performance and cost advantages in existing product lines—and quickly. Manufacturing was pressured to reduce costs and increase the throughput of existing products by stressing reliability and improvements in basic process efficiency. Manufacturing neither wanted to—nor was often able to—invest expensive capital in tooling up for new products or new process concepts.

A NEW, COOPERATIVE MODEL

A new model is now emerging. It does not rely on linear or unidirectional processes. Rather, it contains looping, iterative sequences in which all functions required to advance ideas work together, allowing convergence to commercially viable products or processes. The authors base this model largely on learnings gained from a rapidly increasing number of case histories (see Appendix), involving various situations and experiments in organizational processes. Still in the formative stage, the model will probably undergo revisions. Presented here is the current state of the model, a snapshot of a still-evolving base of knowledge and experience.

The model recognizes that there are three critical phases in the life of a successful innovation:

1. Figuring out what the organization wants to do.
2. Getting started.
3. Getting it done.

Exhibit 3
Innovation Principles and Some Key Application Examples

A. Get business participation—to derive the best thinking on business perspectives, needs, and opportunities.
- Involve the business early.
- Intersperse research, engineering, manufacturing, and marketing.
- Understand major driving forces and forecast customer needs.
- Get early and frequent exposure to potential customers.
- Develop marketing strategy early.
- Have an integrated process to watch and forecast competitor activity.

B. Get broad R&D participation—to get the best technical ideas and approaches.
- Involve experts and nonexperts.
- Involve idea people and do-ers alike.
- Involve basic or applied research and engineering groups.
- Aim to demonstrate both scientific and engineering feasibility concurrently.

C. Involve the people who will do the work—to build commitment and enthusiasm.
- Identify and engage all organizations or functions needed to advance an idea to commercialization.
- Identify and engage the people with the needed skills.
- Estabilsh cross-organizational teams with a high level of work autonomy.
- Ensure prompt, frank communication.

D. Encourage dialogue among all functions—to get the best answers.
- Hold topical workshops to generate dialogue and converge to solutions.
- Ensure dynamic and interactive feedback loops for flow of information among the functions.
- Share knowledge as equal partners do.

Each of the phases can require a distinct organizational process, but within each process, some general principles can be applied to encourage successful innovation. Using the four innovation principles described in Exhibit 3 can greatly reduce or essentially eliminate the effect of organizational boundaries and parochial interests in all phases of the process, increasing the speed and assurance of commercialization.

FIGURING OUT WHAT TO DO

Unless an innovation is mandated at the start or defined by a pressing need or idea, figuring out what to do can be a difficult, frustrating problem. One source of the problem is that research, development, engineering, and business groups often look to each other for the identification of the innovation objective. They encounter the following communications loop: research, development, and engineering ask business what it wants. Business in turn, asks what can they do. Research, development, and engineering may believe they can do anything or may not want to commit themselves. Therefore, they reply by asking business once again what it wants, and the loop continues without an answer.

This loop can be avoided by generating a goal or set of goals derived from an opportunity arena. Such technological opportunities arise in four main forms, described as innovation modes and summarized in Exhibit 4.

> **A new model for innovation is emerging in which all corporate functions required to advance ideas work together.**

The problem-solving mode resembles the business-driven model because it focuses on business needs. The opportunity mode resembles the hand-off model because the driver is opportunity. The cooperative nature of the new model distinguishes these two innovation modes. *How* problem-solving and opportunity occur and are dealt with is new and different.

Included in Exhibit 4 is major innovation, efforts directed at significant technological breakthroughs that competitors and customers alike will recognize as having changed the rules of the game. Major innovation is characterized by very aggressive goals; a high degree of involvement by business and research, development, and engineering representives at all levels; the ability to sustain the effort despite the distractions of other technical demands and changing business performance; and the generation and application of a new science base.

Also included in Exhibit 4 is the strategic option, which includes research programs aimed at securing a future technology-based option to meet a long-term business need or opportunity. In this mode, basic science programs address the key technical hurdles needed to generate the option.

As indicated in Exhibit 4, the longer the time horizon, the more fundamental and strategic the nature of the innovation. The corporation needs a portfolio of innovation modes, the balance determined by the pace of technological change, competitive position, and future business uncertainty or business change directed by senior management. A corporation that favors problem-solving and other incremental activities risks major advances by competitors or innovations from outside the base business. A corporation that favors long-term options risks a loss of competitiveness by insufficiently applying technical solutions to real business needs.

Whatever the portfolio, each innovation-mode component should be strategically aligned with the corporation's business strategies and objectives. Strategic alignment requires a clear understanding of the short, intermediate, and long-term business view, including key uncertainties and how new science and technology can influence or change these uncertainties. Strategic alignment does not require a uniform consensus on what technologies may or may not be in vogue. It does require knowing the most likely boundaries of business interests. Strategic alignment paves the way for a highly empowered workforce, spanning all corporate functions, to apply its full creative potential to the generation of technological innovations. It provides the framework for the cooperative approach.

Given strategic alignment, a set of goals—a mission—can be derived by involving key representatives from all relevant corporate functions. An inno-

Exhibit 4
Technological Opportunities

Innovation Mode	Timeframe	Scope	Customer	Driver	Research*
Problem solving	Short	Incremental	End user, operating unit	Need	Applied
Opportunity	Short to intermediate	Incremental to step-out	To be defined	Opportunity	Basic and applied
Major innovation	Intermediate	Step-out	End user, operating unit	Opportunity	Basic and applied
Strategic option	Long	General, broad	Corporation	Need or opportunity	Basic

Note:
*In this context, applied research is the use of existing scientific knowledge to generate and advance a technological innovation; basic research is the generation of new, fundamental scientific knowledge to provide the base for technological innovation.

vation sponsor group is convened to establish broad goals as a charter for future collaborative activity. The group includes business and research, development, and engineering managers, whose level depends on the scope and importance of the opportunity area. The sponsors help define technical approaches and continue to serve as programs start up and evolve. The sponsors empower an organizing group to engage the people needed to define the best technical approaches and to sharpen and clarify the charter. These people may also propose modifications to the original goals after extensive discussions and analyses.

> Successful innovations happen when someone or some people really want them to and have serious stakes in them.

Exhibit 5 describes the subsequent process elements. The output of the illustrated process can be an innovative problem solution, a promising target for a new technical idea or approach, an extensive research plan to enable a proposed major innovation, or proposed strategic business options with their associated research strategies.

GETTING STARTED

The following sections describe the groups and activities necessary to get started.

Innovation Sponsor Group

Successful innovations happen when someone or some people really want them to and have serious stakes in them. Such people see opportunities for an organization to do more than it has and do it better, and they champion the process of doing so. As the clients for the innovation process, this group of management-level stakeholders in research, technology, and business can access or command resourcing levels of the major functional units that will be involved in carrying an innovation from idea to implementation.

In addition to establishing the charter, the innovation sponsor group demonstrates support, empowers working-level activities, and kicks off an innovation workshop by indicating the stakes each member holds personally, along with statements of the perceived opportunity and the consequences of not realizing the opportunity. Each member commits, in principle, the resources to support the effort. These members ensure constancy of purpose and visibly commit to work to drive worthy ideas through corporate barriers, while establishing and adhering to key milestones toward achieving commercial success.

Innovation Team

The innovation team is a cross-functional group of individuals who create the innovation (i.e., research, engineering, development, field implementation, and marketing). Team membership is self-selected by interest and commitment to the innovation enterprise. Representatives from all functional units are

Exhibit 5
Innovation Work Process

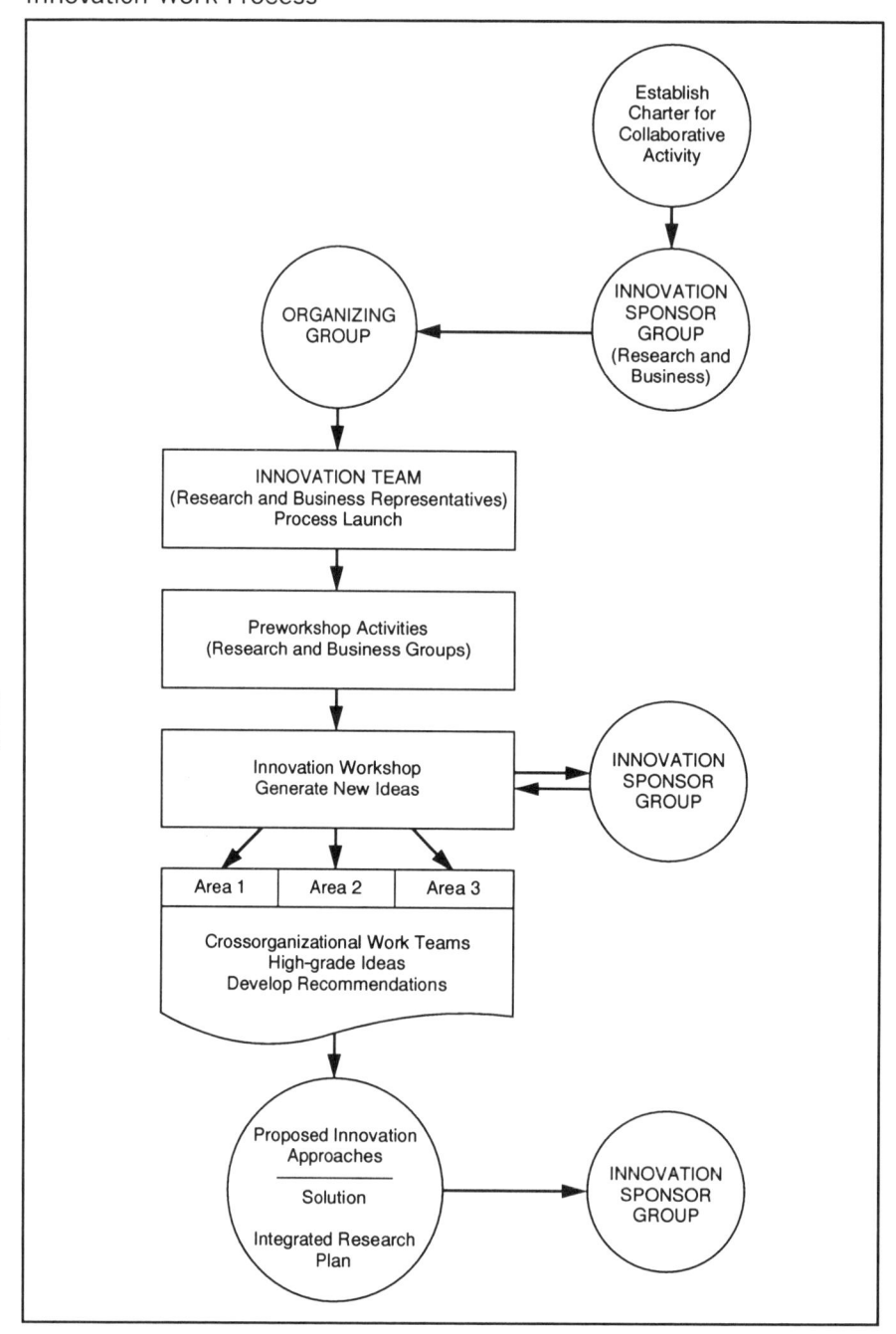

```
                          ┌─────────────┐
                          │ Establish   │
                          │ Charter for │
                          │Collaborative│
                          │  Activity   │
                          └──────┬──────┘
                                 ▼
                          ┌─────────────┐
         ┌────────────┐   │ INNOVATION  │
         │ ORGANIZING │◄──│  SPONSOR    │
         │   GROUP    │   │   GROUP     │
         └──────┬─────┘   │(Research and│
                │         │  Business)  │
                │         └─────────────┘
                ▼
    ┌───────────────────────────────────────┐
    │         INNOVATION TEAM               │
    │ (Research and Business Representatives)│
    │           Process Launch              │
    └───────────────────┬───────────────────┘
                        ▼
    ┌───────────────────────────────────────┐
    │       Preworkshop Activities          │
    │   (Research and Business Groups)      │
    └───────────────────┬───────────────────┘
                        ▼
    ┌─────────────────────────┐     ┌────────────┐
    │   Innovation Workshop   │────►│ INNOVATION │
    │   Generate New Ideas    │◄────│  SPONSOR   │
    └──────┬──────┬──────┬────┘     │   GROUP    │
           ▼      ▼      ▼          └────────────┘
        ┌──────┬──────┬──────┐
        │Area 1│Area 2│Area 3│
        ├──────┴──────┴──────┤
        │Crossorganizational │
        │     Work Teams     │
        │   High-grade Ideas │
        │Develop Recommendations│
        └──────────┬─────────┘
                   ▼
         ┌──────────────────┐
         │Proposed Innovation│
         │    Approaches     │      ┌────────────┐
         │──────────────────│─────►│ INNOVATION │
         │     Solution     │      │  SPONSOR   │
         │──────────────────│      │   GROUP    │
         │Integrated Research│     └────────────┘
         │       Plan        │
         └──────────────────┘
```

NURTURING INNOVATION

necessary to bring an idea to commercialization. The team members may call on additional resources or special capabilities of in-house organizations yet identify strongly with the innovation team. The team takes responsibility for its own direction and decisions and often develops its own leadership. This leadership can change as the program evolves, reflecting changing priorities. It establishes tactical milestones, usually aggressive in accomplishment and timing. The innovation team is responsible to its innovation sponsor group. The team should employ such tools as facilitated brainstorming or idea-growing sessions for fresh ideas.

In modern corporations, opportunities for technological innovation come from both short-term problem solving and long-term technology-based business options.

The team has some similarities to skunkworks, as practiced successfully in many industries. However, it is much broader in the scope of its constituents. Unlike skunkworks, the innovation team is fully aligned with the mainstream organization and does not operate separately or at the fringe of the organization.

Innovation Workshop

The innovation workshop is tailored to the innovation mode (i.e., problem-solving, opportunity, major innovation, strategic option). It is not just another meeting; it is hard work, enjoyable work, the beginning of a different kind of work. It can be a simple one-day meeting for generating new ideas for problem solving, or it can be a sequence of sessions for a new major innovation initiative. Regardless of structure, the innovation workshop is usually an off-site event, yielding importance to and allowing concentration on the subject. The following outcomes can be achieved by the innovation workshop:

- Defining the most promising approaches.
- Developing fragmentary ideas into robust concepts.
- Commitment to action.
- Follow-up.

Working with the innovation team, trained facilitators formulate an agenda including idea-generating activities of proven effectiveness. They coordinate arrangements for the innovation workshop, move toward the desired outcome during the workshop, and plan follow-up activities to move the innovation process forward.

Workshop participants are a collection of different kinds of people—from idea people to those who excel at getting things done. They come from all the organizations (e.g., research, engineering, development, manufacturing, marketing, and law) that will be necessary for generating new ideas, and then moving them toward commercialization. Some workshop participants choose full, direct, innovation team involvement. Others serve only as resources to contribute their technical expertise and ideas to help launch the innovation activity. This composition ensures that the innovation team is comprised of highly motivated, committed individuals. Achievement of the required balance of skills and the release of team members from previous responsibilities is assured by the innovation sponsor group, working through their respective line organizations.

At the end of the workshop, the sponsors and participants select the most promising ideas for technical approaches for further development. They base the selection on the appeal of the ideas and the competitive edge they promise, more than on their initial feasibility or an in-depth analysis. The selected ideas become the subject of follow-up by topical teams, members of which self-select for post-workshop involvement.

Research Plan

In postworkshop activities, the topical teams develop ideas into robust technical approaches, then recommend a plan of action for each approach. In the process, team members anticipate technical hurdles, which become the focus of the research effort. The workshop process emphasizes the generation and development of innovative technical approaches. The innovation sponsors group may elect to support all or part of the teams' recommendations; typically, multiple technical approaches are pursued until a particular approach or set of approaches is chosen.

GETTING IT DONE

This stage of technological innovation involves the same cooperative and mutually supportive approach (in which ideas and solutions emerge from all corporate functions) that characterizes previous steps. Exhibit 6 depicts the innovation loop, an iterative process that leads from the definition of innovation approaches to a commercial product or process. The entry point is the innovation work process shown in Exhibit 6. The exit point occurs after customer response to field testing indicates that the goals and objectives set by the innovation work process have been achieved.

Interim steps involve technology development by the innovation team, rapid prototyping to gauge field performance and determine required improvements, leading to technical approaches to achieve the improvements. The degree of sophistication or readiness of a prototype depends on the specific technological innovation and the complexity and cost of field tests. For large process units that require large investments, prototypes must have a high degree of technical readiness. For example, a lot of small-scale experimentation, along with computer simulations, is necessary to generate a sophisticated first-generation prototype. On the other hand, a relatively small-scale test to determine market acceptance of a new product or service can often employ a relatively primitive prototype.

Exhibit 6
The Innovation Loop

[Diagram: A circular loop showing Innovation Team connecting Innovation Work Process → Develop Technology → Rapid Prototyping → Measure Customer Response → Improvements, back to Innovation Work Process. Results from Measure Customer Response feed into a central Product or Process box.]

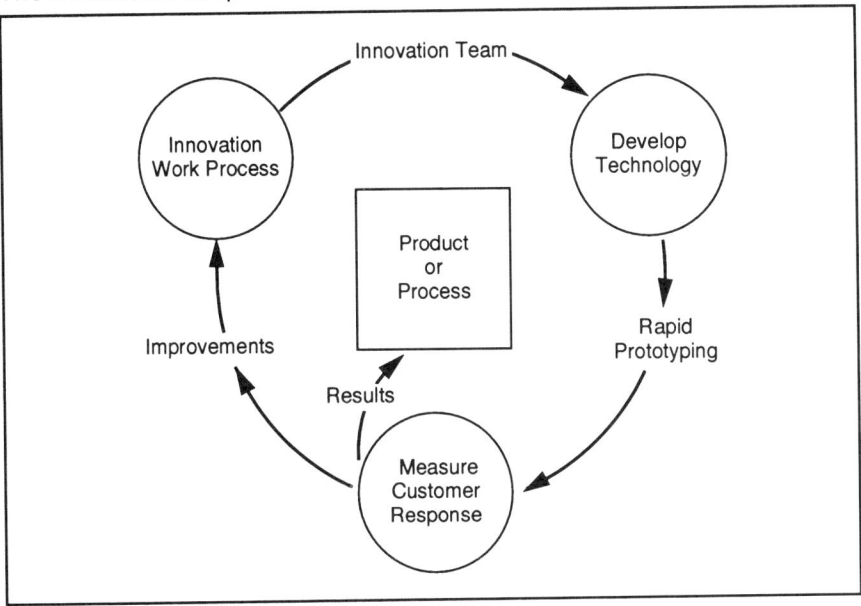

Exhibit 6 does not explicitly show the organizational groups responsible for each process component, because all the groups needed to bring an idea to commercialization are involved in the innovation loop. Also not shown are the critical milestones and checkpoints required by the process. These milestones are established by the innovation sponsor group in concert with the innovation team. They review decisions involving resources and other parameters needed to progress the loop.

Comparing Exhibit 6 with Exhibits 1 and 2 reveals several major differences. First, functions—not organizations—are shown. Second, the source of the potential innovation is neither research, development, and engineering nor the business. The source is shown schematically as the innovation work process, which encompasses inputs and views of all groups. Third, the progression to commercialization is not shown as a direct, unidirectional process, but is represented schematically as a loop.

CONCLUSION

In a process as complicated as innovation, it matters little whether ideas arise from technical units or from the business. In modern industrial corporations relying on a broad scope of technologies for their businesses, opportunities for technological innovation come in forms that range from short-term problem solving to long-term technology-based business options. Ideas that address these

Exhibit 7
Exxon Innovation Experiences (1988-1992)

Workshop	Year	Innovation Mode
Marketing innovation session	1988	Major innovation
Retail concepts workshops	1989	Major innovation
Catalytic separations workshop	1989	Strategic option
Separations opportunities	1990	Opportunity
Computing and communications workshop	1990	Opportunity
Synchrotron beamline session	1990	Opportunity
Marketing technical services workshop	1990	Problem solving
Lubricant dispersants workshop	1990	Major innovation
H_2S detection	1990	Problem solving
Hydrotreating workshop	1991	Opportunity
HF innovation session	1991	Problem solving
Bitumen workshop	1991	Major innovation
Lube degradation workshop	1991	Opportunity
Basic chemicals week	1991	Major innovation
Volume prediction workshop	1992	Problem solving
Heavy oils innovation workshop	1992	Major innovation
Relevant science guidance workshop	1992	Strategic option

opportunities, whatever their form, must grow into technical concepts and advance to commercialization with the cooperation, support, and contributions of all functions. Innovation is developed in a cooperative approach that reflects the corporation's business strategies and objectives. Companies that use this approach empower their workforces to engage in the creative activities typical of small firms. In this way, large companies capture much of the nimble, fast-moving abilities of smaller firms, while retaining their resource advantages and economies of scale.

The principles and techniques presented in this chapter differ substantially from the theory and practice of earlier approaches to innovation. These principles and techniques are largely derived from the case histories listed in Exhibit 7. However, they have also been influenced by such factors as the quality process, which calls for systematizing the design, deployment, and measurement of important organizational systems; a study of key twentieth-century US technological innovations; characteristic practices of small companies that have rapidly and successfully commercialized technological innovations; and Japanese methodologies.

Acknowledgments

The authors wish to acknowledge N.J. Mass for stimulating discussions and ideas on the requirements for major innovation and the many Exxon research, technology, and business managers for their interest, support and contributions to Exxon's innovation effort. Special thanks are due to Dr. R.A. Petkovic.

ACQUIRING AND USING TECHNOLOGY COMPETITIVELY

Henry LaMuth

To compete more effectively, companies must learn to acquire and use technological innovations. This chapter discusses the sources from which companies can acquire new technology or technical ideas and describes how companies can determine their own strengths and weaknesses in developing and using technology.

A company that improves productivity and adheres to quality standards cannot compete unless it also examines its project evolution and development process. New processes are necessary to compete these days, and Japanese companies are leading the shift to them.

Japanese corporations recognize that someone somewhere is always trying to push their products out of the market by outselling them on the basis of quality, price, feature, or support. Consequently, Japanese corporations are redefining their product development cycle by institutionalizing product obsolescence. They are formalizing the product life cycle by using an overlapping development approach. While one organizational group designs a product for manufacturability, a second creates enhancements and upgrades to the product, and a third performs the early development and exploratory work to make the product obsolete in five years or less.

To a large extent, this paradigm holds in the US computer industry, though without a five-year obsolescence plan. The evolution of computer electronics microprocessor and memory chips and peripherals has been swift since the late 1960s, and US digital electronics companies have promoted this rapid change to gain and maintain their worldwide market share in computer-based products.

What is novel is that Japanese corporations appear to want to formalize this approach as a generic industrial model. If this model can be successfully transplanted to other industries, the dynamics of the computer industry will then be replicated in industries that have been less agile technically. Japanese companies have proved, for example, that the machine tool industry was ripe for new technologies. To stay competitive, companies in all industries, regardless of company size, must adapt or fail; they must learn to acquire and use technology more effectively.

This new approach merely recognizes the realities of the marketplace. It formalizes the various phases of product evolution and creates parallel activities that are in competition. Acquiring and using technologies effectively are still distinctly different processes. Nonetheless, Japanese companies are often more open to technology innovation than are US companies, which is one reason

---

HENRY LAMUTH is president of AlphaComm, Inc.

that very large Japanese corporations are as technically agile as smaller, more entrepreneurial US companies.

The techniques and attitudes that encourage optimal technology acquisition and innovation are part of the overall Japanese corporate culture. These attributes and techniques are also prevalent in successful US digital electronics industries. For other US companies to compete in this new development arena, they must lower their barriers to technology acquisition and innovation.

WHAT IS TECHNOLOGY?

Technology is the system by which society provides its members with those things it needs or desires, and it includes the methods or processes a society uses to handle specific technological problems. The practical, industrial, or mechanical arts are included in this definition, as are the applied sciences. Technology is not limited to traditional engineering implementations. Language and writing are technologies. Applied mathematics is technology. Know-how is technology, as is the management of a process. Someone who has solved a difficult problem is correctly said to have engineered a solution.

The optimal solution to even the simplest problems often requires a mixture of expertise and points of view. For example, a program may be held back because the development team cannot get purchase orders processed without excessive delays. These delays might be attributed to corporate pace or excessive justifications and approvals. An observer might deduce that the corporate technologies associated with the purchase process need repair. Both the process and the equipment may need to be replaced, and they are both technologies.

By definition, technology entails practical knowledge. However, having practical knowledge is not the same as solving a problem. A technology can be wrong, or it can be poorly applied. Furthermore, technology innovation is more than passing know-how from one person or organization to another. It is solving a problem by using technology from any source. The problem must, figuratively speaking, reach out for the requisite technology. The technology must also be available for transfer.

Technology innovation typically occurs when problems require new solutions, when the existing ways of solving a problem (or meeting a new requirement) will not work. In other words, new approaches and technologies must be tried, and this is when technology innovation occurs.

WHICH TECHNOLOGIES ARE IMPORTANT?

Technology innovation is inefficient, tentative, or nonexistent when the technologist forces it and it does not arise naturally from a problem. To be effective in dealing with a problem, personnel must be encouraged to be creative. When they are not, the reason is that their companies have a cultural problem, not that they have a technology problem.

The following axioms are true of most technology innovation processes. Most technologies are available to anyone, but key technologies can create a competitive edge. Overall product success requires putting technologies together during technology transfer; product definition and initial design; R&D (including process and support); design, engineering, and production; infrastructure support (e.g., by suppliers and postdelivery support). Leading-edge technologies are one of the last places to look for solutions to problems.

Technology is not limited to traditional engineering implementations.

Most technologies are available to everyone, but critical technologies—or key know-how—are not always easy to get. Most of the technologies required for total quality management (TQM) in a manufacturing organization are easy to get. Some parts of those technologies are management related, and other parts are attitudinal. Needed equipment may have to be custom designed, and that technology element may be unique and key in providing a competitive advantage. However, companies use many of the same consultants, and personnel from different companies read many of the same source materials. Therefore, most technologies used in problem solving are common and waiting for anyone who looks for them.

The many technologies that are brought together during different product life cycle phases do not have to be cutting-edge technologies to achieve a competitive advantage. For instance, massively parallel computer architectures that provide inexpensive supercomputer capabilities use common off-the-shelf processors in novel ways. The whole in such products is clearly greater than the sum of the individual (and common) parts.

Regardless of its origin, a technologically innovative product must be properly supported (as many are not) to succeed. Often technological efforts focus too narrowly on product development. Without field or applications support, each of which requires unique technologies, products will fail. For instance, the lack of adequate software has, to date, limited the use of parallel-architecture computers primarily to a small class of scientific and engineering applications.

WHAT ARE THE BARRIERS TO TECHNOLOGY INNOVATION?

A company can begin to become more effective in technology innovation by auditing its strengths and weaknesses, a process that many organizations skimp on. Because it is nearly impossible to be unbiased during this self-analysis, senior management must be committed to the process, and the input of people outside the organization must be sought.

The goal of this audit is to define and create a culture that optimizes technology innovation. The following activities help in this process:

☐ Questioning every practice and procedure.

- Performing benchmarking, technical audits, and process reviews (e.g., continuous process improvement, TQM, concurrent engineering, and computer-aided acquisition and logistics support compliance).
- Knowing the competition and how it does business.
- Assessing the competition's performance with the same criteria the company applies to itself.
- Identifying standards the competition follows or should follow in domestic and international markets.
- Envisaging the perfect or ideal product life cycle, assuming that financial and accounting issues, proofs of implementation, and corporate barriers are unimportant and that all necessary technologies are mature and available.
- Justifying not using the ideas of Deming, Juran, Taguchi, or Crosby.
- Identifying steps to render products obsolete in the shortest possible time.
- Defining and describing why any action or approach cannot be implemented or changed.
- Identifying how to make the company's products obsolete.

These exercises demonstrate that an analysis is only as valid as its underlying assumptions. To examine its assumptions, a company should ask as many questions as possible, and it should answer no question before identifying and questioning all underlying assumptions.

Organizational tradition is difficult to change. Therefore, a company must ask many questions, even though some may seem foolish. They will seem so only because they examine assumptions that are perceived as unassailable. For example, until it was strongly questioned, quality was thought to add costs to an organization. Providing good service should be similarly examined. Companies should ask themselves: Is it cheaper to keep customers happy or to find a replacement for them?

Technology innovation can occur only in a culture that support and encourages it, but neither the culture nor the innovation develops automatically. Technological problems must be addressed without preconceptions about possible solutions. The organization must honestly evaluate unexpected solutions and technologies and be willinng to abandon current practices.

The technology innovation process also must be well thought out, or the technology that is developed and transferred might make things worse. For example, the factory and office automation initiatives during the 1970s and 1980s were both cases of technologies pushed by vendors rather than demanded by users. The result was partial solutions.

An organization often prefers its own technical expertise to that of outsiders. However, without objective data, a company has no idea how well it is doing in absolute terms. Companies can rarely satisfy all of their technology needs from strictly internal sources.

One way to break down parochial habits is to encourage staff at all levels to access information. Some people may argue that there is no such thing as too much information; others may say that personnel can spend too much time focusing on irrelevant information. What is and is not relevant is a judgment call. The key to identifying relevant technologies is to provide access to information and encourage information acquisition. One way of identifying barriers to the free exchange of information is to establish a continuous process improvement (CPI) team to address the issue.

A company can become more effective in technology innovation by auditing its strengths and weaknesses in this area.

However, before a CPI team advises management on improving information acquisition, senior management must commit to making the sweeping changes. Free access to information and the ability to use that information often collide with established management practices. In too many companies, managers think little of innovation or the hot pursuit of information. A cultural change is necessary in such companies, a change no less sweeping that the one to inculcate a serious respect for quality.

WHERE CAN COMPANIES FIND TECHNOLOGY?

Once an organization accepts that it needs more information, it must acquire it. Japanese companies excel in this endeavor, as do many information-related organizations. In the US electronics and pharmaceutical industries, information acquisition is a highly developed priority. These industries use procedures for keeping abreast of new developments; over time they develop informational data bases that are extensive, current, and easily updated.

The following representative, but not exhaustive, list includes many sources of information:

- Common sources:
    - Vendors.
    - Partners.
    - Subcontractors.
    - Staff.
    - Literature.
    - Competitors.
    - Conferences.
- Less commn sources:
    - Patents and licenses.
    - Customers.
    - Training.
    - Workshops.
- Even less common sources:
    - Patent analysis.
    - Universities and consortia.
    - Small businesses and venture sources.
    - Showcases (i.e., Baldrige winners and CALS).

— Government laboratories and R&D centers.
— Government data bases and publications.
— Shadow R&D teams.
— Contract R&D.

It takes considerable effort to peruse and use these sources, and not all are useful for all organizations. Experience will determine which are best for a given purpose.

US corporations often emphasize the common and less common sources in this list. Large companies also pursue the even less common approaches. Japanese companies generally emphasize all of these sources equally.

The key to information acquisition is to expect and help everyone in the organization to participate. Often, only senior managers or senior technical people are allowed to attend conferences or workshops. Even fewer staff members meet customers or interact with the marketing department. All employees should be exposed to these activities and be encouraged to share what they have learned.

Companies often worry about time and costs associated with letter writing, photocopying, and telephoning and how to control the potential abuses associated with them. These fears are not unfounded, but strong measures to protect against their realization can discourage and quash individual initiative. There are nonintrusive ways to limit the abuse of these activities without limiting their utility.

HOW CAN TECHNOLOGY BE PUT TO WORK?

Once an organization begins to establish a culture consistent with information acquisition, it must ensure that its efforts are properly channeled. Professional R&D organizations always begin a project with literature searches and assessments of how the proposed technologies are practiced elsewhere. The key is targeting the information being gathered. Once the acquisition process begins, it naturally expands to include peripheral data that may later become important.

A common mistake is to gather wrong or incomplete data. Often, this happens because a problem is ill defined or not defined at all. For example, during the 1970s many industry watchers thought that factory productivity would be enhanced by the use of robots. Information about robotics, however, was insufficient. Subsequent studies showed that human factors and quality control, aided by computer systems, were a far more effective solution to production problems. Furthermore, because robots were found to be more effective for some operations than others, manufacturability by robots was required in product designs.

The following guidelines can help companies put technology to work:

☐ Technology should be implemented as a complete systems solution.

- Requirements for products, processes, and support should be defined in detail.
- Technologies (whether available or not) should be identified for the entire product life cycle, including:
  - Prototyping through production.
  - Products, process, and support.
  - Infrastructure and equipment.
- It is wise to simulate the product and production cycle.
- It is important to have the right people with the requisite skills as gatekeepers, problem solvers, mentors, and team players.

The key to information acquisition is to expect and enable everyone to participate.

This list is an eclectic mixture of strategic and operational advice. Its importance is not to describe how to implement a development team but to identify a company's major obstacles to technology innovation.

The development of a product, its manufacturing, sales, and support represents a system that must be defined. A new product may contain new technologies and require new manufacturing and testing procedures and equipment. It may also require a company to redefine its support practices (e.g., repair, replacement, and warranties). The technologies involved in creating a new product, therefore, involve more than just the product itself.

For example, to implement factory automation, a company should research more than just robots or other machines; it must examine the physical layout, warehousing and distribution requirements, and software needs. How staff interacts with and works in a new environment is also critical. A complete life cycle analysis is needed for new designs that, for instance, replace mechanical system with electronic ones.

New products may involve a larger electronic content than their predecessors did. When more embedded processors are required, the software component also increases. Frequently, a product demands a completely new set of engineering skills as it evolves. Detailed attention must also be paid to all aspects of product definition, requirement, manufacturability, and support.

Also important is having the right people on board. More important than their skills is their attitude, which is critical to changing corporate culture. Retraining is also an attitude issue. Staff members who have the right attitude are retrainable, and those with strong intellects can retrain themselves to rapidly become contributing experts in new areas.

Organizations that have become parochial overestimate how skilled their staff members really are. Managers in such companies exclude people with new ideas and attitudes because their corporate culture demands that ideas filter from the top down.

Many problems can arise when management overestimates the capabilities of its staff. For example, a new technology might be touted in the professional literature, and a staff member may be assigned to assess it. An employee who has trouble understanding the technology will probably give it weak marks rather than admit confusion. The company will probably decide not to pursue the technology but to watch and see how it matures and thus perhaps miss an opportunity. A recent example of this phenomenon occurred with fuzzy logic, which Japanese and Chinese companies have had considerable success with in control systems. It is now mature enough for US companies to pay attention to it, but Japanese companies have already introduced products that are capturing a large domestic market share in such areas as air conditioners and environmental controls.

Companies need managers and staff who are not team killers, who are not autocratic, and who are not not-invented-here people. Such personnel distrust their colleagues' capabilities, undermine team confidence, and inhibit action by overmanaging and demeaning team progress. The most brilliant technician can ruin a project by exhibiting team-killer behavior. This all-too-common (and often protected) employee kills innovation. This person also is often the senior expert who stalls the consideration of new technologies.

CONCLUSION

As a society, the US encourages and rewards innovation and is, as a result, very dynamic. Unfortunately, however, our US corporations have often become enclaves that suppress, not harness, creativity.

US companies must encourage people with needs to reach out for technologies that will satisfy these needs. Companies must avoid solutions looking for problems—to do otherwise is to guarantee inefficiency. They must also ensure that they think through their product, support, and service offerings well enough to identify necessary technologies. Successful entrepreneurial organizations focus on a specific product and market and ensure that the entire company participates in the information acquisition and technology innovation process. The free flow of information in a dynamic organization distinguishes it from one that has trouble innovating and competing. In the final analysis, it is not technologies that determine the success of a company or product but how the technologies are used.

MANAGING INNOVATION IN START-UP VERSUS ESTABLISHED ENVIRONMENTS

Russell Horres

Managers in established corporations have been repeatedly surprised by the appearance of technologically innovative product concepts from the design teams of newly formed companies. This study contrasts the dynamics of developing innovative new products in start-up organizations with developing them in established corporations.

The story of technological progress is replete with examples of poorly funded, understaffed entrepreneurial organizations outpacing established market leaders by introducing innovative new products. Well-known examples in the microcomputer field include the fledgling Microsoft Corp. and Apple Computer clearly outproducing industry giants. In the medical products area, a start-up company recently beat industry leaders to the market with infra-red thermometers. How do such start-up ventures outperform industry giants with much larger staffs, facilities, and capital resources?

Established companies face some serious problems that start-up organizations do not in managing innovation in product development. The most significant problem is management's attempt to balance the complex interrelationships among functional departments, each of which in its own way adds requirements to the product attributes, development schedule, design procedures, and manufacturing processes. Individually, each requirement can be rationalized and justified as essential to the success of the project. Any corporation with more than 10 years of development experience can cite ample reasons for the establishment of checks and balances on the innovation process, and functional managers are all too eager to refer to the corporate history.

The first department to enter the process is R&D, which often can cite reasons for restrictive development guidelines, extensive design reviews, and documentation as the right way to develop products. An equivalent factor in a market-driven organization is the product planning function, which reacts to the length of development projects by striving to include every conceivable new feature and by requiring interrelationships with existing product lines from each new product. Product planning seldom appreciates the added complexity and attendant schedule commitments of such requirements. The purchasing function becomes deeply involved in establishing a vendor base that meets a predetermined list of requirements and controlling the sourcing of vendors by requiring conformance of parts to established bills of materials. Operations with established, capital-intensive manufacturing processes and well-defined

---

RUSSELL HORRES is the president of CyberRx, Inc., the second start-up company he has headed.

manufacturing procedures generate additional constraints. Service functions, having experienced difficulties maintaining existing products, develop an additional set of product requirements. The quality assurance and reliability engineering functions, often independent of product development teams, establish product performance standards and corporate quality goals. The cost accounting function, capital approval processes, corporate legal counsel, and, depending on the nature of the business, the regulatory affairs function can all add requirements for conforming to established procedures.

Convention is a powerful force, and no well-schooled project manager is going to commit heresy by attacking the very fabric of business practices. Although many participants bemoan lengthy development cycles, functional areas provide only the appearance of improving the flow of innovative products, each—according to its own perception—finding value-added contributions to the integration of the new-product design effort with existing processes. Although an individual contribution may seem essential, taken altogether, numerous contributions can actually cripple a project, just as the Lilliputians of *Gulliver's Travels* subdued the giant with a myriad of tiny threads.

IN THE BEGINNING

At its inception, all of a start-up corporation's efforts are devoted to the creation of an innovative product. There are no existing products to build, market, or sell; all efforts are focused on design. The concept of departmentalization does not exist, and at a minimum, there is a recognition of overlapping roles among the participants. The acquisition and management of capital to fund these efforts are paramount and supported by all founding members.

In the face of limited resources, much effort must be contracted out during this initial phase. Designers have carte blanche to meet future customers' needs with only broad targets for features, manufacturability, serviceability, and product costs. The development team sources all vendors, procures parts, and defines quality controls and specific manufacturing processes down to the shipping container. Very often, design-team members are also involved in the manufacturing start-up and initial marketing of the concept.

When the new product becomes accepted, R&D funding is cut back drastically as efforts turn to marketing and production. The new company experiences phenomenal growth, tiers of management develop, and departmentalization ensues. Departments establish policies and controls, innovation slows, and the founding entrepreneurs frequently become disenchanted with the pace and leave the Lilliputians to their work.

FUNCTIONAL BARRIERS

The major participants in new product development are marketing, R&D, and manufacturing. Although these roles may be recognized in the new enterprise, the concept of departments and individualized functions do not have to be dealt with. In the established corporation, functional areas create significant barriers to innovative product development. These barriers are manifest in the communication difficulties among the functional groups and their appreciation of the corporate goals and initiatives. Interdepartmental barriers are often deeply rooted, with attitudes passing from generation to generation like a virus.

> At its inception, all of a start-up corporation's efforts are devoted to the creation of an innovative product.

Most contemporary corporations have recognized the limitations of sequential product development and have instituted product development teams with representatives from the participating functions. Concurrent engineering—that is, developing manufacturing processes simultaneously with the product design—is also becoming widely employed. These management processes are intended to help ensure that the resulting product will meet customers' needs, optimize a company's competitive position, achieve the company's predetermined cost and quality objectives, and be integrated smoothly and rapidly into the manufacturing process. Unless care is taken to ensure teamwork and effective communication, these processes will only establish a platform on which to exercise functional imperatives and thereby subvert the very purpose for their existence. Creating the processes is easy; actualizing their promise is a genuine challenge.

Providing a forum for frequent exchange is seldom an answer to reducing barriers. The answer lies in developing an understanding of each function's capabilities, motivation, jargon, and roles and in building respect for the value each function brings to the development process. Individuals representing the various functions must understand the value of teamwork; the team needs to nurture nonconformists to exploit the creativity that such individuals can bring to problem solving.

LEADERSHIP STYLES

Much has been said about the role of leadership styles in managing innovation. Although no right or wrong style can be proved to stimulate innovation, the start-up environment seldom permits the luxury of experimentation. Most often, a directive, controlling style complemented by the charisma and vision of the innovator produces highly effective leadership. The intense focus of the development activity and small size of the start-up corporation result in total alignment of purpose and minimize the role of communication and people skills.

To complement the leadership of the new venture, the start-up organization's team membership must be composed of carefully selected individuals with similar values. Each member must understand his or her role in the venture and be prepared to dedicate a major portion of time to the success of the endeavor. Rigid perceptions about job descriptions are unlikely to be successful in the start-up environment.

RISK TAKERS

By its very nature, a new enterprise attracts more results-oriented individuals. There is little gratification for process-oriented individuals seeking to expand their management abilities by coordinating large numbers of staff members in the new enterprise. On the other hand, the ability to accomplish tasks, make decisions, and reach goals in a highly efficient manner presents significant opportunity for gratification for both types of individuals.

Individuals who choose to associate with a start-up, particularly those who leave the comfort of the established organization, are also risk takers. Contrary to the notion that these individuals are not team players, there are close working relationships within the founding group of many new ventures. Indeed, the pressures to attain profitability and the enormous time commitment often induce stresses in the staff as the team replaces family in its demand for time. Not infrequently, this imbalance results in the loss of founding members. This closeness and peer pressure for commitment often overwhelm even mature managers who have aspired to participate in a new enterprise for years.

The established corporation that pursues innovative product development does not benefit from the personality traits that are characteristic of people who elect to join a new venture. In a more mature company, employees are conditioned to be process oriented and averse to risk, and as team members, they are functionally oriented and career focused. Although the commitment a company gains from these individuals can be great, progress occurs largely on a business-as-usual basis.

MOTIVATION

The subject of motivation cannot be overlooked in examining performance differences between the start-up and the mature corporation. In the start-up, equity participation and the possibility for significant returns are incentives to make a major commitment in time and creativity. Stories of stockroom clerks and secretaries becoming wealthy during the acquisition or public sale of a new venture are legendary and are frequently discussed among team members in start-up organizations. The potential personal reward also provides the rationale to these individuals' families to understand the extraordinary commitment of time and energy that new ventures demand.

The drive for fairness and equity across corporate functions inhibits the established corporation from wielding this power tool. Incentives are limited and seldom effective. For the majority of the employees in these more mature organizations, success means they will be assigned to another team.

FOCUS

A fundamental leverage that the start-up enjoys is focus and dedication to the new product development process. Without the attendant distraction of ongoing operations, each team member can devote substantial effort to the success of the endeavor. There are certainly fewer resources dedicated to the effort, but each resource is more effective because it is not diverted to competing activities.

> By its very nature, a new enterprise attracts more results-oriented individuals.

By contrast, the project development team in the established corporation must be concerned with ongoing production, marketing, and sales. Because the R&D representatives are largely focused on development, they consider requests for assistance on field problems or for customization of existing products a distraction. Other team members face even more significant distractions because the major emphasis of their functions is maintaining the ongoing business of the corporation. Personnel development and training programs are additional time commitments. The impact that these competing activities have on the ability of team members to focus on the project is seldom appreciated.

ANALYSIS PARALYSIS

No single factor has more relevance to the established corporation's lack of responsiveness than its predilection for overanalysis and reanalysis of options. In most cases, the innovation with which the start-up blindsides the established competition has been under consideration for years by a strategic planning group in an established organization but has been viewed as too risky or radical for the market. Each successive pass at an innovative concept seems to become increasingly conservative until the organization finally convinces itself that the product won't be successful.

The most expedient product development actions taken by large corporations are those that respond to competitive product introductions. These initiatives are destined to be late in the marketplace and will result in, at best, adequate second-generation products. A compounding factor is the dilemma a market leader faces in any strategic decision to render an existing, successful product line obsolete. As difficult as this decision is, it will ultimately be mandated by external factors if the company is to remain a leader in the face of competitive innovation.

A variant on the theme of overanalysis is the reevaluation of a project once it is under way. At the first signs of difficulty or schedule delay, risk aversion takes over, more analysis is done—sometimes by a new forecasting team—and the project is canceled. Indeed, the continual turnover of team members is a major cause of schedule delays and goal inconsistencies. US industry's habit of rotating people through assignments without regard to the goal of the assignment must be broken.

POSSIBLE ALTERNATIVES

It is not feasible or necessarily desirable to attempt to simulate the dynamics of managing innovation in start-ups in corporations with established markets.

Risks must be managed more carefully when losses can be significant and product introductions in mature markets demand extensive coordination between corporate functions.

This does not mean that managers must accept an overly bureaucratic process. In addition to systematically evaluating the steps in the innovation process that are not value-added, managers can structure some portion of corporate resources to focus on the technology base of new markets, and entrepreneurial dynamics can be encouraged in new product innovations for these sectors. The most widely recognized and successful structure for these enterprises is the dedicated, freestanding venture organization that has the freedom of a start-up and the financial resources of the established corporation. Although other structures might be successful, few will argue that the new ventures group is not a powerful source for driving innovation.

UNLEASHING CREATIVITY

This chapter illustrates the complex nature of corporate dynamics surrounding the management of product development activities. The need to reduce product development life cycles and increase worldwide competition is placing significant pressure on the product development process. Unless drastic steps are taken to change corporate thinking, policy, and action, the established corporation will continue to be outmaneuvered in the area of innovative product development.

Major US corporations have the capacity to be creative but have locked out this creativity from within through bureaucratic, risk-averse, conservative corporate policies. The areas in which a change of culture would have the greatest impact include product planning, team focus, team member consistency, incentives and rewards for risk takers, tolerance of nonconformity, and reduction of functional barriers.

SCENARIOS FOR INNOVATION STRATEGIES

This chapter presents a case study that illustrates the successful application of the scenario technique.

Horst Geschka
Monique Verhagen
Barbara Winckler-Rub

Scenarios are pictures of the future built under certain plausible assumptions. Alternative assumptions for future development are combined—with the assistance of a computer program (INKA)—in a manner similar to the way in which consistent structures are formed. The scenario technique is an effective planning tool when the time-horizon of planning is long-range and uncertainties are high. This chapter presents a case study that illustrates the successful application of the scenario technique. In addition, alternative assumptions are interpreted and worked out in detail and conclusions are drawn.

DEFINING SCENARIOS

The term scenario was first used in the 1960s by futurologist Herman Kahn in reference to designing pictures of the future regarding strategic issues in a military setting. Scenarios typically represent the framework for future—neither subjective nor normative—plans and strategies. To serve in the formulation of innovation strategies, scenarios must be based on facts and plausible assumptions. Such facts and assumptions are products of the present situation and show possible paths into the future. Scenarios are deemed to provide insight into the future, starting at a given present situation and leading up to a point in time in a future situation, as shown in Exhibit 1. Scenarios are therefore the consistent result of assumptions for which plausible and comprehensive reasons can be given. As a rule, several alternative assumptions are established and developed.

Scenarios can be the answer to the negative experiences attributed to conventional forecasts, which often feign an unjustified and unfounded accuracy. The forecasts concentrate on one or more parameters that, as a rule, must be quantifiable. Most forecasting methods fail to reflect the complexity of the interdependence and the qualitative aspects of a topic. Scenarios, on

---

HORST GESCHKA is the founder of Geschka & Partner Management Consultants. He is also a professor at the Technical University of Darmstadt.

MONIQUE VERHAGEN is at Geschka & Partner Management Consultants.

BARBARA WINCKLER-RUB, an employee at Geschka & Partner Management Consultants since 1989, works mainly in the field of strategic planning with an emphasis on the scenario technique.

Exhibit 1
The Path of Scenarios

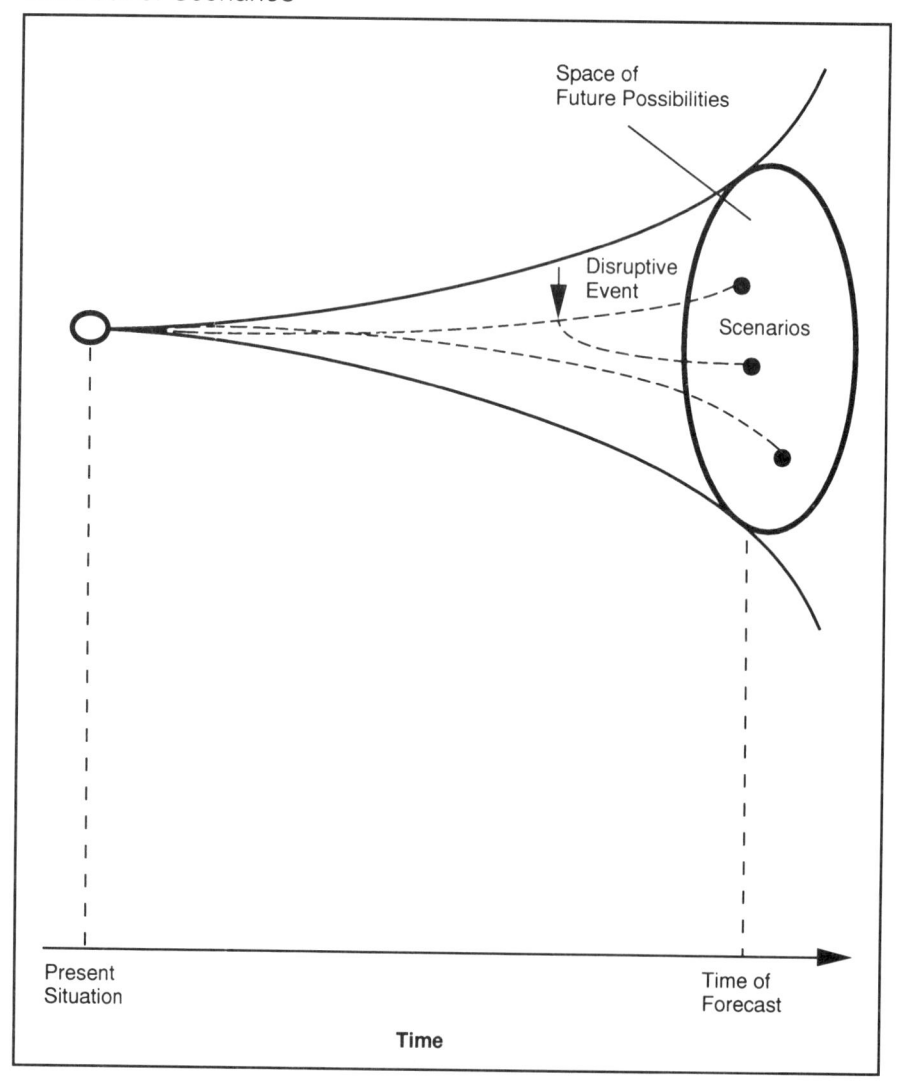

Diagram: funnel-shaped "Space of Future Possibilities" extending from Present Situation to Time of Forecast, with dashed scenario paths leading to Scenarios, and a Disruptive Event marker.

the other hand, emphasize the qualitative aspects of a situation and show alternative possibilities of development, which are systematically devised.

A CASE STUDY

A company that produces sanitary products (e.g., products for water supply and water drainage, including armatures and other components) is considering future strategy. A conventional forecast, based on a quantitative method (e.g., regression analysis), may show that x amount of faucets of type A and y amount of faucets of type B will be needed in the year 2002; qualitative factors of influence are not considered. A scenario, on the other hand, may examine the possible influence of microelectronics on the future development of sanitary products, or the relationship between trends in values and sanitary products, thereby emphasizing qualitative factors. Qualitative factors that may influence the future of the field under investigation, however, will be considered.

To serve in the formulation of innovation strategies, scenarios must be based on facts and plausible assumptions.

SCENARIO DEVELOPMENT

The various methodical approaches for the systematic development of scenarios are as follows:

- *Visions.* Intuitively developed views of the future that are largely normative and partially based on research and facts.
- *Simulation models.* The actual relations are expressed in a model. The analysis results from variations of exogenous variables to obtain new results (forecast values) each time a variable changes.
- *Megatrend scenarios.* Current trends used as indicators and driving forces for future developments. The initial trends are elaborated on by competent researchers and conclusions about future development in the field under investigation are drawn from the overall trends.
- *Scenarios of impacting fields.* Developments in the future are derived from external influences for a defined field under investigation. Alternative scenarios are produced on the basis of alternative assumptions about developments in the future with a high degree of uncertainty.

In the mid-1970s, Battelle of Frankfurt, Germany developed a technique for drawing up scenarios of impacting fields. This technique proposes that a future situation cannot be accurately forecast by extending the correlations of the past if the external conditions of the field under investigation are subject to major changes—as are most fields today.

The external factors of influence must be followed up and their chains of cause and effect investigated. The future situation of a topic is, therefore, derived from the future development of the external fields of influence. However, there are many uncertainties regarding the individual influences; different directions can often be seen. Therefore, alternative assumptions must be put forward for uncertain factors of influence. Once alternative assumptions have been structured into consistent sets, alternative scenarios can be developed.

CHARACTERISTICS OF THE SCENARIO TECHNIQUE

The scenario technique has the following characteristics:

- The starting point is a thorough analysis of the current situation leading to an understanding of the correlations in the field under investigation.
- Meaningful alternative assumptions are made for factors of influence with an uncertain future development.
- Scenarios are developed as path scenarios. The future situation as well as the way leading from the present to the future condition is described.
- The development of scenarios also takes into account disruptive events with an intensive impact on the investigated topic.
- The result is several alternative pictures of the future which are consistent in themselves.

THE SCENARIO TECHNIQUE

The scenario technique, as shown in Exhibit 2, comprises the following eight steps:

- Structuring and defining the topic.
- Identifying and clustering areas of influence.
- Establishing projections and alternative descriptors.
- Forming and selecting alternative consistent sets of assumptions.
- Developing and interpreting the selected scenarios.
- Introducing and analyzing the effect of disruptive events.
- Deriving the consequences for the field under investigation.
- Designing measures.

Structuring and Defining the Topic

First, the topic must be defined as precisely as possible. Determinations must be made as to what belongs to the subject and where the borderline for unconsidered subjects lies. Referring to the example, sanitary products must be defined. Questions such as "What does the group of sanitary products investigated comprise?; What about hot water heaters, soap dishes, decalcification devices, and the like?; Will both sanitary products in private bathrooms and kitchens and those in companies be considered?"; and "What are the common characteristics of sanitary products?" must be answered before an examination of the environment can be performed.

The following is the topic definition and the field of investigation of the case:

- Topic definition: Market and technology of sanitary systems and components in 2002.

Exhibit 2
The Scenario Technique

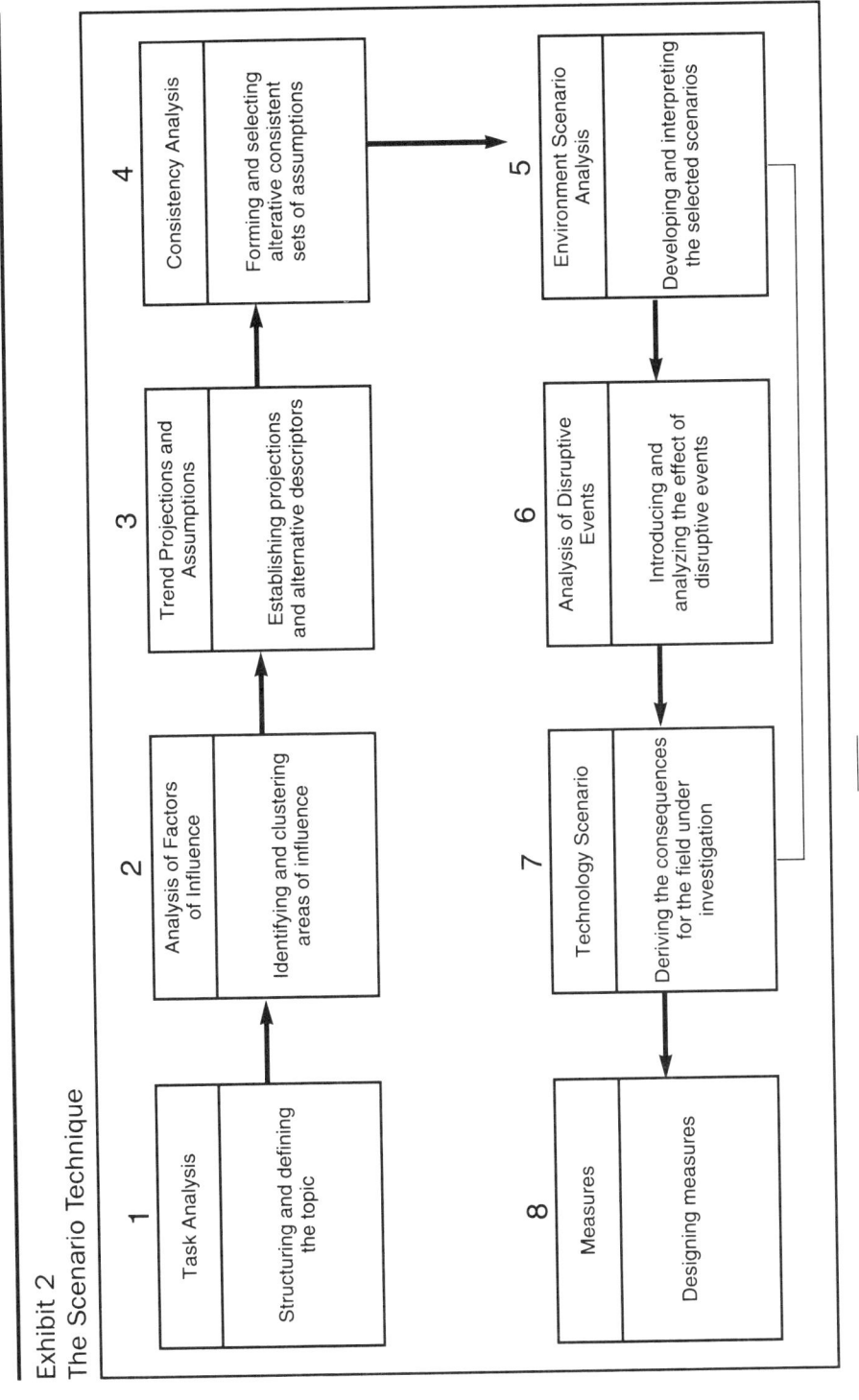

1	Task Analysis	Structuring and defining the topic
2	Analysis of Factors of Influence	Identifying and clustering areas of influence
3	Trend Projections and Assumptions	Establishing projections and alternative descriptors
4	Consistency Analysis	Forming and selecting alternative consistent sets of assumptions
5	Environment Scenario Analysis	Developing and interpreting the selected scenarios
6	Analysis of Disruptive Events	Introducing and analyzing the effect of disruptive events
7	Technology Scenario	Deriving the consequences for the field under investigation
8	Measures	Designing measures

INNOVATION STRATEGIES

☐ Field of investigation: Total product technology of water supply and water drainage, including armatures and other components for home, hotels, and hospitals.

Topic definition requires the collection and analysis of background information on the subject. A definition of the task acceptable to all participants in the project must be determined. Structural characteristics and unsolved problem areas should be highlighted.

Identifying and Clustering Areas of Influence

All areas influencing the field under investigation must be identified, possibly by means of a brainstorming technique. During the application of the brainstorming technique, areas of influence are written on cards (one area per card) by all session participants. The cards are then passed on to other participants. Each participant must read the ideas on the card and try—through association—to further develop the idea or to devise new areas of influence. With this method, 30 to 50 areas of influence can be quickly generated. The areas must be divided into clusters and each cluster given a heading (i.e., field of influence). The fields of influence pertinent to sanitary technology are shown in Exhibit 3.

Establishing Projections and Alternative Descriptors

Descriptors for the established factors of influence must be defined. The transition from the areas of influence to descriptors is shown in Exhibit 4. While areas of influence are the direct result of a brainstorming session, descriptors should be formulated in such a way as to enable the qualitative or quantitative indication of their states and to cover all relevant areas of influence. For qualitative descriptors, index values (i.e., 1992 = 100, 2002 = 120) or brief statements should be used. Usually, 20 to 40 descriptors are elaborated.

The actual state of the descriptors should be determined and assumptions for the target year of the scenario, based on known forecasts and expert knowledge, should be made. For descriptors that have several possible developments, these alternative developments should be identified and labeled as alternative assumptions. These descriptors are called alternative descriptors. The descriptors of the case (alternative descriptors and unambiguous descriptors) and their assumptions are shown in Exhibit 5.

Forming and Selecting Alternative Consistent Sets of Assumptions

To form and select alternative consistent sets of assumptions, a consistency matrix must be completed, as shown in Exhibit 6. The consistency matrix contains all alternative descriptors and their corresponding alternative assumptions. The consistency matrix cannot be completed until a determination has

Exhibit 3
Areas of Influence

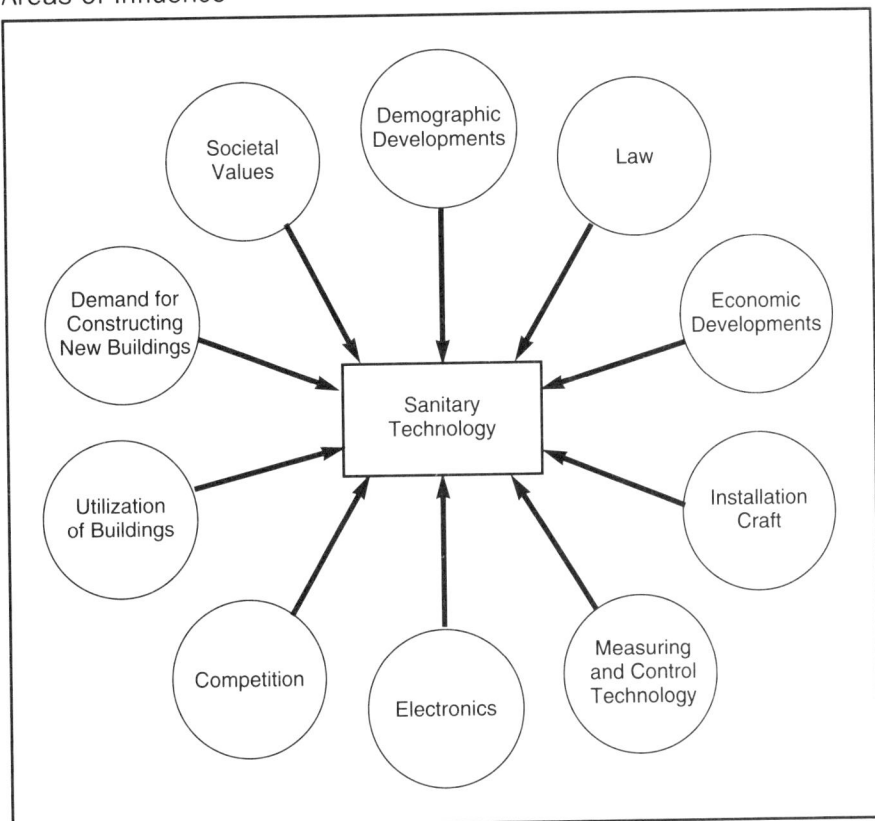

[Diagram: Ten circles arranged around a central box labeled "Sanitary Technology", with arrows pointing inward from each circle. The surrounding circles are labeled: Societal Values, Demographic Developments, Law, Demand for Constructing New Buildings, Economic Developments, Utilization of Buildings, Installation Craft, Competition, Electronics, Measuring and Control Technology.]

been made as to whether two assumptions for two different descriptors fit very well together or are contradictory. If the assumptions fit very well, +3 (the highest value) is entered; if they are contradictory, -3 (the lowest value) is entered. If the assumptions are not correlated with each other, 0 is entered.

After the consistency matrix has been completed, the INKA program performs a consistency analysis that will result in the development of alternative scenarios. The consistency analysis will lead to combinations of assumptions that are highly consistent but at the same time contain very different assumptions. Combinations of assumptions that satisfy these two criteria are the best starting point for the development of scenarios. The algorithm of the consistency analysis and the final proposal for two or three alternative scenarios executed by INKA are composed of two major steps.

Exhibit 4
Areas of Influence Become Descriptors

Factors of Influence	Descriptors
Economic development	Development of GNP
Building market	Demand for building new production plants
Problems with PVC	Prohibition of PVC
More single households	Number of households
Behavior of competition	Diversification strategies
Installation craft	Acceptance of new technology by craftsmen
Application of sensors	Measuring and control technology
Electronics	Influence of electronics on sanitary technology

Range Analysis That Identifies Descriptors That Largely Affect Consistency. The range analysis will present descriptors that will strongly influence the alternative scenarios. These descriptors will therefore serve as the basis for the proposed scenarios. To find these descriptors, the maximum range (R_{max}) and the minimum range (R_{min}) of a descriptor are calculated for each horizontal line of the consistency matrix, as shown in Exhibit 7. The bigger the difference (R_d) between the maximum range and the minimum range, the bigger the influence of the descriptor in question. Only the two or three descriptors that have the biggest impact on the sum of consistency will be used in the second step.

Formation of the Most Consistent Combinations of Assumptions. Basic combinations of the most influential descriptors are set, as shown in Exhibit 8. If two descriptors have the biggest impact on consistency, four initial paths can be fixed; in case of three descriptors, eight initial paths can be fixed.

The other descriptors are added to the initial paths and the assumptions are chosen to facilitate the formation of the most consistent of assumptions. For this purpose, all combinations of assumptions are presented in a decision tree with limited enumeration. A maximum attainable sum of consistency of a partial combination of assumptions will be rejected immediately upon its becoming smaller than the sum of consistency of an already existing established combination. As a result, only those combinations of assumptions that are most consistent are selected and proposals of combinations of assumptions are developed that are simultaneously different and most consistent. Two sets of assumptions that form scenario proposals are shown in Exhibits 9 and 10.

Developing and Interpreting the Selected Scenarios

The assumptions devised for the alternative descriptors from the two or three proposed scenarios in step 4 will be added to the simplest of the selected

Exhibit 5
Descriptors and Assumptions

Descriptors	Assumptions
1. Attitude toward technology	a) Hostility toward technology increases b) Hostility toward technology remains as is
2. Population of Germany	a) Population grows from 78 million to 82 million b) Population remains 78 million
3. Mobility of people	a) Increases by EC and changing East-West relations b) Stabilizes after a period of turbulence
4. Number of households	a) Number of households increases by about 20% b) Number of households increases by only 10%
5. Development of GNP	a) Long-term moderate growth b) Long-term strong growth
6. Volume of investments	a) Volume increases to 150 b) Volume remains at 100
7. Interest rate on mortgages	a) Increases to 15% b) Stabilizes at 9% c) Decreases to about 7%
8. Promotion of property ownership	a) Property ownership is hardly encouraged b) Property ownership is strongly encouraged
9. Construction law	a) Increased enforcement of law b) Relation of law
10. Prohibition of PVC	a) PVC will be prohibited b) PVC recycling rule c) Law on PVC will remain unchanged
11. Rental law	a) Legislator favors tenant b) Legislator favors landlord
12. Demand for building new production plants	a) Demand increases by 20% (120) b) Demand decreases by 10% (90)
13. Differentiation strategies	a) Competition expands the number of products offered b) Competition concentrates on a few products
14. Flexibility of competition	a) Competition invests in additional services b) Competition concentrates on cost leadership c) Competition innovates rapidly
15. Diversification strategies	a) Competition from related markets is a threat b) Competition from related markets is not a threat
16. Use of buildings	a) Desire for flexibility of buildings increases b) Present flexibility of buildings is sufficient
17. Acceptance of new technology by craftsmen	a) Craftsmen have difficulty accepting new technology b) Craftsmen appreciate new technology
18. Measuring and control technology	a) Sensor technology is known but expensive b) Sensor technology is accepted and relatively cheap
19. Electronics[1]	a) Electronics gains importance in sanitary technology

Note:
[1] This descriptor is not ambiguous and will therefore not be included in the consistency matrix.

Exhibit 6
Consistency Matrix

Exhibit 7
Calculation of Descriptors

	1a	1b	2a	2b	3a	3b	4a	4b	5a	5b	6a	6b	etc.
5a	0	0	2	-2	1	2	-1	1	*	*	-2	2	
5b	0	0	-2	2	2	-1	1	1	*	*	2	-2	
R	0	0	4	-4	-1	3	-2	0	*	*	-4	4	

R = Range
R_{max} = (0) + (4) + (3) + (0) + etc. = 21
R_{min} = (0) + (-4) + (-1) + (-2) + (-4) + etc. = -21
Rd = R_{max} - R_{min} = 42

assumptions. As scenarios are developed from the present, the intervals should not be too big (it is easier to make estimations and judgments for a time span of 4 to 5 years as opposed to one to 10 years). For example, if the year of the initial situation is 1992, 2 intervals of 5 years would be chosen.

Another reason for developing scenarios in intervals is to, once again, ensure that all assumptions fit very well together. At each interim, the status of all descriptors will be described and their consistency will be checked. In addition, any reactions to developments in the previous period will be pursued and stated in the next interim. Thus a development structure in the course of time is elaborated, extending from the present to the target year of the scenario. Any inconsistency will be disclosed.

Introducing and Analyzing the Effect of Disruptive Events

A disruptive event is a suddenly occurring incident unforeseen by the trend that may change the direction of the trend. Trend-breaking events can be either catastrophes (earthquakes, nuclear reactor explosions) or positive incidents (political reconciliation, advent of technology).

During the evaluations, the results that have the strongest impact on the scenarios and show a relatively high probability will be highlighted. Next, significant trend-breaking incidents selected in this way will be described in greater depth and introduced into the scenarios. Their effect will be monitored, thus providing new scenario variants.

Deriving the Consequences for the Field Under Investigation

This step is composed of the following two procedures:

- If a defined strategic question exists (e.g., What are the possibilities of going into genetic engineering?), consequences can be derived directly from the

Exhibit 8
Setting of Basic Combinations

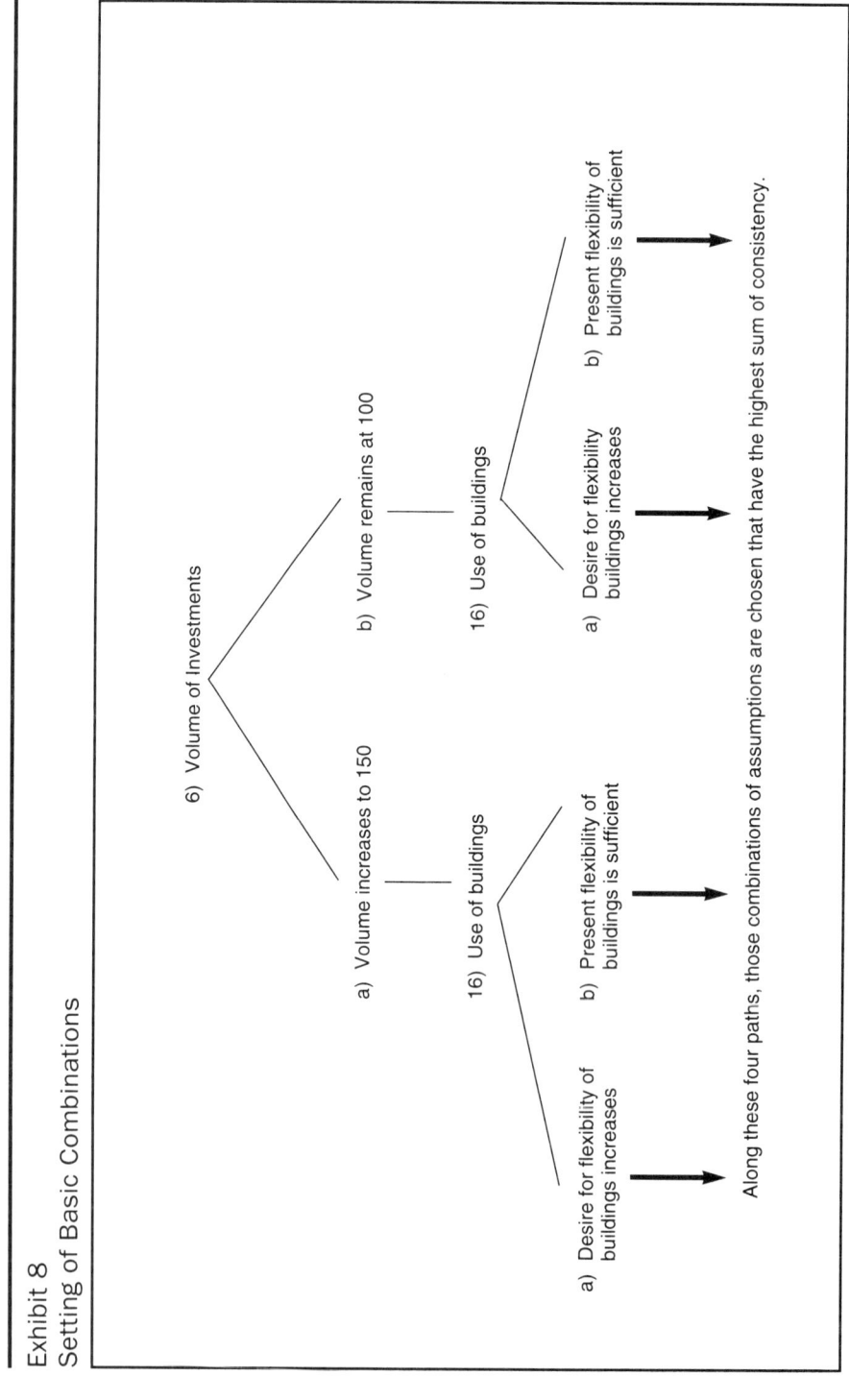

6) Volume of Investments

a) Volume increases to 150
b) Volume remains at 100

16) Use of buildings

a) Desire for flexibility of buildings increases
b) Present flexibility of buildings is sufficient

16) Use of buildings

a) Desire for flexibility of buildings increases
b) Present flexibility of buildings is sufficient

Along these four paths, those combinations of assumptions are chosen that have the highest sum of consistency.

NURTURING INNOVATION

Exhibit 9
Scenario Proposal A

Descriptors	Assumptions	Consistency
1. Attitude toward technology:	a) Hostility toward technology increases	2
	b) Hostility toward technology remains as is	0
2. Population of Germany:	a) Population grows from 78 million to 82 million	3
	b) Population remains 78 million	3
3. Mobility of people:	a) Increases by EC and changing East-West relations	8
	b) Stabilizes after a period of turbulence	3
4. Number of households:	a) Number of households increases by about 20%	8
	b) Number of households increases by only 10%	3
5. Development of GNP:	a) Long-term moderate growth	-5
	b) Long-term strong growth	8
6. Volume of investments:	a) Volume increases to 150	15
	b) Volume remains at 100	-7
7. Interest rate on mortgages:	a) Increase to 15%	3
	b) Stabilizes at 9%	2
	c) Decreases to about 7%	0
8. Promotion of property ownership:	a) Property ownership is hardly encouraged	-1
	b) Property ownership is strongly encouraged	7
9. Construction Law:	a) Increased enforcement of law	-7
	b) Relaxation of law	-6
10. Prohibition of PVC:	a) PVC will be prohibited	10
	b) Law on PVC will remain unchanged	9
11. Rental Law:	a) Legislator favors tenant	0
	b) Legislator favors landlord	7
12. Demand for building production plants (91:100):	a) Demand increases by 20% (120)	-6
	b) Demand decreases by 10% (90)	8
13. Differentiation strategies:	a) Competition expands the number of products offered	-5
	b) Competition concentrates on a few products	4
14. Flexibility of the competition:	a) Competition invests in additional services	1
	b) Competition concentrates on cost leadership	3
	c) Competition innovates fast	-2
15. Diversification strategies:	a) Competition from related markets is a threat	5
	b) Competition from related markets is not a threat	3
16. Use of buildings:	a) Desire for flexibility of buildings increases	1
	b) Present flexibility of buildings is sufficient	15
17. Acceptance of new technology by craftsmen:	a) Craftsmen have difficulty accepting new technology	-6
	b) Craftsmen appreciate new technology	-2
18. Measuring and control technology:	a) Sensor technology is known but expensive	-2
	b) Sensor technology is accepted and relatively cheap	-3
Sum of consistency		112

Exhibit 10
Scenario Proposal B

Descriptors	Assumptions	Consistency
1. Attitude toward technology:	a) Hostility toward technology increases	-4
	b) Hostility toward technology remains as is	4
2. Population of Germany:	a) Population grows from 78 million to 82 million	-1
	b) Population remains 78 million	5
3. Mobility of people:	a) Increases by EC and changing East-West relations	-1
	b) Stabilizes after a period of turbulence	8
4. Number of households:	a) Number of households increases by about 20%	-3
	b) Number of households increases by only 10%	6
5. Development of GNP:	a) Long-term strong growth	8
	b) Long-term moderate growth	-5
6. Volume of investments:	a) Volume increases to 150	-7
	b) Volume remains at 100	14
7. Interest rate on mortgages:	a) Increase to 15%	-7
	b) Stabilizes at 9%	0
	c) Decreases to about 7%	9
8. Promotion of property ownership:	a) Property ownership is hardly encouraged	3
	b) Property ownership is strongly encouraged	-4
9. Construction Law:	a) Increased enforcement of law	-5
	b) Relaxation of law	6
10. Prohibition of PVC:	a) PVC will be prohibited	-4
	b) PVC recycling rule	-5
	c) Law on PVC will remain unchanged	4
11. Rental Law:	a) Legislator favors tenant	-3
	b) Legislator favors landlord	4
12. Demand for building production plants (91:100):	a) Demand increases by 20% (120)	-3
	b) Demand decreases by 10% (90)	7
13. Differentiation strategies:	a) Competition expands the number of products offered	9
	b) Competition concentrates on a few products	-1
14. Flexibility of the competition:	a) Competition invests in additional services	2
	b) Competition concentrates on cost leadership	0
15. Diversification strategies:	a) Competition innovates fast	5
	b) Competition from related markets is a threat	4
16. Use of buildings:	a) Competition from related markets is not a threat	-1
	b) Desire for flexibility of buildings increases	6
17. Acceptance of new technology by craftsmen:	a) Present flexibility of buildings is sufficient	6
	b) Craftsmen have difficulty accepting new technology	-1
18. Measuring and control technology:	a) Craftsmen appreciate new technology	-1
	b) Sensor technology is known but expensive	-1
	c) Sensor technology is accepted and relatively cheap	-1
Sum of consistency		104

impacting field scenarios and ideas for measures to be taken can be developed.
- For tasks of a more general character (e.g., determining a technology concept), working out scenarios for the field under investigation is beneficial. This is also done in all cases in which the scenarios on a specific topic are to be published.

Scenarios emphasize the qualitative aspects of a situation and show alternative possibilities of development, which are systematically devised.

There is a strategic question: How will the market and the product technology of sanitary systems and components develop until the year 2002? Therefore, scenarios for the field under investigation are not devised; consequences are theorized. Consequences can be either intuitive or structured. It is best to begin with the intuitive concept and elaborate the structured methodology later.

Designing Measures

Although the determination of measures is not part of the scenario technique, the reasons for establishing a scenario should be kept in mind. It is useful to involve the participants of the scenario development in the determination of measures; teamwork has proved to be worthwhile also for this step.

APPLICATIONS OF SCENARIOS

Scenarios can be developed for any situation that can be sensibly circumscribed. Heterogeneous issues (e.g., a strongly divisionalized company) are less suitable for this technique. Scenarios can be developed for technologies, market segments, countries or regions, operational functions, or strategic business units. They can also be assigned to various aggregation levels.

In general, scenarios should be considered when uncertainties about future developments are great and effective long-range planning has to be made. These criteria fully apply to tasks in the field of technology or innovation strategy development.

SCENARIOS AND INNOVATION STRATEGY DESIGN

Scenarios cannot generate new technological concepts or inventive principles. Normally, several technological possibilities and lines of development are known. The planner must determine the following:

- What directions to focus on.
- In what area to acquire or build up a knowledge base.
- What application fields to concentrate on first.

INNOVATION STRATEGIES

Scenarios can assist in making such determinations. For example, if it is foreseen that PVC production may be forbidden, a new material to be used as replacement must be considered. As a result:

- Visions of technical solutions may be highlighted (e.g., a total electronic bath system).
- Promising fields of application may become clear (e.g., transfer of the know-how of private sanitary systems to public lavatories).
- Possibilities for combination and integration with other technologies may become evident. If it is known that water consumption will very much increase, the integration of control systems in water supply, helping to save water, may be considered.
- Critical factors and bottlenecks for application, market introduction, and further diffusion can be seen. When end users, once professional craftspeople, become private do-it-yourselfers, both the marketing strategy and the distribution concept must be altered.

Scenarios are often the basis for the formulation of the requirements and specifications of new products. Scenarios also disclose the external driving forces behind R&D programs and projects, namely:

- Customer needs.
- Legislation.
- International competition.
- Substitution trends.
- Interrelated technologies.
- Infrastructure.
- Societal trends.

The following are examples of scenarios developed primarily to serve as an input to the development of innovation strategies:

- Genetic engineering.
- Sensor technologies.
- Telecommunications systems.
- CAD/CAM systems.
- Automobile technology.
- Production technology for a pharmaceutical firm.
- Treatment of wood waste.
- Electrical installation technologies.
- Applications of superconductivity technology.
- Expert systems.

- Home automation.
- Commercialization of satellites.

Scenarios can be used to formulate most of the dimensions of an innovation strategy.

> **The INCA program performs a consistency analysis that will result in the development of alternative scenarios.**

STRATEGIES DERIVED FROM SCENARIOS
Scenarios can be either intuitively or structurally translated in strategy development.

The Intuitive Approach
This entails a thorough reading of the individual scenario, placing oneself completely into the future situation, and then asking the following questions:

- What does it mean for us?
- What are the opportunities and threats resulting from it?
- What problem areas show up?
- What innovations may have good chances of acceptance and diffusion?
- What R&D directions should be followed and intensified?

The Structural Approach
A matrix must first be developed using the descriptors and strategic variables related to innovation (e.g., new products, R&D budget, market segments, plant improvements). Impact scores must then be inserted in the field or matrix to express the strength of the external influence (descriptor) on the individual variable. The strategy evolves by developing guidelines and individual measures for each single strategic variable, taking mainly into account those descriptors with the highest impact scores. The intuitive evaluation allows creativity and is enjoyable. It does not, however, provide a guarantee for all the essential aspects being covered. In comparison, the structural evaluation covers the complete area but is more labor intensive.

THE GOVERNING SCENARIO
Upon formulation of strategy, problems and probabilities similar to those present in other scenarios should be identified. These points will become part of the innovation strategy. Further guidelines and strategic measures are to be oriented to the governing scenario. However, at the same time, flexibility should be a part of the strategy to enable a changeover to another scenario. As a rule, the more probable scenario is the governing scenario. Companies with a dominating market position can also choose as the governing scenario the path of development promising the greater success. In this case, however,

the market power and creativity must be applied to the impacting fields such that the favored scenario will actually come to fruition.

If the governing scenario is one under which influencing factors determine a declining market, the development of that scenario should be carefully examined and analyzed to determine any signal indicating such a trend. The establishment of a monitoring system for new markets to enter may be beneficial.

INDEX

A

Aligning R&D with business objectives
   about   3–15
   case study   63–83
   creating partnerships   209–213
   gaining participation of non-R&D staff   26–28
   problems in   209
   role of R&D personnel   201–202
Applied research
   applied scientist's role   345–347
   profile of an applications researcher   347–349
   role in R&D   348–351

B

Business managers
   gaining their support   111–112, 126–127
Business units
   forming cross-disciplinary teams with   241–248

C

Chargebacks
   case study   173–180
   for R&D services   173–180
Competencies
   core technological   75–80, 85–97
Concurrent engineering
   case study   365–379
   competitive analysis   371
   customer input   370–371
   design goals   374
   obstacles   367–369
   people problems   377–378
   product testing   376
   project planning   374–375
   rapid prototyping   375–376
   regulatory environment   371–372
   setting up a team   372–373
   transfer to production   376–377, 423–432

Continuous improvement
   in customer-driven R&D   140–141
Consultants
   and scientific networks   485–495
   choosing   116–120
   technology management   32–33
Core competencies
   and self-managed teams   236–238
Core technological competencies
   and corporate retrenchment   99–104
   case study   103–104
   developing action plans   95–97
   life expectancy estimates   90–95
   methodology for evaluating   88–90
   selecting   86–87
Cost accounting
   case study   173–180
   for R&D   173–180
Critical success factors
   and self-managed teams   234–236
Cross-disciplinary teams
   difficulties of   246–248
   for product development   24–248
   managing   243–246
   product development with   465–474
   technology transfer with   465–474
Customer focus
   in selecting R&D projects   134–139, 208–209

D

Downsizing
   case study   99–104

E

Employee development
   and self-managed teams   236–238
   incentives   131
   recognizing R&D personnel   181–193
   recruiting   127–128
   rewarding personnel   181–193
   strengthening technical skills   116
Evaluating R&D productivity
   about   149–171
   developing an R&D evaluation system   217–219

F

Focus groups
   analysis and reporting 337
   and project selection 134–139
   getting the most out of 337–339
   moderating a group session 337–339
   organizing 334–337
   recruiting respondents 336
   study design 335–336
   value to an R&D manager 339–340

G

Goals
   examples of R&D strategic 113–114

I

Information technology
   acquiring and using 511–518

L

Linking R&D
   to business objectives 3–15, 17–23
   to business planning 35–43, 110
   to improve R&D productivity 159–164

M

Manufacturing
   creating R&D partnerships with 209–213
   input into R&D project development 433–452
   transferring products to 423–432
Marketing
   R&D's role in 110–111
   strategic 110–111
Multinational companies
   identifying laboratory sites for 144
   managing R&D in 143–148
   R&D strategies for 148

O

Operating units
   creating R&D partnerships with 209–213

   cross-disciplinary teams 241–248, 465–474
   differences with R&D in time horizons 210–211
   differences with R&D in technology evaluations 211
   lack of interest in building bridges 211–212

P

Personnel
   emotional conflicts 198
   encouraging creativity 197–198
   gaining support for process redesign 393–394
   motivation guidelines 155–163
   organizational structure 191–204
   performance appraisals 187–188
   recognition of performance 181–193
   recognition of team performance 184–195
   rewards for 181–193
Planning
   aligning R&D with business objectives 3–15, 35–43, 209–213
   developing an R&D evaluation system 217–219
   developing a technology plan 17–23
   establishing control 215–217
   integrating business and R&D 35–43
   leadership in 26–28
   problems in 209
   strategic technological 25–33, 45–61, 63–83
   technological forecasting 21–23
Process
   redesign 383–399
Product development
   case study 291–298, 353–364
   controlling 314–317
   cross-disciplinary teams and 241–248, 464–474
   customer input in 134–139, 208–209
   design 299–309
   evolution 401–409
   improvement 311–319
   incentives 131
   planning problems 209
   preparing for technology transfer 453–463
   problems 281–288
   risks 311–319

Product development process (continued)
  scenarios for innovation   513–520
  with cross-disciplinary teams   241–248
Product development process
  a model for innovation   497–509
  auditing   292–294
  best organization for   273–279
  case studies   273–299, 353–364
  cross-functional teams and   465–474
  developing an R&D evaluation
      system   217–219
  evolution   373–381
  generating new ideas   311–319
  improvements   311–319
  manufacturing input during   433–452
  preparing for technology
      transfer   453–463
  problems   311–319
  technology transfer during   411–421
  tracking   294–298
Project assurance
  checklist   276
Project management
  controlling project variables   223–229
  developing an R&D evaluation
      system   217–219
  planning   209, 215–219
  project control   217–219
  project planning   215–219, 224–226
  project control   215–219
  reporting   226–229
  scheduling   224–226

Q

Quality
  and team structure   232
  management program   259–269
  total quality management   259–269

R

R&D administration
  centralized   122–123
  decentralized   123–124
  strengthening   107–120
R&D infrastructure
  cross-disciplinary teams and   241–248
  effect of R&D organization on
      productivity   161–164
  R&D administration   107–120
  R&D strategy development   107–120
  self-managed teams   233–234

R&D investment
  and company goals   4–5
R&D management
  budgets   131–132
  centralized organizations   122
  cross-disciplinary teams and   241–248,
      465–474
  decentralized organizations   123
  developing an R&D evaluation
      system   217–219
  employee incentives   131
  in multinational settings   143–148
  in the technology base   213–218
  influence on productivity   159–164
  microcontrolling R&D   9–15
  motivating personnel   195–203
  partnerships with operating
      units   209–213
  planning documents   7–8
  process redesign   383–399
  recognizing and rewarding
      personnel   181–193
  strategic planning   25–33, 107–120
  strengthening   107–120
  total quality management
      program   259–269
R&D planning
  and R&D operations   110–114
  controlling project variables   223–229
  documents   7–9
  establishing control   215–219
  problems   209
R&D productivity
  controlling project variables   223–229
  evaluating   149–171
  improving   150–159
  measuring   150–159, 164
R&D project
  cross-disciplinary teams and   241–248,
      465–474
  customer input in   134–139, 208–209
  design   299–319
  developing an R&D evaluation
      system   217–219
  evaluation techniques   289–298
  evolution   401–409
  improvement   311–319
  manufacturing input   433–452
  mix   125–126
  preparing for technology
      transfer   453–463
  problems   209, 281–288
  reporting   226–229

R&D project (cont'd)
  reviews   220
  risks   311–319
  scenarios for innovation   525–542
  technology transfer during   411–421
R&D project management
  controlling project variables   223–229
  problems   209
  process redesign   383–399
  reviews   220
  setting requirements   223–224
R&D strategies
  developing one   107–120
  for multinational companies   148
Redesigning
  case study   383–399
  R&D processes   383–399
Reporting
  effective   128–131
  on R&D projects   226–229

S

Scheduling
  R&D projects   224–226
Scientific network
  case study   485–495
  establishing   485–495
Senior management
  winning support of   111–112, 126–127

Start-up firms
  managing innovation in   519–524
Strategic planning
  case studies   45–61, 63–83
  creating an R&D plan   25–33, 107–112
  developing for R&D   107–120
  generic outline   7
  linkage to R&D projects   8

T

Teams
  cross-disciplinary   241–248, 465–474
  leading   249–258
  motivating   249–258
  recognition of   184–185
  self-managed   231–240
Technological innovation
  a new approach   437–449
Technology
  acquiring for new product development   511–518
Technology transfer
  during a project   411–421
  for small companies   475–481
  laying the groundwork   453–463
  to manufacturing   423–432
  using cross-functional teams   465–474